Baltic Iron in the Atlantic World in the Eighteenth Century

The Atlantic World

Europe, Africa and the Americas, 1500–1830

Editors

Wim Klooster
Clark University

Benjamin Schmidt
University of Washington

VOLUME XIII

Baltic Iron in the Atlantic World in the Eighteenth Century

By

Chris Evans
Göran Rydén

BRILL

LEIDEN • BOSTON
2007

Cover Illustration: Leufsta *herrgård*. The *herrgård* in the mid-eighteenth century, viewed from the north, from the direction of the lower forge. To the right is the *bruksgatan*, leading to the church and, just beyond it, the *bruk* office. Courtesy of Jernkontoret.

This book is printed on acid-free paper.

Library of Congress Cataloging-in-Publication Data

A C.I.P. record for this book is available from the Library of Congress

ISSN 1570-0542
ISBN 978 90 04 16153 5

CONTENTS

PREFACE

This is a book about iron making and the international trade in iron during the eighteenth century. Iron, it is argued, was the very stuff of modernity. The nails and bolts into which it was hammered fastened the new urban fabric of Georgian Britain in place and held tolerably rigid the creaking sailing ships that carried ever greater volumes of commodities across the oceans. Indeed, iron became omnipresent in the eighteenth-century Atlantic world. Iron—and its alloy steel—was there in the precision instruments of Enlightenment science, just as it was in the shackles that restrained enslaved Africans as they made their way to the plantations of the New World.

The greater availability of iron in the eighteenth century is usually seen as a consequence of Britain's Industrial Revolution. It was not. Iron became ubiquitous in Britain, and those parts of the Atlantic world with which Britain traded, because of imports from the Baltic. That link between the Baltic and the Atlantic is our theme and the basis for a wider argument about Atlantic history.

This project began in the mid-1990s through a chance meeting of the two authors. A preliminary conversation outside a pub near the National Archives in London revealed that we shared some assumptions about historical causality and historical practice. (We also discovered a shared conviction that *Fullers London Pride* is the world's most thirst-quenching beer.) We were soon convinced that the story of Britain's iron industry and that of Sweden should properly be told as a single, intertwined story. Such was our theoretical conclusion; the difficulty lay in finding materials with which to demonstrate the point. Eventually, we found our answer in Somerset Archives. An exploratory trip in 1998 had us examining the business papers of Graffin Prankard, an early eighteenth-century Bristol merchant. Prankard's letters were dotted with mysterious hieroglyphic squiggles—mysterious, that is, to those unacquainted with Swedish industrial history. To Göran Rydén, the symbols that Prankard scratched into his letter books were instantly recognisable; they were the brand marks stamped onto bars of Swedish iron.

Prankard, it transpired, was a major importer of Swedish iron. Better still, he was an avid buyer of iron from the estates of Charles De

Geer, Sweden's leading ironmaster. And the De Geer ironworks were, like Prankard's business, very well documented. We had our empirical link—one that bridged the North Sea and that shed light on both Atlantic commerce and life in Swedish iron making communities. Finding an appropriate way of presenting our findings, however, posed a new challenge. We took inspiration from another shared passion: baroque music.

The fugue, it seemed to us, was an appropriate metaphor through which to arrange our material. The way in which the different musical parts were held in dialectical tension was a model to which we aspired when organising our analysis. Yet there were, of course, many different styles of the baroque. Johann Sebastian Bach's *Die Kunst der Fuge*, which we have come to think of as the soundtrack to our labours, can be heard in many different ways. Jordi Savall's rendition, for example, in which wind instruments augment the strings, stands in contrast to more traditional performances. History comes in forms that are equally various. There are national styles and these are not always congruent. Historical writing in Britain is often literary in form; narrative is preferred to analytical exposition. In Sweden the reverse is true. Full and systematic analysis in the German style is the model to be followed. Our aim in this book is not to reconcile the two, but to exploit the strengths of each tradition.

Ours has been a closely coordinated collaborative venture. The research questions emerged from lengthy joint discussion; much of the archival research was undertaken in the company of one another; and the completed text is the result of drafts and counter-drafts that have been swapped back and forth numberless times. Although the last touches to the text have been made by Chris Evans as the native English speaker in our partnership, this is in every way a joint project.

Needless to say, we have incurred many debts. Heading the list of creditors are two accomplished historians: Åsa Eklund and Owen Jackson. Åsa's *licentiate* thesis at the University of Uppsala, undertaken under the supervision of Göran Rydén, showed us what could be done in tracing patterns of commerce between Stockholm and Bristol. Owen, who was employed at the University of Glamorgan in 2000–2001 as a research assistant funded by the Economic and Social Research Council, took up the challenge. He laboured long and hard on Graffin Prankard's account books, converting the often confusing contents into a body of data that was usable for historical analysis. We thank them both.

We are grateful too to the archivists, librarians, curators and picture librarians who have been unfailingly helpful to us in the preparation of this book. It is invidious to single out any one individual, but we hope our professional colleagues will not take it amiss if we mention Yngve Axelsson at Jernkontoret in Stockholm, who has been tireless in locating illustrations for us. Beyond academe, particular thanks go to Lena and Peter Bergström, the owners of Gammelbo *herrgård*, who so hospitably allowed us to explore the archive in their attic. We thank also those who generously alerted us to information we would otherwise have missed—some very distant, like Mrs Lee Paltridge of Perth, Western Australia, who shared her knowledge of her Shallard ancestors with us.

Several of our colleagues have read our text in manuscript. We have profited greatly from the comments of Norry Laporte, Leos Müller, Rolf Torstendahl, and Anne Kelly Knowles, the last of whom, with her professional expertise as an editor, quite apart from that of a historical geographer, made us re-think our whole approach to the use of illustrations. The conversation of our friends and colleagues is also reflected in this book, even though they may not recognise it.

At Brill, in Leiden, we would like to thank Boris van Gool who has assisted us in all the practical matters of turning our manuscript and illustrations into a readable book.

We would also like to thank the bodies that have funded our research.

- Chris Evans gratefully acknowledges the generosity of the Economic and Social Research Council, whose award (no. R000223109) allowed the basic gathering and processing of data on the British side to go ahead in 2000–2001, and the Leverhulme Trust, whose award of a Study Abroad Fellowship in 2002–2003 allowed him to enjoy the hospitality of the Department of Economic History at the University of Uppsala, where the bulk of the text was written. Chris Evans has also been privileged to hold a Caird Short-term Fellowship at the National Maritime Museum in 2001 and a visiting fellowship at the Winterthur Museum and Library in 2005. Throughout, he has enjoyed the support of the Faculty of Humanities and Social Sciences at the University of Glamorgan.
- Göran Rydén wishes to thank the *Axel och Margret Ax:son Johnsons Stiftelse för Allmännyttiga Ändamål* and the *Wilhelm Ekmans fond för tryckbidrag* at the University of Uppsala.

Our greatest debt, however, one we will never redeem, is to our parents—Audrey and Kenneth Evans and Margareta and Kjell Rydén. This book is dedicated to them.

LIST OF MAPS, FIGURES, ILLUSTRATIONS,* AND TABLES

Maps

Figures

* The colour illustrations can be found between pages xvi and 1.

Illustrations

Tables

ABBREVIATIONS

Angerstein	Torsten Berg and Peter Berg (eds), *R.R. Angerstein's illustrated travel diary, 1753–1755: industry in England and Wales from a Swedish perspective* (London, 2001)
BCA	Birmingham City Archives
BL	British Library
CCL	Cardiff Central Library
ET	Eric Touscher
FJ	Francis Jennings
GA	Glamorgan Archives, Cardiff
GP	Graffin Prankard. All letters ascribed to Prankard in the text are taken from three of his surviving letter books in Somerset Archives: DD/DN 424 (June 1728 to March 1732), DD/DN 425 (April 1732 to November 1734), and DD/DN 426 (December 1734 to October 1738).
GS	Georg Swebilius
JJDG	Jean Jacques De Geer
KB	Kungliga Biblioteket, Stockholm
LDG	Louis De Geer (1705–1758)
NLW	National Library of Wales, Aberystwyth
NMW	National Museum of Wales, Department of Industry
RA	Riksarkivet, Stockholm
SA	Somerset Archives, Taunton
Schröder	'Dagbok rörande Handel, Näringar och Manufakturer m.m. Uti Danmark, Holland, England, Frankrike och Tyskland. Under verkställde resor, Åren 1748–1751 förd af Samuel Schröder' (Kungliga Biblioteket, X:303). Samuel Schröder's travel journal occupies two manuscript volumes, which are cited as Schröder I or II.
SML	Science Museum Library, London
SS	Samuel Shore
TNA	The National Archives, London
ULA	Landsarkivet i Uppsala
UUB	Uppsala Universitetsbibliotek

CHAPTER ONE

THE WAREHOUSE OF THE WORLD COMMERCE AND PRODUCTION IN THE EARLY MODERN ATLANTIC WORLD

In the Great Warehouse

				cwt.qr.lb	@	£.s.d.
Iron Crows				58.0.4	21/–	60.18.19
	Old					
Brewer's Squares				274.2.15	12/6	0.10.0
Broads Short				67.0.24	15/6	52.1.10
	Swedish			01.1.25	21/3	1.11.3
Steel Blister'd				11.1.16	23/–	14.10.10
	Rolled			3.0.6	25/–	3.16.4
	German			0.2.6	44/–	1.9.1
	Faggot			0.3.0	25/–	0.18.9
Rod Iron				62.1.0	17/6	54.12.6
Outside Rods				7.0.0	16/–	5.12.0
Rod wire				1.2.18	21/–	1.4.10
Strong hoops				2.1.1	17/6	1.19.6
Rolled plate				1.1.4	27/–	1.17.1
Mill hoops				5.1.6	16/–	4.4.10
Coach & Chaize	Tyre			3.3.22	22/–	4.6.10
Cart Tyre				1.3.0	20/–	1.15.0
	.....					
Hoes Barbados						
	Narrow	No. 0	24 & 1	)		
		1	185 & 8	)		
		2	391 & 8	)	8/–	268.17.4
		3	70 & 9	)		
	Broad	No. 1	86	)		
		2	95 & 10	)	12/–	145.0.0
		3	59 & 10	)		
[Hoes] Jamaica		No. 0	72 & 8	)		
		1	138 & 11	)	12/–	141.10.0
		2	24 & 3	)		
[Hoes] Carolina		No. 0	88 & 11	)		
		1	46 & 8	)	13/–	93.19.7
		2	9	)		
	....					

In the Nail Warehouse

Flat heads	4lb	42m	1/6	3.3.0
	9	71	2/8	9.9.4
	11	94	3/2	14.17.8
	14	23	3/9	4.7.2
Sharp	9	36	2/8	4.16.0
	10	52	2/11	7.11.8
	11	213	3/2	33.16.1
	14	9	3/9	1.16.11
	18	3	4/8	0.16.6
	20	15	5/2	3.17.6
	21	47	5/4	12.15.4
	22	15	5/7	4.5.9
Clasp	7	210	2/4	24.10.0
	20	48	5/3	12.12.0
	21	48	5/5	13.2.0
Flat points	7	110	2/4	12.16.8
	11	211	3/2	33.8.2
	14	22	3/9	4.5.4

....

In the Bar Iron Warehouse

Russia Iron	95.3.7	13/–	62.5.7
Swedish squares	3.3.7	17/–	3.4.10
Old Iron	34.1.1	11/–	18.16.10
Pig hogs	12.0.7	15/6	9.6.11
Short Broads	14.0.0	15/6	10.17.0
Swedish ditto	1.2.15	21/3	1.14.8
Thimble Iron	21.1.1	17/–	18.1.5
Boltstaves	98.1.24	19/–	93.10.9
Scrap steel	15.2.24	14/–	11.0.0

....

The Warehouse

The list above is part of an inventory of goods stored in the warehouses of Crowley Hallett & Co at Deptford in 1751. The warren of shops, cellars and garrets occupied by Hallett and his partners was crammed with a great miscellany of iron and steel articles.[1] Each entry in the inventory veiled untold processes and transactions, for the nails, hinges,

[1] BL, Oriental and India Office Collections, MSS Eur F 218/115.

chains or shovels that had come to a temporary rest in the shuttered darkness of Hallett & Co's warehouses had arrived by the most circuitous and varied of routes. The blister steel in the Great Warehouse had been shipped from the North East of England, where iron from the Swedish county of Uppland underwent conversion in cementation furnaces; and it was iron from the Basque Country that lay ready for the smiths in the Company's anchor shop beside Deptford Creek. Some commodities had already been carried thousands of miles. The Russian bar iron, for example, had been smelted and refined amid the *taiga*; the distance from Russia's Ural frontier to the Thames was so great, and the logistical difficulties so acute, that two years were required for the journey. Other goods were being held in readiness for shipment across the Atlantic. The destination of different plantation hoes, each dedicated to a specific form of tropical agriculture, was announced by their description in the inventory. 'Hoes Barbados' were distinguished from 'Hoes Jamaica' and from 'Hoes Carolina'.

Hallett & Co's warehouse was home to a range of rather prosaic goods. Whip saws and poll axes had none of the finesse that polished Hanoverian consumers looked for in foliated Sheffield plate or japanned *objets*. These were working tools, not ivory-handled table cutlery. The Caribbean-bound hoes had a severe practicality to them, suggesting little of the sugar or coffee that would be served in metropolitan salons. Much has been written in recent years about the role of exotic groceries and tropical timber in enriching the material culture of eighteenth-century Britain; but for all that, it was the humble tools that lay oiled and wrapped in Deptford that were of fundamental importance for Britain's Atlantic empire. They were at the commercial hinge that joined Baltic Europe—the Europe of rye bread and herring, of tar and potash—to the western ocean. The English merchantmen that heaved their way south to the Canaries and then westward on the trade winds to the Antilles were held together by hemp from Riga and by bolts and hoops beaten out of Swedish iron. And every one of the steeled machetes swung by enslaved Africans in Jamaica originated in ore that had been hauled from the giant mine at Dannemora, 60° north of the equator. In short, the westward advance of British capitalism drew strength from a northern hinterland that was rich in the mineral and vegetable resources that Britain lacked. It is that relationship that supplies the main theme of this book: how trade between the Baltic and Britain—more particularly, the trade in iron—contributed to the world economy in the eighteenth century.

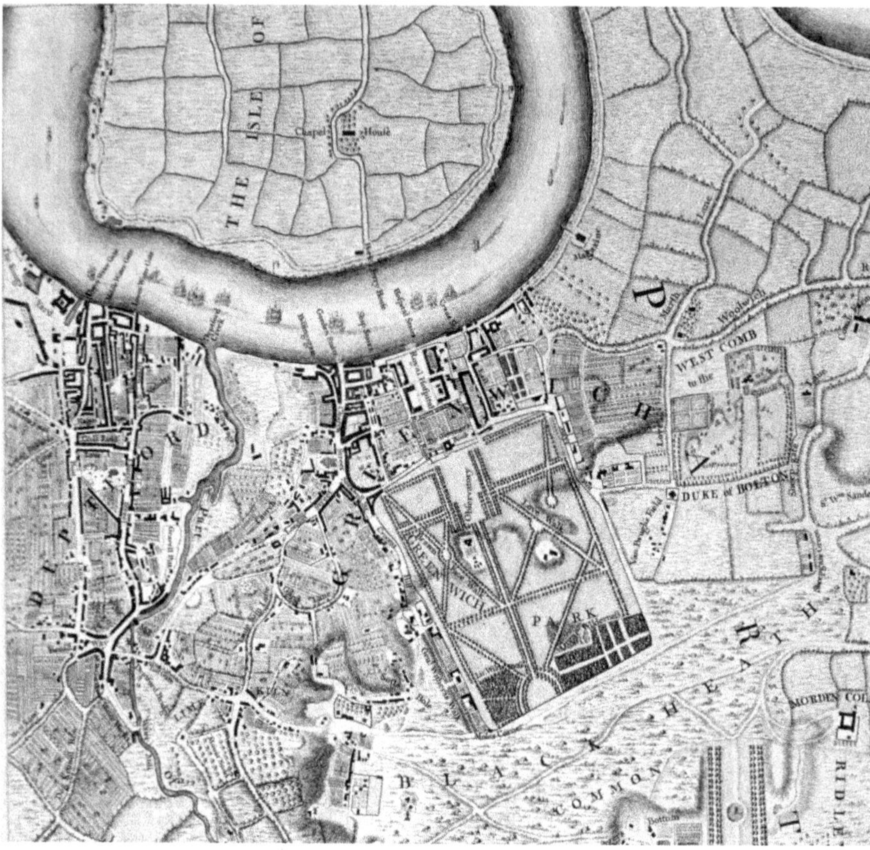

Map 1.1. Deptford and Greenwich in the 1740s, as shown in John Rocque's *Cities of London, Westminster, the Borough of Southwark, with the Country near Ten Miles around* (1746).

Courtesy of the Guildhall Library, London.
Caption: Deptford, three miles below London Bridge, was an important staging post for the export of English ironmongery, where many hardware merchants maintained warehouses. With its anchor shops and a naval base ('The King's Yard'), Deptford was host to a lively maritime economy. A little further downstream was the imposing Royal Naval Hospital at Greenwich, and just beyond, the warehouses of Theodosia Crowley.

Illustration 1.1. The Royal Dockyard, Deptford in the 1750s.

Courtesy of the National Maritime Museum, London.
Caption: This 1757 painting by John Cleveley the Elder shows the *Cambridge*, a 100-gun man-of-war, being floated from the dry-dock in which she was built. The Great Storehouse of the Yard stands adjacent. The warehouses of Crowley Hallett & Co were just downstream, where Deptford Creek emptied into the Thames.

In all, Crowley Hallett and his partners had goods valued at £13,000 stored at Deptford and at locations in the City of London. At Greenwich, a few hundred yards to the east, a still larger magazine could be found. The Thameside depot of Theodosia Crowley, the aunt of Crowley Hallett, was unrivalled for the range of hardwares that it housed, for 'the Lady Crowley' was Britain's foremost manufacturer.[2] When inventoried in 1728, upon the death of Theodosia's husband John, the goods that lined the racks and gangways of the Greenwich warehouse were appraised at £48,115. The variety was such as to tax the descriptive powers of Mrs Crowley's clerks. Over 80 types of file were manufactured at one of her factories on Tyneside, and 154 varieties of nail emerged from the complex of workshops she owned at neighbouring Winlaton.

The inventory was a device for bringing order to this tumult of goods. It categorised and labelled, fixing on metallic objects designations that would be as well understood by the storekeepers who retailed them in tidewater Virginia as by the artisans in south Staffordshire who had hammered them out. The inventory also enumerated, assigning weight and measure to the contents of the warehouse. Above all, the inventory imposed synchrony on articles that had been fabricated at various points in time and that were at different stages in their life-cycles as commodities. For the sake of analytical clarity it immobilised goods that were in transnational flux.[3]

Although the inventory is an ancient way of handling data, one that can be found on the Sumerian clay tablets that are the earliest forms of human inscription, it had particular appeal for the early eighteenth-century European—still more the British—mind. It was a tool of enormous utility for a society in which the 'exchange of forms of mobile property' had a new salience, threatening, as many contemporaries saw it, the eminence of land as the embodiment of wealth. It was a means of mapping out a 'world of moving objects' in which novel commodities proliferated and freshly minted financial instruments hastened the circulation of goods.[4] Yet the zest for listing and quantifying that was so

[2] For the Crowley family see M.W. Flinn, *Men of iron: the Crowleys in the early iron industry* (Edinburgh, 1962).

[3] See the discussion in Jack Goody, *The domestication of the savage mind* (Cambridge, 1977), chapter 5.

[4] Quotations taken from J.G.A. Pocock, *Virtues, commerce, and history: essays on political thought and history, chiefly in the eighteenth century* (Cambridge, 1985), p. 109.

marked a feature of Augustan England masked a conceptual difficulty. How was wealth to be defined? And were certain forms of property more fruitful than others? How, to extend the question, was new wealth generated? And how best retained?

Daniel Defoe (*c.* 1660–1731) supplied one answer with a hydraulic metaphor:

> ...an estate's a pond, but trade's a spring: the first, if it keeps full, and the water wholesome, by the ordinary supplies and drains from the neighboring grounds, it is well, and it is all that is expected; but the other is an unexhausted current, which not only fills the pond and keeps it full, but is continually running over, and fills all the lower ponds and places about it.[5]

The agrarian world, in other words, was stable and secure, but that was 'all that [could be] expected'. Landed property was stately but it was not dynamic. The generation of new wealth depended upon trade and the protean energies that it embodied. For Defoe, as for the mercantilist writers of the seventeenth century, it was axiomatic that foreign trade was the key to economic aggrandisement. If the shipment of British goods to overseas markets outweighed the influx of foreign-made goods onto the domestic market then the nation's wealth would grow. Foreign merchants would have to make good their deficit by shipping bullion to their British counterparts—a satisfyingly tangible settlement of accounts. Such a desirable state of affairs should be enforced, wherever possible, by appropriate legislation. The Navigation Acts of the 1650s did just that. Overseas trade was to be conducted in British or colonial-made bottoms, the colonies were to be the exclusive preserve of British exporters, and the most valuable colonial goods could only be shipped to foreign markets via British ports.

It was the necessity of monitoring trade that made the listing and enumerating of goods such an important practice in the early modern period. Mercantilist thinkers believed that wealth was a finite substance; it comprised products of the natural world, which were, of necessity, fixed in extent. It followed, then, that enrichment, whether of nations or individuals, would be the result of capturing a larger share of the wealth that a benign providence had put at humankind's disposal. Circulation should therefore take priority over production, and little importance was to be attached to consumption. In mercantilist theory

---

[5] Daniel Defoe, *The complete English tradesman* (1738; first edn 1725) p. 322.

it was the movement of goods that merited the most intense scrutiny, not the method of their fabrication or the final uses to which they were put.[6] It was this that accounted for the salience of the inventory as an intellectual and organizational device in the mercantilist age.

But this is to treat mercantilism as a static mode of thinking when it was not. Although circulation and exchange were ever the preoccupations of mercantilist thought, by the end of the seventeenth century there was a keener appreciation of production. The workshop, so to speak, was encroaching upon the warehouse. Debate over the 'balance of trade', which early theorists such as Thomas Mun and Edward Misselden had considered largely in terms of the inflow and efflux of specie, gave way to discussion on the effective exploitation of labour. Later seventeenth-century controversialists such as Nicholas Barbon and Sir Josiah Child placed more emphasis on the role of commerce in providing employment. A vigorous promotion of trade would boost manufacturing activity, which would in turn encourage the growth of population. A large population, industriously employed, was identified as a central component of national wealth.[7]

The later mercantilists took a more capacious view of trade, one that extended beyond the act of exchange to include aspects of production. Defoe presented trade as a complex and ambiguous phenomenon. 'Trade', he wrote in his *Plan of the English commerce* (1728), 'like Religion, is what every Body talks of, but few understand: The very Term is dubious, and in its ordinary Acceptation, not sufficiently explain'd'. Defoe ventured a clarification that explicitly yoked production to exchange:[8]

> The general heads of Home-Trade are best contain'd in the *two* plain and homely Terms *Labouring* and *Dealing*. 1st *The Labouring Part*, this consists of Art, Handicraft, and all Kinds of Manufactures; and those who are employ'd in these Works, are properly called *Mechanicks*; they are employ'd, generally speaking, about the first Principles of Trade, (*viz*) the Product of the *Land* or of the *Sea*, or of the Animals living on both: In a Word, the ordinary Produce of the *vegetative* and *sensative* Life; such as Metals, Minerals and Plants, the immediate Produce of *Vegetation*, or

[6] Lars Magnusson, *Mercantilism: the shaping of an economic language* (1994), pp. 68–80.

[7] Joyce Appleby, *Economic thought and ideology in seventeenth-century England* (Princeton, 1978), pp. 112 and 154ff; Magnusson, *Mercantilism*, pp. 134–38. See also Julian Hoppit, 'Political arithmetic in eighteenth-century England', *Economic History Review*, XLIX, 3 (1996), 516–40.

[8] Daniel Defoe, *Plan of the English commerce. Being a complete prospect of the trade of this Nation, as well home as foreign* (2nd edn, London, 1737), pp. 2ff.

> such as Flesh, Skins, Hair, Wool, Silk &c. grown with, and produc'd by the Animals as the Effect of *sensitive* Life.
>
> 2. *The Dealing Part*; this consists of handing about all the several Productions of Art and Labour, when finish'd by the Hand of the industrous Mechanick, and made useful to Mankind; conveying them from Place to Place, and from one Country to another, as the Necessity and the Convenience of the People call for them; and that upon such Terms and Conditions of Delivery, as they can best agree about among themselves, and this is Trade…

'One vast Piece of Machinery'

'Dealing and Manufacturing', Defoe concluded, 'comprehends all Trade.'[9] Malachy Postlethwayt took up the theme. The entry on 'Manufacturers' in his *Universal dictionary of trade and commerce* (1751) began with a conventional genuflection to the bounty of nature, but Postlethwayt moved on to advocate a closer attention to manufacturing.

> We begin to be now convinced, that we are nearly as much enriched by the labours of our fellow-creatures, as by the productions of the earth; and, if we have reason to rejoice at the abundance which nature, from year to year, produces for us, we may reap no less reasonable satisfaction from all the variety of employments in human society, and especially by means of our manufactural arts. The first proof of this have been taken from numberless kinds of business, which our servants and the very meanest labourers perform for us; not in our houses only, but from one end of the earth to the other: what they are doing on the banks of Newfoundland, at Potosi, at Mocha, or in the island of Amboyna, concerns us no less than the being decent in our apparel and habitations. Let us consider the reason we have to esteem artizans of every kind for their industry, and find new motives, from the numberless services they do us, to rectify our way of thinking concerning them.[10]

Joseph Massey, writing in 1760, concurred. Production and trade were functionally integrated: '[t]he various Branches of our Manufacture and

---

[9] Defoe, *Plan*, p. 3.

[10] Malachy Postlethwayt, *The universal dictionary of trade and commerce: with large additions and improvements* (4th edn, London, 1774), *sub* 'Manufacturers'. Postlethwayt alluded to four of the major centres of world trade: the great cod fishery of the Grand Banks, the immense Andean silver mine at Potosi, the coffee-growing hinterland of the Arabian port of Mocha, and the East Indian spice island of Amboina.

Trade, when nationally considered, may aptly enough be compared to one vast Piece of Machinery'.[11]

A willingness to look upon manufacturers and artisans in a positive light could also be found among contributors to the *Encyclopédie* (1751–1766), the *magnum opus* of the French Enlightenment. Production had a pivotal position in the intellectual universe that the encyclopédists defined. They did not share the mercantilists' reverence for trade; their concern was to bestow 'a new dignity on craft and technology'.[12] Denis Diderot, editor-in-chief of the enterprise, boasted of the unusual lengths to which his contributors had gone in their pursuit of knowledge:

> We addressed ourselves to the most skilful artisans of Paris and the kingdom: we took the trouble to go into their workshops, to question them, to write under their dictation, to develop their thoughts, to draw from them the terms proper to their professions...[13]

The encyclopédists were generally respectful of the craftsmen whose practices they described. The expertise of artisans should be acknowledged, Diderot thought, and the self-regard that it bred in workmen tolerated as 'the only means to obtain from them more perfect products'.[14] Yet the *Encyclopédie* was as prescriptive as it was descriptive. The illustrative plates that accompanied the *Encyclopédie* appeared to fulfil Diderot's claims for the work as a repository of concrete, useful knowledge, but the plates presented a vision of workshop practice that was, despite the detailed depiction of tools, abstract and deracinated. Operatives were shown in postures that were curiously lifeless, gesturing towards the implements with which they laboured rather than wielding them. The workshop as envisioned by the encyclopédists was far removed from the clutter, noise and noisomeness of the actual *atelier*. The project of the *Encyclopédie* was to critique the hierarchies of the *ancien régime*. This was most clearly the case with respect to the aristocracy and the Church, but the encyclopédists also detected obscurantism in the workshop where, they asserted, craft mystery and artisanal conviviality

[11] Joseph Massey, *Representation concerning the knowledge of commerce as a national concern* (1760), quoted in Hoppit, 'Political arithmetic', 521.

[12] C.J. Koepp, 'The alphabetical order: work in Diderot's *Encyclopédie*', in S.L. Kaplan and C.J. Koepp (eds), *Work in France: representations, meaning, organization, and practice* (Ithaca, 1986), p. 239.

[13] Denis Diderot, 'Prospectus to *Encyclopédie*', quoted in Koepp, 'Alphabetical order', p. 248.

[14] Denis Diderot, 'Arts' in *Encyclopédie*, quoted in Koepp, 'Alphabetical order', p. 240.

stood in the way of rational, productive labour—hence the conceptual concern with the division of labour as a way of reducing human toil to a scientifically irreducible core, shorn of the drinking, joshing, feasting, cruel horseplay and camaraderie that encrusted workshop routine in the Paris of Louis XV.

The *Encyclopédie* devoted 5,000 words to the common pin, an item that 'undergoes eighteen operations before it gets into the shops'.[15] In doing so, the *Encyclopédie* anticipated Adam Smith, who famously extolled pin making in the opening pages of *The wealth of nations*. This 'very trifling manufacture', as Smith described it, exemplified the 'increase in the productive powers of labour' brought about by the division of labour. 'One man draws out the wire, another straights it, a third cuts it, a fourth points it, a fifth grinds it at the top for receiving the head...the important business of making a pin is, in this manner, divided into about eighteen distinct operations'.[16] An opulent nation, Smith declared, was one that had carried the division of labour to its furthest extent. Yet the division of labour in manufacturing industry was not arbitrary; it was governed by the extent of the market. An extensive market allowed for the subdivision of tasks, whereas in the 'lone houses and small villages which are scattered about in so desert a country as the Highlands of Scotland, every farmer must be butcher, baker and brewer for his own family'. By linking the division of labour to the extent of the market Smith posited a firm and mutually beneficial relationship between trade and production. The two marched *pari passu*.[17]

'In the stages that preceded capitalist society', Karl Marx averred, 'it was trade that prevailed over industry; in modern society it is the reverse.'[18] From his vantage point in the nineteenth-century industrial

---

[15] Quoted in Philipp Blom, *Encyclopédie: the triumph of reason in an unreasonable age* (2004), p. 144.

[16] Adam Smith, *An inquiry into the nature and causes of the wealth of nations* (1776: Indianapolis, 1981), p. 15.

[17] It was once common to speak of a Smithian revolution in economic thought. Recent authorities are more circumspect, allowing for greater continuity between Smith and the mercantilist writers against whom he set himself. See Appleby, *Economic thought*, pp. 94, 182, 202 and 271ff, and Magnusson, *Mercantilism*, pp. 1ff. Studies linking mercantilism and the Enlightenment are in short supply. For one important exception see John Robertson, 'The Enlightenment above national context: political economy in eighteenth-century Scotland and Naples', *Historical Journal*, XL, 3 (1997), 667–97, which identifies political economy, a study of 'human betterment', as being at the very core of the Enlightenment project (673).

[18] Karl Marx, *Capital: a critique of political economy*, vol. III (Harmondsworth, 1981), p. 448.

world, Marx could be confident in this assertion. For the economic theorists who preceded him the situation was less clear-cut. The relationship between commerce and production was problematic and whether one should be assigned priority over the other a matter of dispute and puzzlement. In the eighteenth century, as high mercantilist theory fell from favour, models that emphasised the primacy of commerce gave way to those that acknowledged the wealth-generating capacities of manufacturing industry. Enlightenment thinkers dwelt more upon the reciprocity of trade and production and hesitated about elevating one over the other.

This book takes the analytical ambivalence towards trade and production in early eighteenth-century economic discourse as its point of departure. That ambivalence should not be viewed as a sign of intellectual bewilderment; rather, it should be seen as reflecting a reality whose features were hybrid and transitional, in which trade and production were intermelded in such a way that it would be impossible to speak of one as dominant. When contemporaries spoke of the 'Iron Trade' they spoke of trade as Defoe defined it: a process that comprehended 'Dealing *and* Manufacturing' [italics added]. When a group of eighteenth-century ironmasters declared that the 'Iron Trade is beyond all dispute for Imployment of Hands & on all other Accts the second in ye Kingdom', acknowledging the seniority only of woollen textiles, they were defining the sector as extending far beyond blast furnaces and forges.[19] Those capital-intensive installations gave work to fewer than 1500 men nationally at the mid century; far larger numbers, tens of thousands more, were employed in the making of hardware. All of them, nailers and scythe grinders as much as blast furnace keepers, were seen as members of the iron trade. Equally, the capitalists who were concerned in the iron trade rarely restricted themselves to a single facet of production. Crowley Hallett and his partners were exemplars in this respect. They were international merchants who imported iron from Stockholm and shipped hardware to the New World; they were industrialists who owned blast furnaces, forges and slitting mills; and they were wholesale ironmongers. They roved across the boundaries that would later demarcate 'primary processing', 'secondary manufacturing', and 'wholesale distribution'.

---

19 Sheffield Archives, SpSt 60487.

The Swedish traveller Samuel Schröder showed similarly scant regard for sectoral boundaries when he compiled his 'Notes on the English Iron Trade' in 1749. His starting point was a survey of bar iron making at English forges. That led to a discussion of the market for bar iron in Britain, and how imports from Sweden, Russia, Spain and the American colonies swirled about it. A description of the uses to which malleable iron was put in British manufacturing followed. Finally, Schröder addressed the marketing of British-made hardware domestically and internationally. Samuel Schröder's approach prefigures that taken in this book. Our aim is to range across sectoral and national frontiers, and by so doing disinter a commercial web that once joined the Baltic and Atlantic worlds. In this, forges in Siberia and *Bergslagen* (the iron making region of central Sweden) were yoked to the metal ware manufacturing districts of the English Midlands, and the articles wrought up in Dudley or Wednesbury were consigned to places as varied as the Bight of Biafra and the Carolina Lowcountry.

Iron Histories

By adopting this wide-angle perspective on the making and marketing of iron and iron wares we depart from established historiographical practice in both Britain and Sweden. Writing on the British iron industry has been surprisingly sparse given the centrality that the industry had for the Industrial Revolution. Major studies have been few and far between, dwarfed by those devoted to textiles and comfortably out-numbered by those concerning coal. T.S. Ashton's classic study of 1924, *Iron and steel in the Industrial Revolution*, set the tone for much of what followed.[20] A magisterial treatment of its subject, Ashton's attention to technological change provided a template from which his successors were reluctant to depart. The origins of coke smelting developed a sub-literature all of its own, whilst the most important synoptic addition to the canon, Charles K. Hyde's *Technological change and the British iron industry 1700–1870* (1977), cleaved to Ashton's priorities, its methodological superiority notwithstanding.[21]

---

[20] T.S. Ashton, *Iron and steel in the Industrial Revolution* (Manchester, 1924; revised edition 1963).

[21] See R.A. Mott, 'Abraham Darby (I and II) and the coal-iron industry', *Transactions of the Newcomen Society*, XXXI (1957–59), and *idem*, 'The Coalbrookdale group

The historiography of British iron making has been obdurately supply-sided. Revisions to Ashton have been made by those who wished to reassess the take-up of new technologies (in the case of Hyde), and by those who disputed Ashton's gloomy assessment of the iron industry in the last decades of the charcoal era (in the case of Flinn and Hammersley), but little has been done to explore the use of iron.[22] Very little notice has been taken of the fact that most of the iron consumed in the British Isles between the 1720s and the 1790s would have been of Swedish or Russian provenance; massive import penetration, a matter of great concern for contemporaries, has gone largely unremarked by historians. Indeed, the market for malleable iron in Britain has gone virtually unexplored.[23] Insofar as explorations have been made, they have been oblique and indirect, embedded in studies of the entrepreneurial organisation of the iron industry in the charcoal era. By the late seventeenth century the British iron industry comprised a number of interlocking partnerships, each controlling a network of blast furnaces, forges and processing mills, and each with members engaged in the hardware trades. These were meticulously, not to say exhaustively, investigated in the mid-twentieth century. The best-known of the partnerships, that centred on the Foley family in the West Midlands, was unravelled by B.L.C. Johnson in the 1950s.[24] Arthur

---

Horsehay works: Part I', *Transactions of the Newcomen Society*, XXXI (1957–59), 271–87 and 'Part II', XXXII (1959–60), 43–56, and more recently Nancy Cox, 'Imagination and innovation of an industrial pioneer: the first Abraham Darby', *Industrial Archaeology Review*, XII, 2 (1990), 127–44.

[22] M.W. Flinn, 'The growth of the english iron industry 1660–1760', *Economic History Review*, XI (1958), 144–53, and G.F. Hammersley, 'The charcoal iron industry and its fuel', *Economic History Review*, XXVI (1973), 593–613. For an overview see J.R. Harris, *The British iron industry 1700–1850* (1988).

[23] Gross domestic consumption of bar iron is calculated in Peter King, 'The production and consumption of bar iron in early modern England and Wales', *Economic History Review*, LVIII, 1 (2005), 1–33, but the functioning of the market is not attended to.

[24] B.L.C. Johnson, 'The Stour valley iron industry in the late seventeenth century', *Transactions of the Worcestershire Archaeological Society*, XXVII (1950), 35–46; *idem*, 'The charcoal iron industry in the early eighteenth century', *The Geographical Journal*, CXVII (1951), 167–77; *idem*, 'The Foley partnerships: the iron industry at the end of the charcoal era', *Economic History Review*, VI (1952), 322–40; *idem*, 'The iron industry of Cheshire and Staffordshire, 1688–1712', *Transactions of the North Staffordshire Field* Club, LXXXVIII (1953–54), 32–55. See also B.G. Awty, 'Charcoal ironmasters of Cheshire and Lancashire, 1600–1785', *Transactions of the Historical Society of Lancashire and Cheshire*, CIX (1975), 71–124; R.G. Schafer, 'Genesis and structure of the Foley "Ironworks in Partnership" of 1692', *Business History*, XIII (1971), 19–38; and P.W. King, 'The Vale Royal company and its rivals', *Transactions of the Historical Society of Lancashire and Cheshire*, CXLII (1992), 1–18.

Raistrick performed the same service for the scarcely less important partnership of the Spencer family in Yorkshire.[25] Together, Johnson and Raistrick presented clear evidence of the seamlessness of primary production, secondary processing and marketing—evidence that the *industry* was, in fact, a *trade*. The iron industry was characterised less by vertical integration within firms—for these 'firms' were amorphous, shifting entities—than by a constant exchange of raw materials and semi-processed goods between loosely connected business associates.[26] Yet the insights of Johnson and Raistrick remained implicit in the empirical material that they presented. Nor was much attention given to the role of Baltic iron in the web of exchange that they described, salient though it was. Indeed, the recognition that British iron making and metalware manufacturing was but part of an international division of labour, as seemed plain to Samuel Schröder in the 1740s, has not been taken up by subsequent historians.[27]

The historiography of the Swedish iron industry is quite different, not least in its scale. Iron making plays a central role in the narrative of Swedish national development from the sixteenth century to the present; its history has therefore been accorded lavish attention. That said, Swedish studies, like their British counterparts, tend to be limited

[25] A. Raistrick, 'The South Yorkshire iron industry, 1698–1756', *Transactions of the Newcomen Society*, XIX (1938–39), 51–86; A. Raistrick and E. Allen, 'The south Yorkshire ironmasters, 1690–1750', *Economic History Review*, IX (1938), 168–85. See also G.G. Hopkinson, 'The charcoal iron industry in the Sheffield region, 1500–1775', *Transactions of the Hunter Archaeological Society*, VIII (1961), 122–51.

[26] 'Although there was vertical integration within each of the partnerships, considerable traffic took place between them, as well with independent forge and slitting-mill masters. Iron at all stages of manufacture entered into this trade.' Johnson, 'Foley partnerships', p. 331. Curiously, the most overt application of the approach developed by Johnson in the 1950s came in a work devoted to the significance of transport in European industrialisation, not the British iron industry *per se*: Rick Szostak's *The role of transportation in the Industrial Revolution* (Montreal, 1991). 'Iron, as a producer goods industry, needs special treatment', Szostak announced; 'proper coverage requires that one looks at the uses to which iron was put' (p. 91). This led Szostak to reconstruct a production chain that began with the smelting of ore and terminated with the marketing of metal wares. This, in turn, was the basis for conjectures about the relationship between improved transport organisation, price levels and the extent of the market for metalwares. The analysis of the latter may not have been wholly convincing, but the line of inquiry was suggestive, despite its brevity and reliance upon secondary literature.

[27] The issue was broached by Marie B. Rowlands in her important study *Masters and men in the West Midland metalware trades before the Industrial* Revolution (Manchester, 1975) but not pursued at any length. It does not feature at all in David Hey's *The rural metalworkers of the Sheffield region: a study of rural industry before the Industrial Revolution* (Leicester, 1972).

in scope. As in Britain, entrepreneurship and technological development have been foregrounded. This 'top-down' bias stems from the format in which the literature first appeared, that of the company-sponsored history. Many steel combines commissioned official histories in the years after 1945 and these, naturally enough, dwelt upon the foundation and descent of their various constituent works, many of which had been in existence for two or three hundred years. The best-known of these company histories, and the only one to transcend the limitations of the genre, was *Fagerstabrukens Historia*, published in five volumes between 1957 and 1959.[28] Its authors did not restrict themselves to the institutional history of the five steelworks that had merged to form *Fagerstabruk* in the 1920s; they ventured an overview of the Swedish iron industry as a whole, one that was to stand as an orthodoxy for the remainder of the twentieth century.[29]

In 1987 Karl-Gustaf Hildebrand, one of the *Fagerstabrukens Historia* authors and doyen of Swedish industrial history, revisited the themes that he had first explored three decades earlier. His book, issued in English in 1992 as *Swedish iron in the seventeenth and eighteenth centuries: export industry before the industrialization*, was a brilliant summation of Swedish iron making in the preindustrial era. There were, nonetheless, striking lacunae. Very little was said, for example, about the world of work. However, Hildebrand's restatement of the orthodoxy came at a moment when research on the premodern iron industry, which had known a period of quiesence, was reviving. A new generation focused upon the very issues that were underplayed in the established literature, namely labour and everyday life in iron making communities (*bruk*).[30] The work of Anders Florén on the making of bar iron and metalwares at Jäders

---

[28] *Fagerstabrukens historia* (Uppsala, 5 vols, 1957–59). The key volumes were: K.-G. Hildebrand, *Del I. Sexton- och sjuttonhundratalen* (1957); A. Attman, *Del II. Artonhundratalet* (1958); and S. Montelius, G. Utterström, and E. Söderlund, *Del V. Arbetare och arbetareförhållanden* (1959).

[29] The reason for the enduring influence of the *Fagerstabrukens historia* authors, particularly Hildebrand and Attman, apart from the intrinsic value of their scholarship, is that parts of the first two volumes in the series were published separately in the 1980s: Artur Attman, *Svenskt järn och stål* (Stockholm, 1986) and K.-G. Hildebrand, *Svenskt järn. Sexton- och sjuttonundratal. Exportindustri före industrialismen* (Stockholm, 1987). The latter, which was revised extensively to take in research completed since its first publication, has been translated into English (see below). For a general treatment of Swedish iron making see E.F. Heckscher, *Sveriges ekonomiska historia från Gustav Vasa. Andra delen. Det moderna Sveriges grundläggning* (Stockholm, 1949), chapter 6.

[30] See Maria Ågren (ed.), *Iron making societies: early industrial development in Sweden and Russia, 1600–1900* (Oxford, 1998) for a presentation in English of these new trends.

bruk between 1640 and 1750 led the way. Taking his cue from the model of proto-industrialization propounded by Jürgen Schlumbohm, Florén explored how workers in the forges and workshops of the *bruk* were gradually deprived of their autonomy.[31] Control over the labour process shifted. It did so, at least in part, in response to developments on the market for metalwares. The market, Florén suggested, was a historical variable in grievous need of investigation.[32] The need to understand Swedish iron in the context of an international market had been acknowledged by Hildebrand when he had made a provisional appraisal of overseas markets in the 1950s, but in a valedictory survey in 1997 he could still lament the paucity of research on distribution and marketing: 'what is needed is many more studies in the history of iron from a consumer perspective'.[33]

The British and Swedish historiographies of iron mirror one another in one important respect. The British have been steadfastly incurious about the Baltic iron that flooded their domestic market in the

[31] Jürgen Schlumbohm, 'Relations of production—production forces—crises in proto-industrialization', in Peter Kriedte, Hans Medick, and Jürgen Schlumbohm, *Industrialization before industrialization* (Cambridge, 1981), pp. 94–125.

[32] Anders Florén, *Disciplinering och konflikt. Den sociala organiseringen av arbetet. Jäders bruk 1640–1750* (Uppsala, 1987). Florén returned to the issue of the market at greater length in a subsequent study of the iron trade in the southern Netherlands: *Vallonskt järn. Industriell utveckling i de södra Nederländerna före industrialiseringen* (Uppsala, 1998).

[33] K.-G. Hildebrand, 'Gammalt och nytt i det svenska järnets historia. En översikt över fem årtionden', *Dædalus 1997. Svenskt järn under 2500 år. Från gruvpigor och smedsdrängar till operatörer* (Stockholm, 1997), pp. 1–30. For Hildebrand's pioneering work in this area see his 'Foreign markets for Swedish iron in the eighteenth century', *Scandinavian Economic History Review*, VI (1958), 3–52. Because iron was such a significant part of Sweden's export trade in the eighteenth century it has also been studied by historians concerned with trends in external trade and shipping. See in particular Heckscher, *Sveriges Ekonomiska Historia*, pp. 644–91; Staffan Högberg, *Utrikeshandel och sjöfart på 1700-talet. Stapelvaror i svensk export och import 1738–1808* (Lund, 1969) and Kurt Samuelsson, *De stora köpmanshusen i Stockholm 1730–1815* (Stockholm, 1951). None of these older studies, however, with the partial exception of Heckscher, concerned themselves with international markets in the way that Hildebrand did in 1958. In recent years, however, a new generation has addressed these issues more directly. Leos Müller has analysed the sale of Swedish iron on the Amsterdam market in the seventeenth and eighteenth centuries in his *The merchant houses of Stockholm, c.1640–1800* (Uppsala, 1998) and explored the links between the development of the American market and Swedish shipping policy in the nineteenth century in his *Consuls, corsairs, and commerce: the Swedish consular service and long-distance shipping, 1720–1815* (Uppsala, 2004). Åsa Eklund has studied the regional distribution of Swedish iron on the British market in her 'Iron production, iron trade and iron markets. Swedish iron on the British market in the first half of the eighteenth century', (Licenciate thesis, Department of Economic History, University of Uppsala, 2001).

eighteenth century, while the Swedes have shown little taste for tracking the routes taken by iron from *Bergslagen* once it had been shipped from Stockholm and Gothenberg. This book seeks to step into this historiographical no-man's-land and open a dialogue between the two national literatures, one that can enrich both. In doing so, we draw upon a conceptual construct first deployed by Immanuel Wallerstein and his World-System school, that of the global commodity chain (GCC).[34] At its simplest, a GCC can be defined as a 'network of labour and production processes whose end result is a finished commodity'.[35] But what is to be gained by tracing an apparently common-sense sequence of events? Firstly, there is much analytical value in following a GCC along its entire length, appraising each link or node that it contains, for this allows an analysis of economic activity that crosses conventional sectoral boundaries. The 'sequential stages of input acquisition, manufacturing, distribution, marketing, and consumption' are considered as a whole. The GCC model is also to be commended for highlighting issues of 'territoriality' and 'governance'. Global commodity chains, by their very nature, traverse national frontiers and thereby pose questions about why certain functions are spatially distributed in the way that they are. The dispersal or concentration of activity has to be accounted for. So too does the concentration of authority—for decision-making and profit extraction are powers that are spread unevenly, often very unevenly. Finally, the GCC model is valuable for its willingness to relate different levels of social organisation one to another. A GCC comprises 'sets of interorganizational networks clustered around one commodity or product, linking households, enterprises, and states to one another within the world-economy'.[36]

Conceiving of the world economy in the early modern era as a set of interlaced global commodity chains is helpful. It lends shape and direc-

---

[34] In what follows we draw upon the discussion of recent developments in GCC theory in the editors' introduction to S. Reimer and A. Hughes (eds), *Geographies of commodity chains* (London, 2003), and Peter Dicken, Philip F. Kelly, Kris Olds and Henry Wai-Chung Yeung, 'Chains and networks, territories and scales: towards a relational framework for analysing the global economy', *Global Networks*, I, 2 (2001), 89–112.

[35] Thomas K. Hopkins and Immanuel Wallerstein, 'Commodity chains: construct and research', in Gary Gereffi and Miguel Korzeniewicz (eds), *Commodity chains and global capitalism* (Westport CT, 1994), p. 17.

[36] Gary Gereffi, Miguel Korzeniewicz and Roberto P. Korzeniewicz, 'Introduction: global commodity chains', in Gereffi and Korzeniewicz, *Commodity chains*, p. 2.

tion to commercial networks that might otherwise remain analytically amorphous and vapid. But that is not to say that GCC theory supplies automatic answers to the questions that it poses. On the contrary, there appears to be no satisfactory method of accounting for shifts in the territoriality of production or the overturn of governance structures. The GCC model, despite foregrounding dynamism as a distinguishing feature of the capitalist world economy, does not provide a theoretically grounded explanation of that dynamism, other than by invoking abstract and empirically questionable macro-level phenomena such as Kondtradieff waves. For that reason, our use of the commodity chain concept is largely as an organising metaphor; it allows us to explore the multiple transactions and physical transmutations that *inter alia* took metallic matter from *Bergslagen* ore pits to the rice fields of the Carolinas. We are not committing ourselves to the more prescriptive features of world-system theory, those that see the eighteenth-century world economy as irrevocably structured in concentric socio-geographical zones that turned around Amsterdam or London.[37] The volume of recent research that argues for a multi-centred world economy, one in which European domination was not destiny foretold, precludes that.[38] The commodity chains that passed through Basra, Surat or Molucca were multi-directional; they did not converge on London or Amsterdam.

---

[37] For a survey of world-system theories see Fernand Braudel, *Afterthoughts on Material Civilization and Capitalism* (Baltimore, 1977); Immanuel Wallerstein, *The modern world-system: capitalist agriculture and the origins of the European world-economy in the sixteenth century* (New York, 1974) and *The modern world-system II: mercantilism and the consolidation of the European world-economy, 1600–1750* (New York, 1980); Giovanni Arrighi, *The long twentieth* century (1994).

[38] See K.N. Chaudhuri, *Trade and civilisation in the Indian Ocean: an economic history from the rise of Islam to 1750* (1985); Sushil Chaudury and Michel Morineau (eds), *Merchants, companies and trade: Europe and Asia in the early modern era* (Cambridge, 1999); Andre Gunder Frank, *ReOrient: global economy in the Asian age* (1998); Jack A. Goldstone, 'Efflorescences and economic growth in world history: rethinking the 'Rise of the West' and the Industrial Revolution', *Journal of World History*, XIII, 2 (2002), 323–89; Kenneth Pomeranz, *The great divergence: China, Europe, and the making of the modern world economy* (2000); John K. Thornton, *Africa and Africans in the making of the Atlantic world 1400–1800* (Cambridge, 1998). See also, for scepticism about the role of overseas trade in European development, Patrick O'Brien, 'European economic development: the contribution of the periphery', *Economic History* Review, XXXV (1982), 1–18.

Trade in the Early Modern World

What was the significance of trade in the pre-industrial age? A word of warning from Ferdnand Braudel should be kept in mind. Early modern peasants, he claimed, 'lived in their villages in an almost autonomous way, virtually in an autarchy', alongside yet separate from a market-orientated economy. There were 'two universes, two ways of life foreign to each other'.[39] Trade affected only a fraction of Europe's people, four-fifths of whom were engaged in agriculture in 1700. Most produced little in the way of surplus, and those that did usually saw it diverted into the pockets of 'unproductive aristocrats and rulers'.[40]

By 1700, however, change was afoot. Braudel's distinction becomes too emphatic. The market economy did not float, like a film of oil, atop a sea of self-subsisting peasant households. The phenomenon of proto-industrialization depended precisely upon the integration of peasant households into long-distance trading networks, upon structural affinity not repulsion between the two spheres. Many families devoted slack periods in the agricultural calendar to the making of textiles, metalwares or wooden goods. Merchants who were equipped with the financial resources and *savoir faire* that most farming households lacked would market the finished goods, enabling peasant communities to augment their subsistence. In such ways were peasant weavers from Silesia to Ulster harnessed to international markets.[41] Proto-industrialization is one example of what Jan de Vries has identified as the 'industrious revolution' of the seventeenth and eighteenth centuries: namely, a redivision of household labour in which individuals engaged with the market economy, seeking money wages as a means of increasing household income and domestic comfort.[42] Why, some Europeans asked themselves, grind corn if you can earn the money to buy bread?

---

[39] Braudel, *Afterthoughts*, pp. 5–6.

[40] Sheilagh Ogilvie, 'The European economy in the eighteenth century', in T.C.W. Blanning (ed.), *The eighteenth century: Europe 1688–1815* (Oxford, 2000), p. 95. See also Jan de Vries, *European urbanization 1500–1800* (London, 1984) for the relatively low levels of urbanization at the start of the eighteenth century.

[41] Sheilagh Ogilvie and Markus Cerman (eds), *European proto-industrialization* (Cambridge, 1996).

[42] Jan de Vries, 'The industrial revolution and the industrious revolution', *Journal of Economic History*, LIV (1994), 248–71. See also Jan de Vries, 'Between purchasing power and the world of goods: understanding the household economy in early modern Europe', in John Brewer and Roy Porter (eds.), *Consumption and the world of goods*, London 1993, 85–132.

Embracing the 'industrious revolution' was not something that could be done arbitrarily or unilaterally, however. There were preconditions. 'Industrious' households required a context, that of bouyant markets, an advanced social division of labour, and—as is implied by the two foregoing conditions—a relatively high level of urbanization. North-western Europe met those conditions. Demand grew prodigiously. Europe underwent a substantial rise in population in the eighteenth century; the 81 million inhabitants of 1700 had become 123 million by 1800. Admittedly, there was not a corresponding rise in the level of urbanization—that scarcely rose at all—but what appeared as pan-European urban stagnation masked major regional disparities. While parts of Southern Europe experienced urban decay there was a dramatic growth of the non-agricultural population in the north and west of the continent. Britain was the outstanding example. Its population (taking in that of Ireland) rose from nine to sixteen millions during the eighteenth century. One-fifth of that population lived in towns with more than 10,000 inhabitants, double the European average, by 1800.[43]

This tilt to the north and west will be familiar to readers of Braudel, who long ago described the early modern European economy as having successive focal centres, each marking a gradual shift from the Mediterranean to the Atlantic. In brief, the Venice of 1450 was surpassed by the Antwerp of 1550. War and political upheaval in the 1570s and 1580s may have extinguished Antwerp's greatness, but Antwerp's eclipse merely facilitated the rise of Amsterdam and the inauguration of Holland's Golden Age.[44] Dutch hegemony over international trade was lengthy, stretching across the seventeenth century. Yet capitalism abhors fixity, so after 1713 it was London that rivalled and then surpassed Amsterdam as the organising centre of European and, for that matter, global commerce.[45]

---

[43] De Vries, *European urbanization*, pp. 36–39.

[44] Braudel, *Afterthoughts; idem, Civilization and capitalism: 15th–18th century. Volume 3: The perspective of the world* (Berkeley, 1992). For the rise of the Dutch see Jonathan Israel, *The Dutch Republic: its rise, greatness, and fall 1477–1806* (Oxford, 1995), and Jan de Vries and Ad van der Woude, *The first modern economy: success, failure, and perseverance of the Dutch economy, 1500–1815* (Cambridge, 1997).

[45] See the summary of developments in Patrick O'Brien, 'Inseparable connections: trade, economy, fiscal state, and the expansion of empire', in P.J. Marshall (ed.), *The Oxford history of the British Empire. Volume II: the eighteenth century* (Oxford, 1998), pp. 60–63.

These developments reflected the rise of an Atlantic economy in the sixteenth and seventeenth centuries. The landfall of Spanish adventurers in the Caribbean in 1492 opened a radically new chapter in the history of both the Old World and the New. The southward thrust of Portuguese navigators along the coast of Africa was no less epochal: it took them to India and, less advertantly, to Brazil. Europeans had hitherto existed on the bleak western margins of a Eurasian economy whose pulse was set in China, India, and the Levant. With the opening up of the Atlantic basin an entirely new arena was added to the world economy and the once peripheral Europeans now found a role of their own. The Americas furnished the precious metals that allowed the Portuguese (and then the Dutch) to buy their way into the Asian spice trade. More importantly, the New World offered an environment in which a range of exotic commodities could be produced on a massive scale. Some, like chocolate, were entirely novel. Others, like sugar and coffee, had been obtainable from the Levant, but only in limited quantities. Now they could be grown in large volumes under the auspices of European planters. Other articles, such as deerskins and beaver pelts from North America's boundless forests, proved to be excellent substitutes for expensive Old World commodities.

Pelts and hides were supplied by Indian trappers who adapted their existing hunting patterns to fit in with European demand, but most New World products depended upon intensive plantation agriculture.[46] The labour demands of such a system were formidable: far too high, in fact, to be met by native populations, especially after Old World pathogens brought about demographic collapse in the decades after the first European contacts. Nor could free European migrants be induced to undertake the gruelling labour of sugar harvesting in sufficient numbers. The labour needs of the plantations could only be satisfied through force. Coercing Native Americans or Europeans proved impractical: the former were too few or too elusive, whilst the supply of European candidates for forced labour—condemned criminals

---

[46] A vast literature addresses this issue, from which it is invidious to select just a sample, but see Robin Blackburn, *The making of New World slavery: from the baroque to the modern, 1492–1800* (1997) for a synoptic interpretation; Ira Berlin, *Many thousands gone: the first two centuries of slavery in North America* (Cambridge MA, 1998) for developments in what was to become the USA; and Herbert S. Klein, *African slavery in Latin America and the Caribbean* (New York, 1986), for developments to the south.

or prisoners of war—was fitful and unpredictable.[47] Africa offered an alternative. Not only were there zones of sub-Saharan Africa that were densely populated, but slavery was widely recognised as a legal condition. Before 1500, however, the *trade* in slaves was relatively limited, and most slaves were women and children employed domestically, not adult males engaged in collective agriculture. Moreover, the busiest export routes ran east and north, to the Islamic world and the Indian Ocean.[48] Insofar as Europeans engaged in the trade, as the Portuguese did in the late fifteenth century, they did so on a small scale. The Luso-Hispanic incursion into the New World changed all that. The African slave trade was transformed in orientation and intensity: the Atlantic supplanted the Indian Ocean, and the steady outflow to the east became a quickening torrent to the west. As the Dutch, French and English established their own transatlantic settlements in the seventeenth century the slave trade gathered pace. At the start of the sixteenth century the Portuguese were shipping about 2,000 captives annually; by the start of the eighteenth century European slave traders were consigning nearly 36,000 a year.[49]

By 1700 it was possible to speak of an integrated Atlantic system. It was characterised by the ecological transformation of those parts of the Americas that were fit for plantation agriculture, and it depended upon massive infusions of African labour. Slaves were procured by trading European or Asian-made textiles, metalwares and fancy goods with African merchants along the Guinea and Angolan coasts. In the African interior the insatiability of American demand prompted the emergence of predatory political formations—states for which slave gathering was a *raison d'etre*. In the Americas, slavery flourished best in tropical or semi-tropical zones, but its influence was also felt in more temperate regions of the New World where European settlers grew prosperous by growing foodstuffs to nourish slaves. Farmers in the Delaware valley, for example, were not paragons of homespun self-sufficiency; they sold their surpluses to Philadelphia factors for shipment to the sugar islands of the Caribbean. Similarly, New Englanders traded salted

[47] For the problems of enslaving Native Americans see Klein, *African slavery*, pp. 41, 83–84, and Alan Gallay, *The Indian slave trade: the rise of the English empire in the American south, 1670–1717* (New Haven CT and London, 2003).

[48] Ronald Segal, *Islam's Black slaves: the history of Africa's other Black diaspora* (2002).

[49] Herbert S. Klein, *The Atlantic slave trade* (Cambridge, 1999), pp. 208, 210.

cod for sugar and molasses, processing the latter into rum.[50] Europe's role in this widening web of commerce was twofold. Firstly, it was the principal source of the manufactured goods that flooded into both the African and the American segments of the Atlantic economy. Secondly, Europe was the principal market for New World commodities. Indeed, the material culture of Europe—its very style of life—was transformed: exotic groceries enlivened Europe's palate, Atlantic hardwoods introduced variety into its domestic interiors, and tropical dyestuffs such as indigo extended its colour spectrum.[51]

But above all, Europe was at the centre of a process of hemispheric capital accumulation. European planters repatriated the fortunes that their African chattels had produced; European manufacturers enjoyed the profits that the sales of their goods in African and colonial marts had generated; and European merchants and brokers took a disproportionate share of the earnings to be had in shipping, insuring, and handling the cargoes, animate and inanimate, that were carried back and forth across the Atlantic. It is the extent and significance of this capital accumulation that has driven the historiography of the Atlantic world forward in recent years. That Atlantic enterprise was of central importance for the emergence of the modern world was the thesis, masterfully expressed, of Eric Williams, whose *Capitalism and slavery* (1944) posited a firm, direct relationship between the slave economies of the Caribbean and industrialization in Britain. Enslaved Africans, Williams argued, had not merely added to the wealth of their owners, they had generated the new investment capital that made the Industrial Revolution possible. Such an argument, advanced by a West Indian anti-colonial intellectual, did little for the imperial *amour propre* of most British historians. As a result, the Williams thesis was subjected to sustained criticism, with most critics focusing upon the extent to which repatriated profits from the Caribbean were ploughed into the new technologies of Britain's Industrial Revolution. Since the empirical difficulties of demonstrating that planters tended to sink their wealth in the woollen mills of the West Riding were considerable, the critics felt that the inadequacy of the Williams thesis had been sufficiently exposed.

---

[50] John J. McCusker and Russell R. Menard (eds), *The economy of British North America* (Chapel Hill NC, 1985).

[51] Maxine Berg, *Luxury and pleasure in eighteenth-century Britain* (Oxford, 2005).

Since the 1970s, however, a new Atlantic historiography has given fresh force to Williams's hypothesis about slavery and industrialization. For one thing, there has been a vast growth in the study of slavery in the Atlantic world. Impressionistic or catch-penny accounts of the slave trade have been superceded by systematic investigation and statistical rigour.[52] The new knowledge that has emerged has done nothing to diminish the ethical enormity of the 'Guinea Trade' but much to document its far-reaching ramifications. The slave trade was a colossal enterprise, involving the transportation of approximately 12 million Africans between the 1440s and the 1870s. It absorbed a vast amount of shipping, commanded the labour of thousands of maritime workers, and summoned into existence skein-like supply networks that carried goods to Liverpool or Nantes, Ouidah or Calabar, Havana or Charleston. These supply chains stretched for thousands of miles; some, as we shall see, began in places such as Gammelbo, a tiny iron making community in central Sweden.

An appreciation of the slave trade's tentacular reach has changed the terms of debate about Atlantic slavery's relationship to the economic development of Europe. This is no longer conducted on the narrow ground of whether the fruits of plantation slavery were invested in new industries (or squandered in conspicuous consumption). It is the logic of Atlantic slavery *as a system* that now seems central to European—and above all British—advancement. The slave trade was a powerful force for transoceanic integration. The infamous triangular trade, by its very nature, brought different parts of the Atlantic littoral into permanent contact with one another. Moreover, the slave system encouraged the spread of market relationships around the entire Atlantic zone. The demand for food on the sugar islands (where every available acre was devoted to raising cane) stimulated farmer-settlers along the Delaware and Susquehanna rivers to grow grains and legumes for export, as has already been noted. Likewise, African farmers along the lower Niger harvested more and more yams in response to the demands of the slave captains at Bonny and Calabar for bulk carbohydrate to sustain their captives during the Middle Passage. By expanding the number of circum-Atlantic agriculturists who produced for the market the slave system drove up the number of potential consumers for manufactured

---

[52] David Eltis, Stephen D. Behrendt, David Richardson, and Herbert S. Klein (eds), *The transatlantic slave trade: a database on CD-ROM* (Cambridge, 1999).

goods. Nowhere was this clearer than in British North America, which took just 6 per cent of English exports in 1700–01, but 32 per cent in 1797–98.[53] The surge of English goods could only be paid for by exporting American crops and commodities to the Caribbbean, or by building the ships—a New England specialism, this—that triangulated the ocean. In these ways, the Atlantic slave system became more than an adjunct to the economic life of Europe; by extending, diversifying and integrating markets around the Atlantic basin it became the mechanism through which Europe's economy was transformed.[54]

Our understanding of this complex system of transoceanic exchange and reciprocation has had a long and troubled gestation. Advocates of the 'Atlantic' as a historical subject *sui juris* have had to contend with older, nationalist historiographies that concentrated on the exploits of Portuguese or British empire-builders, or the prehistory of the United States. Atlantic histories that self-consciously transgress imperial boundaries have been slow to emerge.[55] Indeed, the first attempts to build a pan-Atlantic historiography foundered. The notion of an 'Atlantic civilization' had been broached by Jacques Godechot and Robert Palmer in the 1950s. Did the fact that revolutionary upheavals struck both America and France in the late eighteenth century justify talk of an Atlantic or a 'democratic' revolution, stemming from a common transatlantic experience? Godechot and Palmer thought so, but their suggestion was not always warmly received in the age of the Cold War; for too many it appeared as an ideological cover for NATO. Nor did the notion of a broadly-based 'Atlantic Revolution' appeal to those French historians who were temperamentally committed to the uniqueness of 1789.[56] Other historians took inspiration from Fernand

[53] Kenneth Morgan, *Slavery, Atlantic trade and the British economy, 1660–1800* (Cambridge, 2000), p. 19. The later figure refers to exports from Britain as a whole, not just from England. It should also be noted that Britain's Caribbean possessions, home to 450,000 slaves in the later eighteenth century (and an elite of super-wealthy planters), absorbed 25 per cent of British exports in their own right in 1797–98.

[54] Joseph E. Inikori, *Africans and the Industrial Revolution in England: a study in international trade and economic development* (Cambridge, 2002).

[55] For surveys of the historiography see Bernard Bailyn, *Atlantic history: concept and contours* (Harvard, 2005), and Horst Pietschmann, 'Introduction: Atlantic history—history between European and global history', in Horst Pietschmann (ed.), *Atlantic history: history of the Atlantic system 1580–1830* (Göttingen, 2002), pp. 11–54.

[56] Jacques Godechot, *France and the Atlantic Revolution of the eighteenth century* (1965); Robert R. Palmer, *The age of the democratic revolution* (2 vols., 1959–64). The debate continues to resound, with the latest contribution a firm denial that Anglo-America and France

Braudel, whose vision of the Mediterranean as a single, indivisible civilisation had been dazzlingly expressed in his study of the Mediterranean world in the age of Philip II.[57] Huguette and Pierre Chaunu's monumental *Séville et l'Atlantique (1504–1650)*, published in 11 volumes between 1955 and 1959, owed a plain debt to Braudel and the *Annales* school. Nevertheless, what the Chaunus had embarked upon was a statistical investigation of Spain's trade with her overseas possessions, reliant upon a single national archive. Vast though it was, their work could not match the transnational sweep or thematic range of Braudel's. Nor was it clear that the Atlantic, an ocean that was scene to some of the most dramatic changes in the early modern world, lent itself to the conceptual vocabulary of the *Annalistes*. Braudel's preference for *l'histoire immobile*, in which the pace of chance was glacial, seemed ill-suited to the turbulent western ocean.

The dynamism of the Atlantic economy in the eighteenth century was eminently suited, however, to the intellectual agenda set by globalization in the closing decades of the twentieth century. Flux and turmoil, ever-changing transnational production networks, cultural hybridity, religious syncretism, and movements of peoples and goods that overspread national boundaries: these were as characteristic of the early modern Atlantic as they are of the contemporary global economy.[58] It has been this that has given force to contemporary scholarship on the Atlantic world and led to talk of a full-blown Atlantic History paradigm.[59] Proponents of the new Atlantic history are far less mindful of imperial structures and far more sympathetic to the notion of a single, culturally fluid, polyglot and spontaneously integrated Atlantic world. Here was a multitude of peoples, an infinity of things, a babble of tongues, and an inexhaustible medley of faiths and beliefs. These interwove, trespassed upon one another, and multiplied fruitfully in ways that were essentially

---

were yoked together in a common Atlantic Enlightenment: Gertrude Himmelfarb, *The roads to modernity: the British, French, and American Enlightenments* (New York, 2004).

[57] Fernand Braudel, *La Méditerranée et le monde méditerranéen à l'époque de Philippe II* (1949). An English translation was not issued until 1972, which did much to mute Braudel's influence on Anglo-Saxon scholarship.

[58] The literature of globalization is so vast as to defeat any attempt at citation, but if one place of reference has to be given it should be Manuel Castells's magisterial *The information age: economy, society and culture* (3 volumes, Oxford, 1996–1998).

[59] The literature is so fast-moving that it is best monitored online: for regular updates see the 'Bibliography in Atlantic History' at www.fas.harvard.edu/~atlantic/atlantbib.html. The emergence of a specific paradigm is mooted in S.D. Smith, 'The Atlantic History paradigm', *New England Quarterly* (March 2006), 123–33.

ungovernable. The ocean was not a blank space to be quartered and divided by imperial administrators; it was a fluid environment inhabited by traders, refugees, slaves, sailors, scientists, and religious seekers who habitually evaded mercantilist regulation and bypassed state edict.[60] As a result, there is now a plurality of Atlantics—black, green, red, proletarian, Quaker and Calvinist, and criminal—jostling for historical attention.[61] Likewise, a medley of methodologies and conceptual tools compete for historical business. Should the history of the ocean be Trans-Atlantic, Circum-Atlantic, or Cis-Atlantic?[62]

The old imperial divisions have not been erased from historical scholarship—indeed, the dialogue between the Anglo-Saxon Atlantic and the Iberian Atlantic remains stumbling and irregular[63]—but the study of empire has taken new paths. Some historians, taking their cue, perhaps, from modern commentators who have asked what future the nation state has amid the surge and counter-eddy of globalization, have been led to ponder the origins of those nation states, such as Great Britain and the United States, that rose from the Atlantic cauldron.[64] The question of how an archipelago without dynastic, confessional, ethnic or linguistic homogeneity came to be moulded into a 'United

[60] For a programmatic statement see David Hancock, 'The British Atlantic world: co-ordination, complexity, and the emergence of an Atlantic market economy, 1651–1815', *Itinerario: European Journal of Overseas History*, XXIII (1999), 107–26. For examples of work in this vein see Robin Law and Kristin Mann, 'West Africa in the Atlantic community: the case of the Slave Coast', *William and Mary Quarterly*, 3rd ser. LVI (1999), 307–34, and Ira Berlin, 'From Creole to African: Atlantic Creoles and the origins of African-American society in mainland North America', *William and Mary Quarterly*, 3rd ser. LIII (1993), 251–88, and some of the essays in Peter A. Coclanis (ed.), *The Atlantic economy during the seventeenth and eighteenth centuries: organization, operation, practice, and personnel* (Columbia SC, 2005).

[61] Gail D. MacLeitch, '"Red" labor: Iroquois participation in the Atlantic economy', *Labor: Studies in Working-Class History of the Americas*, I, 4 (2004), 69–90; Peter Linebaugh and Marcus Rediker, *The many-headed hydra: the hidden history of the revolutionary Atlantic* (2000); Gwenda Morgan and Peter Rushton, *Eighteenth-century criminal transportation: the formation of the criminal Atlantic* (Basingstoke, 2004).

[62] These are the approaches identified by David Armitage in 'Three concepts of Atlantic history', in David Armitage and Michael J. Braddick (eds), *The British Atlantic world, 1500–1800* (Basingstoke, 2002), pp. 11–27.

[63] J.H. Elliott, *Empires of the Atlantic world: Britain and Spain in America 1492–1830* (New Haven CT and London, 2006) provides a comparison of two imperial experiences.

[64] J.G.A. Pocock, 'British history: a plea for a new subject', *Journal of Modern History*, XLVII, 4 (1975), 601–28; *idem*, 'The limits and divisions of British history: in search of an unknown subject', *American History Review*, LXXXVII (1982), 311–36; David Cannadine, 'British history: past, present—and future?', *Past and Present*, 116 (1987), 169–91.

Kingdom' between the sixteenth and eighteenth centuries has attracted huge interest.[65] Likewise, the 'British-ness' of the Anglophone colonies in the New World has provoked prolonged debate. To what extent, it has been asked, did denizens of Britain and her Atlantic colonies share a common mentality? Were there not political vocabularies, religious affinities, and cultural practices that spanned the ocean? If so, little reliance can be placed upon the old historiographical reflex of viewing the history of colonial North America as no more than a prelude to the American Revolution, with independence as the pre-given outcome.[66]

To focus in this way upon the American Revolution—as an event in British history, stemming from ideological disputes that were distinctively British—is to return to the question of empire. It is to ask afresh about the role of the state. The tendency of the new Atlantic historiography is very much to downplay imperial structure. Two examples, from two very different points on the historiographical compass, can illustrate that. David Hancock presents a vision of the Atlantic economy as a scene of ebullient commercial endeavour, swarming irresistibly over hapless officialdom. Although not unmindful of the Atlantic's cruelties, this is an essentially positive portrayal that emphasises collaboration and mercantile networking as unifying the ocean. Human agency takes precedence over structure.[67] Peter Linebaugh and Marcus Rediker offer an alternative model, although for them, as for Hancock, the early modern Atlantic was a place of restless activity, driven by the aspirations of human actors. The difference lies in the human actors selected for study. Linebaugh and Rediker deal with the wretched and the outcast, not the sleek merchants analysed by Hancock; their Atlantic is a place riven by ferocious class struggles. The ocean was convulsed by repeated multi-ethnic rebellion as slaves, indentured servants and

[65] See Linda Colley, *Britons: forging the nation, 1707*–1837 (1992); Steven G. Ellis and Sarah Barber (eds), *Conquest and union: fashioning a British state, 1485–1735* (1995); Brendan Bradshaw and John Morrill (eds), *The British problem, c. 1534–1707* (1996); Laurence Brockliss and David Eastwood (eds), *A union of multiple identities: the British Isles, c. 1750–c. 1850* (1997).

[66] For two very different yet landmark additions to the literature see Ian K. Steele, *The English Atlantic, 1675–1740: an exploration of communication and community* (Oxford, 1986) and T.H. Breen, 'An empire of goods: the anglicization of colonial America, 1690–1776', *Journal of British Studies*, XXV (1986), 467–99.

[67] Hancock, 'The British Atlantic world'; *idem*, *Citizens of the world: London merchants and the integration of the British Atlantic community, 1735–1785* (Cambridge, 1995).

maritime proletarians of various lands and racial hues joined together to repudiate Atlantic capitalism.[68]

Yet by emphasising Atlantic insurgency Linebaugh and Rediker give fresh prominence to the state, for it was state-organised coercion that was brought to bear on the masterless men, maroons, and runaway mariners who threatened the process of capital accumulation. Capital was accumulated best in an orderly environment; those who disrupted the process, be they antinominian pirates or mutinous slaves, were to be gibbeted. This reminder of the sanguinary nature of the eighteenth-century state is salutary, for the state had an inescapable presence in the Atlantic trading system, as a well-established literature attests.[69] Mercantilist regulation was the resort of every European power. It prescribed, for good or ill, the course of commercial development. Imperial bureaucracies specified the ports through which trade would run; customs officers defined what was contraband. Legal codes, such as Sweden's *Produktplakat*, granted privileges to national shipping fleets and preferential employment to native seamen. One of the reasons that large merchant fleets were smiled upon was that they were seen as a nursery for the seamen who were to man naval vessels in time of war. As one English commentator explained, 'without that naval force which trade produces, we shall be constantly exposed to the insults and invasions of our neighbours'.[70] Indeed, success in trade was intimately connected with naval aggression. Commercial profits were underwritten by military protection. Maintaining a fleet was expensive, but increases in trade led to an increased tax base, which laid the basis—at least for those states that could harvest the revenues efficiently—for a renewed cycle of imperial advance.

The state was of critical importance. If that was true as a general proposition, it applied *a fortiori* in the case of the international iron

[68] Linebaugh and Rediker, *The many-headed hydra*. For a critique, see Nicholas Rogers, 'Archipelagic encounters: war, race, and labor in American-Caribbean waters', in Felicity A. Nussbaum (ed.), *The global eighteenth century* (Baltimore, 2003), pp. 211–25, and Arne Bialuschewski, 'Between Newfoundland and the Malacca Strait: a survey of the golden age of piracy', *The Mariners' Mirror*, XC, 2 (2004), 167–86.

[69] John Brewer, *The sinews of power: war, money and the English state, 1688–1714* (1989); Daniel A. Baugh, 'Maritime strength and Atlantic commerce: the uses of "a Grand Marine Empire"', in Lawrence Stone (ed.), *An imperial state at war: Britain from 1689 to 1815* (1994), pp. 185–223; O'Brien, 'Inseparable connections'.

[70] Charles Davenant, *An essay upon ways and means* (1695), quoted in William J. Ashworth, *Customs and excise: trade, production, and consumption in England 1640–1845* (Oxford, 2003), p. 87.

trade, for the context in which iron was produced and exchanged in Europe and the Americas was defined by the rise and fall of empires. The ascendancy of Britain in the early eighteenth century coincided with the dismemberment of Sweden's Baltic empire and the aggrandisement of Sweden's nemesis, the Russia of Peter the Great. These contrasting imperial trajectories were of major consequence for the makers and traders of iron: new centres of demand opened up and new supply chains were laid down.

Baltic Transformations

Sweden's emergence as a great power in the seventeenth century was rooted four-square in mining and metal processing. Swedish copper and iron had been exported since the middle ages, but the shipments that went to Danzig and other cities of the southern Baltic in the sixteenth century were rather modest. From the 1620s they underwent a revolution in scale and scope. Iron exports, which had averaged little more than 3000 tons per annum in the late 1620s, leapt to 11,000 tons in 1640, then to 18,000 tons in 1650, and 27,000 tons in 1680. Their destination changed too. Swedish iron now passed through the Sound in large volumes, bound for the Dutch Republic, the gravitational centre of North European commerce.[71]

This startling escalation was a matter of policy. The Swedish state entertained territorial ambitions that could only be fulfilled if the poor and sparsely populated kingdom of Gustav II Adolf (Gustavus Adolphus) could exploit its latent mineral wealth more effectively. For their part, a group of Amsterdam-based merchants were alert to the advantages that preferential access to Swedish copper and iron would give them. The Dutch Republic in its Golden Age, with its busy shipyards and bustling towns, consumed iron on a grand scale. Yet the Thirty Years' War (1618–1648) disrupted the supplies of German iron that usually came down the Rhine, and jeopardised the flow of metalwares from the Spanish Netherlands. Iron was needed in the capital-rich Netherlands; iron was to be had in capital-poor Sweden. This realisation spurred the intervention of Louis De Geer, Willem de Besche and other Dutch merchants in the 1620s, heralding a transformation of Sweden's

[71] Hildebrand, *Fagerstabrukens historia*, pp. 35–59.

industries. The Dutchmen were awarded wide-ranging privileges by the Swedish state, allowing them to establish a network of processing plants. The newcomers were therefore able to take control of Sweden's copper resources (which were Europe's richest), set up cannon foundries at a time when endemic warfare made the gun trade especially lucrative, and to re-direct Swedish iron exports westwards. The greatly increased export revenues that accrued to the Swedish state enabled Gustavus Adolphus to make his sensational entry into the Thirty Years' War in 1628 and humble the Catholic-Habsburg cause in Germany. It was this twin military-industrial initiative that ushered in Sweden's 'Age of Greatness' (*stormaktstiden*).[72]

The transfiguration of Swedish iron making in the mid seventeenth century involved more than an influx of Dutch capital. It was based upon a profound alteration in the social matrix of iron production. Traditionally, iron making was the work of peasant miners (*bergsmän*) who smelted ore at communally owned furnaces and then refined the pig iron into crudely shaped lumps of *osmund* iron. It was this *osmund* iron that was exported to Danzig, Lübeck, and other commercial centres to the south. And it was forge owners in Danzig and elsewhere who had the *osmund* iron drawn out into bars, the form that malleable iron took as an international commodity. Changes imposed by the Swedish state from the 1620s onwards were intended to improve the quality of iron made in *Bergslagen*, and to ensure that the production of bars—the high value-added part of the production process—was carried out in Sweden. A new social division of labour was introduced. Henceforth, *bergsmän* were restricted to the smelting of ore, while the refining of pig iron was entrusted to a new class of professional ironmasters (*brukspatroner*). The *brukspatroner*, deploying greater capital resources and a more specialised workforce, were charged with improving the quality of output. The export of the finished bars was to be the province of international merchants based in specified trading centres. Every town in Sweden was allotted its particular place in an ordered urban hierarchy (the so-called *Stapelstadssystem*). Heading the hierarchy were 24 towns, the *stapelstäder*, through which overseas trade was to be channelled; below

[72] For the relationship between the Swedish state and Dutch entrepreneurs see M.-B. Nergård, *Mellan krona och marknad. Utländska och svenska entreprenörer inom svensk järnhantering från ca 1580 till 1700* (Uppsala, 2001). See also Göran Behre, Lars-Olof Larsson, and Eva Österberg, *Sveriges historia 1521–1809. Stormaktsdröm och småstadsrealitet* (Stockholm, 2001), pp. 190–94.

them came the *uppstäder*, towns that were restricted to internal trade. Of the *stapelstäder*, two were of commanding importance: Gothenberg, founded in 1624, was the outlet for iron from the western county of Värmland, whilst the output of the older mining areas to the north of lake Mälaren was funnelled through Stockholm. The entire production process, from forest clearings to the Stockholm quayside, was policed by a special state agency, the *Bergscollegium* (Board of Mines), founded in 1649.[73]

The establishment of the *Bergscollegium* coincided with the close of the Thirty Years' War. The peace of Westphalia, sealed in 1648, confirmed Sweden as the arbiter of Northern Europe. Sweden's provinces on the eastern shore of the Baltic were extended and consolidated; new territories in Northern Germany were acquired. Riga and Bremen were Swedish cities; the marshy delta on which St Petersburg would one day be built was an as yet insignificant corner of the Swedish province of Ingermanland. It was a striking vindication of the Swedish state's distinctive blend of military mobilisation and industrial *dirigisme*. Yet Swedish power, for all its martial lustre, was insecure. Despite a considerable increase in numbers during the seventeenth century, Sweden remained thinly populated; Charles XII (d. 1718), the last of the great warrior kings, had no more than 1.5 million subjects.[74] Sweden's enemies, on the other hand, were numerous. Were they to combine—as the Russians, the Poles, and the Danes did in 1699—the consequences might be severe. The Swedish crown lacked the manpower to compensate for repeated battlefield losses, so when Charles XII's principal field army was annihilated at Poltava in the Ukraine in 1709 the curtain fell on *stormaktstiden*. The Treaty of Nystad, which concluded the Great Northern War in 1721, brought humiliation. The Baltic empire was lost, ceded for the most part to Russia.

If the collapse of Swedish power was dramatic, so was the near simultaneous advance of Britain. The 'Glorious Revolution' of 1688 had turned Britain, a peripheral actor in European affairs in the 1670s,

---

[73] Anders Florén and Göran Rydén, *Arbete, hushåll och region. Tankar om industrialiseringsprocesser och den svenska järnhanteringen* (Uppsala, 1992).

[74] There was, as will be seen in chapter 3, continuing growth in population, urbanization and agricultural productivity, before and after the end of *stormaktstiden*. See J. Myrdal, *Jordbruket under feodalismen 1000–1700* (Stockholm, 1999); S. Lilja, *Tjuvehål och stolta städer. Urbaniseringens kronologi och geografi i Sverige (med Finland) ca. 1570-tal til 1810-tal* (Stockholm, 2000); and C.-J. Gadd, *Den Agrara Revolutionen 1700–1870* (Stockholm, 2000).

into a key protagonist of Louis XIV's France. The Nine Years' War (1689–1697) and the War of Spanish Succession (1702–1713), fought to thwart Bourbon expansionism, brought about a major overhaul of the British state. Naval and military expenditure grew stupendously, sustained by new methods of public finance that allowed the British state to bring national resources to bear with an unmatched efficiency.[75] The outcome was extremely favourable. The British benefited massively from the Treaty of Utrecht, not least by the granting of the *asiento*, the exclusive right to supply slaves to Spain's American empire. But this triumphal extension of British commercial might in the Atlantic was accompanied by a thickening of Britain's links to the east, to the Baltic. The long struggle against France deepened an already existing dependence upon Swedish *matériel*.

Heightened military demand for malleable iron added to an already extensive civilian market, one buoyed by urban expansion and by slow but sure industrial growth. Then there was demand from colonial markets, which, although starting from a far lower base, grew prodigiously as plantation agriculture in the West Indies and British North America intensified. All of this required a greater volume of iron, but British ironmasters were unable to respond. Their industry was hobbled by a seemingly insurmountable supply problem: the volume of charcoal available for smelting and refining was strictly limited. Despite the careful husbanding of coppice woods, the industry had hit a production ceiling through which it could not break. In the fifty years between 1660 and 1710 the make of bar iron in England and Wales hovered around 13,000 tons. Since the demand for malleable iron moved relentlessly upwards in the same period there was a shortfall that had to be made good with iron from overseas. In 1660 imported iron already amounted to 57 per cent of domestic production. By 1680 imports had achieved virtual parity (96 per cent), and by 1700 foreign imports were equivalent to 127 per cent of domestic output.[76] English consumers had traditionally looked to Spain for additional supplies of iron, but the Basque iron industry was not capable of meeting a surge in demand on this scale, hence the turn to the Baltic.

[75] Brewer, *The sinews of power*; D.W. Jones, *War and economy in the age of William III and Marlborough* (1988).

[76] Calculated from figures presented in King, 'The production and consumption of bar iron', p. 23.

When Swedish iron first appeared on West European markets in the 1620s and 1630s it was funnelled through Amsterdam, the headquarters of Louis De Geer and the other Dutch merchants who had revamped Sweden's metallurgical industries. From the 1660s, however, the locus of the European iron market swung westwards, to England. At first this trade was managed by Dutch and Scottish factors, but by the late 1670s English merchants, mostly Londoners, had assumed control.[77] In 1700 the English market took 44 per cent of Stockholm's iron exports, and the Scottish market a further 5 per cent. Less than 25 per cent went to the once dominant Dutch Republic.[78]

By the 1720s Swedish imports to the British Isles were running at over 15,000 tons annually, edging past the output of Britain's own forge sector. Yet demand continued to spiral upwards, straining even Swedish capacity, so British merchants sought a fresh source of supply. They found it in Russia. Russian iron, shipped from St Petersburg, came to augment and eventually surpass Swedish iron. The rise of an export-orientated iron industry in Russia, as in Sweden, had its origins in military ambition. Peter the Great could not hope to supplant Sweden as the Baltic's leading power while Russia remained dependent upon its great rival for so strategic a material as iron. Thus, the first of a string of state-owned metal works was established in the Urals in 1699, far from the older centres of Russian metallurgy. It was a fateful first step. Seventy-one iron or copper works were established on the Russian empire's eastern edge in the first half of the eighteenth century.[79] The impact of Russian bar iron on the international market was at first muted, but after the Peace of Nystad exports from Siberia began in earnest. Iron from the Urals began to trickle into British ports in the late 1720s, mounting slowly in volume in the 1730s, and taking off in the middle years of the century.

By the 1730s, then, the British iron market had become an arena in which a variety of different irons contended. The locally made product was confronted by Swedish imports, Russian iron had made its debut, and Spanish iron retained a small but significant market share. Bar iron was an ostensibly prosaic material, but the brands that were offered

[77] Sven-Erik Åström, *From cloth to iron: the Anglo-Baltic trade in the late seventeenth century. Part 1: The growth, structure and organization of the trade* (Helsinki, 1963).

[78] W.S. Unger, 'Trade through the Sound in the seventeenth and eighteenth centuries', *Economic History Review*, XII, 2 (1959), 217.

[79] Ågren, *Iron-making societies*, p. 7.

for sale in Britain could be differentiated by subtle variations in their chemical composition and physical properties. Such differences were not apparent to the untrained eye, but the brand marks and identifying symbols stamped on each bar were quite legible to experienced iron merchants. Such marks indicated the distant forge from which the bars had been brought; they also spoke of the very different social environments in which they had been manufactured.

The British iron industry was clearly capitalist in its mode of operation. The indirect method of iron making—the two-stage process involving a blast furnace at which ore was smelted and a forge at which the outcome was refined—had been introduced to the British Isles in the 1490s. It had been taken up by entrepreneurial landowners who saw an opportunity of putting ore and timber on their estates to good account. The landowning elite had the capital to invest in the costly infrastructure, but they lacked the inclination to oversee production. So, by the middle of the seventeenth century, there was a clear tendency for landowners to hire their plant out to an emergent group of professional ironmasters. These ironmasters took charge of every part of the production process: they hired charcoal makers, they set miners to work digging ore, they employed furnacemen and forgemen, and they disposed of the final product. In Sweden, on the other hand, the state consciously parcelled out the production process among distinct social groups. *Bergsmän* and *brukspatroner* each had their allotted tasks, and the sale of iron on international markets was reserved for specialised merchants. Unlike Britain, where ironmasters obtained their inputs for cash, the Swedish system rested in large part on non-monetary exchanges. Peasant leaseholders paid rent to *brukspatroner* in the form of charcoal, and forgemen were often paid in kind. Regulation by the state was designed to optimise the use of forest resources. The *Bergscollegium* imposed production maxima on *bruk* to ensure that *bergsmän* and *brukspatroner* did not compete for charcoal. It was a strikingly successful strategy. The number of forges at work never fell below 400 between the middle of the seventeenth century and the middle of the eighteenth.

In Sweden smelting and refining were kept functionally distinct and spatially dispersed, scattered across an archipelago of production sites in *Bergslagen*. The policy in Russia was quite different. Gigantism and centralisation were preferred, with furnaces, forges, and processing facilities being gathered together. At Ekaterinburg, Peter the Great's showpiece industrial settlement, the works comprised a blast furnace, two forges, and a rolling mill, as well as shops for the making of anchors, sheet iron, steel, and wire.[80] All of this took place within the confines of a single feudal jurisdiction. In the Urals, as in Britain, all parts of the production process were carried out at the behest of ironmasters. But whereas British ironmasters were capitalists who obtained their material and labour inputs on the market, Russian ironmasters were feudal landlords who relied almost exclusively upon extra-economic compulsion. The labour requirements of the Urals iron industry were met by serf labour. Forgemen and artisans, as well as charcoal burners and forestry workers, lacked judicial freedom.

Counterpoint in the 1730s

The presence of Baltic iron on British markets was a matter of deep anxiety for imperial administrators in Whitehall. Each new projection of British power in the Atlantic seemed to highlight British vulnerability in the Baltic. Swedish and Russian iron were required if the translatlantic demand for metalwares was to be satisfied. Worse, from a strategic point of view, was the reliance of Britain's merchant fleet on Russian hemp or Swedish tar. Most deplorable of all was the dependence of the Royal Navy on Baltic supplies. It was this that prompted the British Parliament, from the very start of the eighteenth century, to counterpoise the Atlantic to the Baltic. Given adequate encouragement, could not tar and pitch be obtained from the pine forests of Carolina? That was the aim of legislation passed in 1705 to award a bounty on imports of naval stores from the American colonies.[81] And might not similar legislation be enacted to loosen Britain's ties to Sweden and Russia by promoting colonial iron smelting? Such proposals were actively canvassed in the 1730s and 1740s. The extension of iron production on Europe's eastern frontier should, it was argued, elicit a riposte from across the Atlantic. The toil of serfs in Siberia should be answered by the efforts of enslaved Africans at furnaces in Maryland and Virginia.

---

[80] Anders Florén, 'Social organization of work and labour conflicts in proto-industrial iron production in Sweden, Belgium and Russia', in Catharina Lis, Jan Lucassen and Hugo Soly (eds), *Before the unions: wage earners and collective action in Europe, 1300–1850* (*International Review of Social History*, supplement 2, 1994), p. 97.

[81] See D.G. Kirby, 'The Royal Navy's quest for pitch and tar during the reign of Queen Anne', *Scandinavian Economic History Review*, XXII (1974), 97–116.

As this suggests, writing the history of the trade in iron in the eighteenth century cannot be done as an exercise in Baltic history; it can only be done as a contribution to the history of the Atlantic world, for the need for bar iron in the manufacturing districts of Britain was driven upwards by the deepening demand for metalwares around the Atlantic basin. Nor can the history of the commerce in iron be written as a history of trade in the commonly accepted sense of the term; it must be a history of 'trade' as contemporaries understood it—as an amalgam of 'Dealing and Manufacturing' in Defoe's words, or as Joseph Massey's 'one vast Piece of Machinery'. We must linger in the workshop, not just the warehouse.

Our aim is to extend the reach of Atlantic history, to register how market signals from the western ocean reverberated deep into the European continent. The flow of exotic goods from the Atlantic world into the coffee houses and domestic parlours of bourgeois Europe is well attested to by historians; the impact of oceanic trade on production networks is less familiar.[82] Yet the demands of Atlantic commerce mobilised producers far beyond the hinterlands of the great westward-facing ports. Historians are increasingly aware that weavers in landlocked Silesia, say, felt the tug of Atlantic demand as they bent over their looms making cloth destined for American or African users. We should recognise that forgemen in *Bergslagen* felt the same centripetal pull, so too their counterparts in far distant Ekaterinburg. This had important consequences for the ways in which working life was played out in iron making communities in northern Europe, as we shall see.

The waxing of Atlantic ties also had important intellectual repercussions in northern Europe, notably so in the Swedish case. Swedish intellectuals and state officials—categories that overlapped considerably—were obliged to rethink how the economy and society they knew was configured. The more export-orientated that economy became, the less plausible a closed, cameralist conceptualisation of the Swedish state became. And as the focus of Swedish exports shifted steadily westward in the eighteenth century, so Swedish thinkers and policy makers were compelled to think of their own iron industry as part of an over-arching 'iron system' that girdled the northern hemisphere. An understanding

[82] For an exception, see Klaus Weber, 'The Atlantic coast of German trade: German rural industry and trade in the Atlantic, 1680–1840', *Itinerario: European Journal of Overseas History*, XXVI, 2 (2002), 99–119.

of this international dimension affected the development of economic science in Sweden deeply. The Atlantic experience, in other words, was an important component of the Swedish Enlightenment.[83]

An appreciation of the transnational 'iron system' that emerged in the eighteenth century is pertinent for historians of industrialization generally. One of the most significant debates of recent years over the transition to industrial society has concerned energy use. E.A. Wrigley distinguishes between 'organic' economies, which depend upon vegetable matter for energy, and 'mineral-based energy' economies whose needs are met by fossil fuels. Pre-industrial economies were by definition 'organic', and because of that their growth was constrained. Virtually everything necessary for the sustenance of human life—foodstuffs, raw materials, and fuel—came from the land, so the productivity of agriculture set strict limits on growth. Because the area of cultivable land was finite, any significant rise in the demand for food, for raw materials, or for energy would press too hard on the soil. More industrial materials (like flax, leather hides or wool), more construction materials (principally wood), more energy (wood once more), and more food could not all be had from the same fixed acreage of land. Growth might occur, as Adam Smith suggested, through a more elaborate division of labour or through other efficiency gains in the spheres of production and exchange, but ultimately growth would peter out.[84]

The only escape from the ever-diminishing returns of 'organic' economic growth, Wrigley has argued, was through substituting mineral energy, obtained from coal, for vegetable energy. Put simply, burning coal removed the need to keep large areas of land forested, and if woodlands were no longer needed as a source of fuel then the space they occupied could be devoted to other productive uses. This was the secret of the British Industrial Revolution: the abundance of coal and the deployment of coal-based technologies allowed the British

---

[83] There is little discussion on the links between the Enlightenment and economic discourse in Sweden. For a preliminary treatment see Göran Rydén, '"Det Andra" som det kommersiella och det industriella—Svenska tankar om handel och produktion under 1700-talet', in Maths Isacson and Mats Morell (eds), *Industrialismens tid. Ekonomisk-historiska perspektiv på industriell omvandling under 200 år* (Stockholm, 2002), pp. 37–56. For Swedish economic thought in the eighteenth century generally see K. Petander, *De nationalekonomiska åskådningarna i Sverige. Sådana de framträda i litteraturen* (Stockholm, 1912); Heckscher, *Sveriges ekonomiska historia*, pp. 812–26; and Lars Magnusson, *Äran korruptionen och den borgliga ordningen* (Stockholm, 2001).

[84] E.A. Wrigley, *Continuity, chance and change: the character of the Industrial Revolution in England* (Cambridge, 1988).

economy to slip the bounds of the 'organic' economy and erupt into mineral-fuelled growth.

There was no more spectacular instance of coal-fuelled expansion than that of the British iron industry in the closing decades of the eighteenth century, but it was not obvious that coal-burning technologies would be decisive until very late in the century. Before then ironmasters, merchants and policy makers applied themselves to stretching the boundaries of the 'organic' economy. They did so by acting and thinking globally. Was the British-dominated seaborne trade in bar iron of the early eighteenth century anything less than a raid on the vegetable energy stocks of Sweden and Russia? And was not the promotion of smelting in colonial America by the British authorities inspired by similar motives? It is of course true that some ironmasters in Britain were busily engaged in the development of coal-based technologies, but others were more committed to evading the dilemmas that energy shortages brought about. Their solution was not technological but organisational. It lay in extending an international division of labour so that the energy demands of metals fabrication were dispersed across the northern hemisphere, not concentrated in the wood-depleted British Isles. Only in the 1780s did the technological fix, one that unleashed an immense growth in output, emerge triumphant.

But this is to leap ahead in our story. Our starting point is the 1730s, when Baltic domination of the British market appeared to be irreversible. Our analytical approach, appropriately enough for the late baroque age, is contrapuntal. We follow the French musician François Roberday who opined, when writing of fugues, that 'the Parts being all together, and yet distinguished from one another, may the more easily be examined separately and the relationship they each have to one another more easily be seen'.[85] But for us, commodity chains take the place of the intertwined musical parts. Each can be followed independently, yet each takes its full meaning from the wider Atlantic fugue in which it is situated. Each may, for the purposes of analytical clarity, be accorded priority for a time, but, in the manner of the fugue, none retains its eminence for long. Counting houses, shipping lanes,

---

[85] François Roberday quoted in Jordi Savall, 'J.S. Bach's Musical Testament', in the booklet accompanying the CD *Die Kunst der Fuge* (AliaVox: AVSA9818, 2001). For a discussion of Johann Sebastian Bach in relation to scientific developments in the seventeenth and eighteenth centuries see Christoph Wolff, *Johann Sebastian Bach: the learned musician* (Oxford, 2000), pp. 1–11.

and artisanal workshops are each given their moment of contrapuntal exposure, but none is awarded primacy.

Chapter two picks out a selection of the commodity chains that girded the iron Atlantic. The outcome is a tour of the Atlantic economy from the perspective of the Baltic. Of the commercial nodes that are visited, some have conventional Atlantic coordinates (Bristol, Calabar and Charleston); others (Stockholm and St Petersburg) do not, but—considered as way stations of the international iron trade—they should be seen as part of the Atlantic littoral. Two well-documented actors in this commerce provide the empirical foundation for our analysis: one is Charles De Geer, *brukspatron* at Leufsta, Sweden's largest ironworks; the other is a Bristol iron merchant named Graffin Prankard. For a few brief years in the mid-1730s the connection between Leufsta and Prankard's Bristol warehouse was one of the principal axes of the international 'iron system'. We examine that connection in detail and track the commodity chains that span off, east and west, from the Leufsta-Bristol axis.

Chapter two, then, is a panorama of the ocean at a particular moment in time. The canvas is broad and the depiction detailed. Indeed, chapter two is almost a book in itself. Chapters three and four provide a more orthodox historical narrative. They follow the fortunes of the iron trade in the northern seas from the mid-eighteenth century to the mid-nineteenth century. Chapter three describes how between the 1740s and the 1760s merchants, manufacturers and policy makers in both Sweden and Britain sought to alter the institutional framework within which iron was traded. Chapter four reveals how those efforts were subverted by the sweeping technological changes that revolutionized iron production in Britain between the 1760s and the 1790s. Those changes ended Britain's reliance on Baltic imports, save in a few niche markets. In so doing, they reconfigured the Atlantic economy. Russia was excluded so comprehensively that she ceased to be a major iron exporter in the nineteenth century. Swedish iron was also expelled from the British market in the post-Napoleonic years, but not from the Atlantic world as a whole. Indeed, Sweden found a new point of entry into Atlantic commerce, one that did not rest upon British mediation, which allowed Swedish producers to extend the life of the iron Atlantic into the 1830s and 1840s.

CHAPTER TWO

THE TOPOGRAPHY OF THE EARLY MODERN IRON TRADE, *C.* 1730

Bristol

Bristol in the early eighteenth century was Britain's premier west coast port. It was a trading city of some antiquity, whose medieval prosperity had been based upon the barter of English cloth for Gascon wines. Although the wine trade had been brought low by the loss of Bordeaux and other Gascon ports to the French at the end of the Hundred Years' War, Bristol merchants staged a slow recovery in the sixteenth century by re-directing their trade to the south, to Iberia and the western Mediterranean. A further, fateful re-orientation took place in the seventeenth century as Bristolian seamen spanned the Atlantic to bring sugar and tobacco from England's New World possessions.[1]

The shifting horizon of Bristol's commerce—from the Bay of Biscay, to the Mediterranean, to the Chesapeake and Caribbean—reflected a search for high-value products that could be exchanged for English cloth, fish, and agricultural produce. In this respect, the opening up of the New World provided an epochal opportunity for Bristol's trading community. Here were abundant, fertile lands, readily appropriated by European adventurers, that could be dedicated to the cultivation of crops that delivered a narcotic or stimulant effect, and that therefore commanded a premium price on European markets. Bristol had obvious geographical advantages for conducting a transatlantic commerce. Her position at the mouth of the Severn gave her command of the Atlantic approaches. Admittedly, shipping often had to ride at anchor in Kingsroad before entering the Avon, but once the tidal surge commenced ocean-going vessels could wend their way through the Avon Gorge to Bristol's quays, half a dozen miles upstream.

---

[1] David Harris Sacks, *The widening gate: Bristol and the Atlantic economy, 1450–1700* (Berkeley CA, 1991).

The city's importance for the Atlantic economy was derived not merely from its westward aspect, but from the richness of its hinterland. The Severn was the greatest navigable waterway in the British Isles, and its tributaries stretched deep into the English Midlands and Wales. One contemporary adumbrated the connections: 'by the [Warwickshire] Avon she draws unto herself commodities from Warwickshire; by the help of the Teem, she receives those of Herefordshire and Shropshire; the Wye also brings her some part of the tribute of the former of those counties, and of Radnorshire; and if there be anything left in Herefordshire and Shropshire, the Lugg drains them both; Monmouthshire and the adjacent parts of Wales send their supplies by the Uske....'[2] The busy trade that this 'metropolis of the west' carried on with Ireland might also have been mentioned.

The city had 20,000 inhabitants at the opening of the eighteenth century. It was England's second largest city and her second-ranked port. Bristol was 'the greatest, the richest, and the best port of Trade in Great Britain', Daniel Defoe proclaimed in the 1720s, 'London only excepted'.[3] Fastidious visitors were apt to dwell upon the filth and squalor of the closely packed streets, but none disputed Bristol's commercial vibrancy. The city's trading community, from the merchant elite to petty traders, looked to the west for commercial gain: 'all men that are dealers, even in shop trades, launch into adventures by sea, chiefly to the West India plantations and Spain. A poor shopkeeper that sells candles will have a bale of stockings, or a piece of stuff, for Nevis, or Virginia'.[4]

The city's position, on a tongue of land between the rivers Avon and Frome, made for a long quayside, a mile in extent, all told. The Quay along the lower reaches of the Frome, where ships from the West Indies, North America, and Europe tied up, lay to the west. On the eastern side of the city was The Back, where coasting traffic and the trows that plied the Severn came to moor.

---

[2] Quoted in David Hussey, *Coastal and river traffic in pre-industrial England: Bristol and its region, 1680–1730* (Exeter, 2000), p. 3.

[3] Daniel Defoe, *A tour through the whole island of Great Britain* [1724–26], ed. P.N. Furbank and W.R. Owens (New Haven CT, 1991), p. 181.

[4] Quoted in Jacob M. Price and Paul G.E. Clemens, 'A revolution of scale in overseas trade: British firms in the Chesapeake trade, 1675–1775', *Journal of Economic History*, XLVII, 1 (1987), 3.

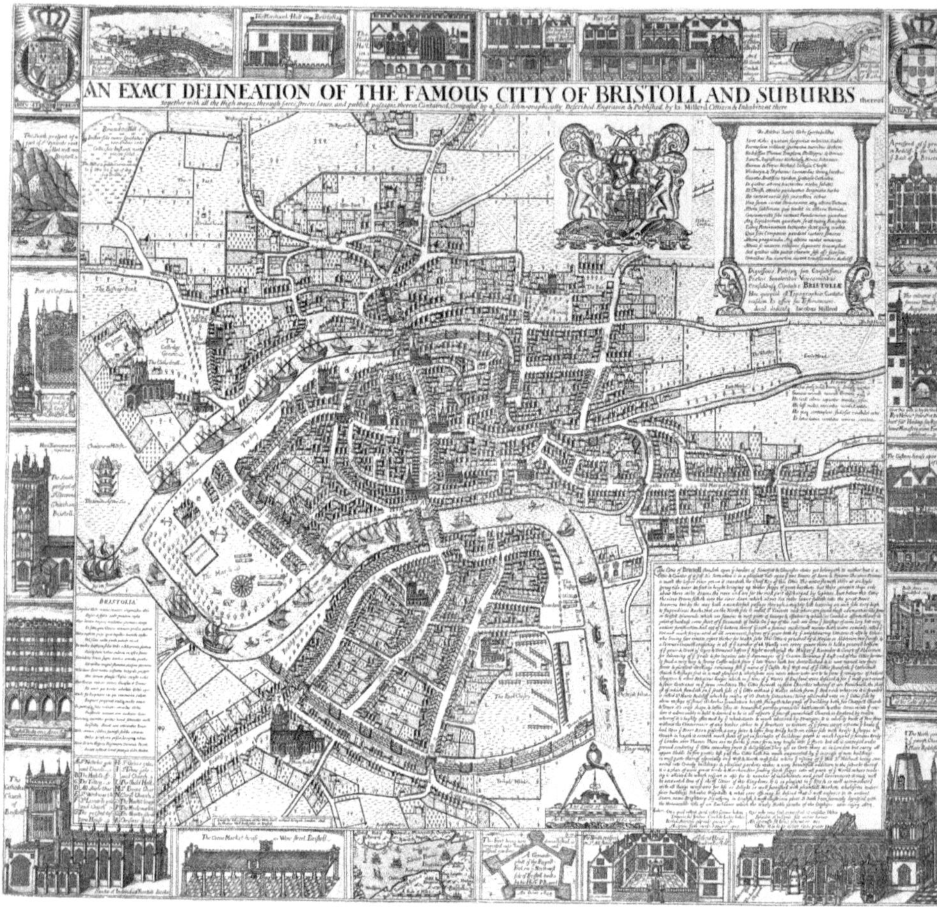

Illustration 2.1. *An Exact Delineation of the Famous City of Bristoll and Suburbs* (1671).

Courtesy of Bristol City Museum and Art Gallery.

Caption: James Millerd's map of Restoration Bristol has the confluence of the Avon ar the Frome on its left margin where a sailing ship is shown heading downstream towards tl Severn. The Quay stretches away in a straight line northwards; the Avon winds eastwar past the medieval core of the city.

This was a city waxing rich on transatlantic trade.[5] Tobacco had been imported to Bristol since the mid seventeenth century, making the port the second most important destination for Chesapeake tobacco after London. The tobacco trade was not the most dynamic part of the city's commercial portfolio in the eighteenth century, it has to be said. Bristol merchants did not pioneer new methods of collecting tobacco in the Chesapeake as their rivals in Glasgow did, nor did they pursue lucrative re-export markets in Europe with the same ardour, preferring to cleave to established but slow-growing domestic sales. Nevertheless, leading tobacco merchants such as Lyonel Lyde or Thomas Chamberlayne were major figures in the city.

It was sugar rather than tobacco that defined Bristol's commerce in the eighteenth century. Whereas Glasgow, Liverpool and Whitehaven successfully challenged Bristol in the tobacco trade, no outport was able to overhaul Bristol in the field of sugar importation until the very end of the century. Bristol merchants had specialised in the trade ever since the take-off of Caribbean sugar production in the third quarter of the seventeenth century, and a good many became planters in their own right, acquiring estates in the West Indies. These sugar merchant-planters were fabulously wealthy, made so by the strong demand for sugar and spin-offs such as rum. A depression in the trade in the 1730s and a sluggish recovery in the 1740s took the edge off the planters' prosperity, but the generation before the American Revolution saw renewed, indeed unparalleled, success.

The sugar trade supported another critical branch of Bristol's Atlantic commerce, the trade in African slaves. Bristol merchants were legally excluded from the 'Guinea Trade' until 1698, but once they were admitted to what had formerly been a monopoly of London's Royal African Company they thrived. The expansion of Bristol's slave trade was particularly swift in the years of peace that followed the end of the War of Spanish Succession. London was eclipsed as a slaving port. By the late 1720s over forty slave ships a year were clearing Bristol, with capacity to carry over 12,000 captives to the slave marts of Jamaica or

---

[5] Kenneth Morgan, *Bristol and the Atlantic trade in the eighteenth century* (Cambridge, 1993). See P.G.E. Clemens, 'The rise of Liverpool, 1665–1750', *Economic History Review*, XXIX (1976), 211–25, and T.M. Devine, *The tobacco lords: a study of the tobacco lords of Glasgow and their trading activities, c. 1740–1790* (Edinburgh, 1975) for the other major west coast ports, and Christopher J. French, '"Crowded with Traders and a Great Commerce": London's dominion of English overseas trade, 1700–1775', *The London Journal*, XVII (1992), 27–35, for a reminder of the capital's enduring importance.

Barbados. Bristol slavers never surpassed the meridian they achieved *c.* 1730, and by 1750 Liverpool had overtaken her southern rival to assume the position of Britain's leading slave port. Even so, slaving remained one of the central pillars of Bristol's international commerce long after the city's supremacy in the trade had passed.[6]

For sugar and tobacco to be imported—and for the slave trade to flourish—export goods had to be found. Metalwares from the West Midlands and textiles from the West Country and further afield figured very prominently here, as will be seen, but export goods were also manufactured in the city itself and its immediate environs. Bristol was a major centre of glass manufacturing. Both window glass and bottles were made, the latter often being employed in the bottling of local cider, beer, and Hotwells mineral water for export to north American and Caribbean consumers.[7]

Bristol was also the seat of Britain's copper and brass trades. The Bristol Brass Wire Company's works at Crew's Hole, two miles south of the city, was one of Europe's largest, with 49 copper smelting furnaces operating in the early 1750s. The Warmley works of Champion & Co did not smelt copper on quite such an extensive scale, but it was as important as a centre of brass manufacture. In fact, the Avon valley between Bristol and Bath was stiff with non-ferrous smelting works, batteries, and rolling mills.[8] All these works were intimately connected with the export trades, for copper and brass wares played a crucial role in the Guinea trade. Copper rods were widely used as a trade good in West Africa, as were brass manillas and 'Guinea kettles'.[9] Indeed, prominent brass manufacturers such as Thomas Coster were also active slavers. Other export trades grew in symbiosis with plantation agriculture in the New World. Factories making clay tobacco pipes, for example, were clustered in the west of the city. Clay was shipped

[6] See David Richardson, 'Slavery and Bristol's "Golden Age"', *Slavery and Abolition*, XXVI (2005), 35–45, and the materials edited by the same author in *Bristol, Africa and the eighteenth-century slave trade to America. Volume 1: The years of expansion 1698–1729* (B[ristol] R[ecord] S[ociety] Publications, XXXVIII, 1986), *Volume 2: The years of ascendancy 1730–1745* (BRS Publications, XXXIX, 1987), *Volume 3: The years of decline, 1746–1769* (BRS Publications, XLIII, 1991), and *Volume 4: The final years* (BRS Publications, XLVIII, 1996).

[7] Cleo Witt, Cyril Weeden, and Arlene Palmer Schwind, *Bristol glass* (Bristol, 1984).

[8] Joan Day, *Bristol brass: a history of the industry* (Newton Abbot, 1973).

[9] See Reinhold Angerstein's extensive comments on the local non-ferrous sector in *Angerstein*, pp. 136–45.

Illustration 2.2. *The South East Prospect of the City of Bristol* (1734).

Courtesy of Bristol City Museum and Art Gallery.
Caption: Samuel and Nathaniel Buck's *The South East Prospect of the City of Bristol* (1734) shows the heart of the city crowded with the towers and spires of its medieval churches. The eastern suburbs of the city, on the other hand, are studded with tall, smoking glass kilns. In the left distance the river Avon is seen snaking towards the gorge through which it made its way to the sea.

in 'from Wales, the Isle of Wight and Poole in Dorset' and the completed pipes were sent in 'huge batches' to the American Colonies.[10] Other export goods such as gunpowder, which was made at mills on tributaries of the Avon such as the Chew, played a critical role in the Indian trade along the Appalachian frontier, securing deerskins for European consumption.

Bristol was then an industrial as well as a commercial centre, and it was not just export industries that were significant. The processing of colonial produce also loomed large. Sugar refineries, fifteen of which were at work in the early eighteenth century, were the most visible expression of this, but other, less celebrated examples could be cited, such as the distilleries that produced turpentine from pine resin brought in from the Carolinas. Port industries that supported Bristol's merchant fleet were also much in evidence: anchor forges, sailcloth factories, and ropewalks.[11] The dynamic relationship between New World slavery, plantation agriculture, maritime endeavour, and domestic manufacturing was well understood by contemporaries. The Bristol merchant John Cary, writing in the 1690s, hailed African slaves as the means 'whereby our Plantations are improved, and 'tis by their Labours such great Quantities of *Sugar, Tobacco, Cotton, Ginger*, and *Indigo* are raised, which being bulky Commodities imploy great Numbers of our Ships for their transporting hither, and the greater number of Ships imploys the greater number of Handecraft Trades at home'.[12]

'They trade with every place on earth', the Swedish traveller Reinhold Angerstein reported upon visiting Bristol in 1754. He alluded principally, of course, to the Atlantic trade, but Angerstein was aware that Bristol merchants traded to the east as well, to the Baltic. Indeed, he knew that Bristol's shipbuilders relied upon the vast timber resources of the Baltic littoral; that they looked to Riga for masts, and to Danzig for planking. He knew also that Bristol's ropewalks were processing hemp from Russia.

---

[10] *Angerstein*, pp. 131, 146. See also R.G. Jackson and R.H. Price, *Bristol clay pipes: a study of their makers and their marks* (Bristol, 1974).

[11] It was estimated that 800 tons of hemp was consumed in Bristol's ropewalks every year in the 1730s: SA, DD/DN 425, GP to Robert & Patrick Mackey, 28 October 1738.

[12] John Cary, *An essay on the state of England in relation to its trade* (Bristol, 1695) quoted in Richard B. Sheridan, 'The formation of Caribbean plantation society, 1689–1748', in P.J. Marshall (ed.), *The Oxford history of the British Empire. Vol. II: The eighteenth century* (Oxford, 1998), p. 399.

Most of all, he knew that products from his homeland played a strategic role within Bristol's commerce. Quite apart from the Swedish tar and pitch that was used so extensively in the port's shipyards, Swedish bar iron found multiple uses in the city and its hinterland. Anchorsmiths made use of it, and slave merchants shipped it to the Guinea coast as a trade good. Above all, Swedish iron was funnelled into Bristol's hinterland, to be converted into steel or wrought up into articles of hardware. Much of the Swedish iron that was landed at Bristol passed into the English Midlands, there to be manufactured into a variety of tools and gadgets, huge volumes of which would then be exported to British North America and the West Indies.

This book explores these processes. It shows how the rise of the Atlantic economy brought about major shifts in the Baltic economy. Naval stores and iron became the most important articles to pass westward through the Sound, overshadowing the grain shipments from Danzig that had defined Baltic commerce in the sixteenth century. Swedish iron came to dominate European export markets in the seventeenth century in large part because of the stimulus of the Atlantic world, mediated as it was through the Dutch and then, more massively from the second half of the seventeenth century, the British market.

The commodity chains that flowed from the Baltic through Britain and on into the wider Atlantic world were woven into complex, braided patterns. A great mass of commodities moved from east to west, first to be wrought into manufactured articles, then to be disgorged into the Atlantic basin. Indeed, a good deal of Baltic wood, hemp, flax, and iron was embodied in the sailing ships that triangulated the western ocean and in which the reciprocating commodity flows of tropical produce were borne back to Europe. These commodity flows, for all their protean diversity, were curbed and moulded by government action: by the regulations of the Swedish and Russian states, then by Britain's Navigation laws.

Making sense of this cat's cradle of commercial exchange presents a formidable problem. It was a transoceanic web with no sure beginning or end. It existed in a state of constant flux, despite the best efforts of European states to impose some sort of fixity upon it. For this reason we have selected a number of different vantage points from which we hope to be able to make this network of iron production, trade and consumption more comprehensible. Bristol is the first of these. As Britain's premier west coast port in the early eighteenth century, Bristol played a key role in connecting the Atlantic economy to the Baltic. If Bristol is the central vantage point, the central actor in this study is the Bristol merchant Graffin Prankard (d. 1756), a man who traded to the east as well as to the west, and who specialised in iron and metalwares. We focus upon Prankard's warehouse as an organising centre through which a variety of Baltic commodities were funnelled to manufacturing centres in the Midlands and the West Country. Many of the articles that were fabricated in the industrial villages that sprawled across the south Staffordshire coalfield or dotted the slopes of the Mendips were returned to Prankard to await shipment westwards. Prankard's depot provides a lens through which such manufacturing activity can be scrutinised, and through which the trade routes that extended east and west can be viewed.

A second vantage point is the *bruk* at Leufsta in Uppland, the seat of Charles de Geer (1720–1778), a descendant of the great Louis De Geer. Leufsta was the largest and most renowned *bruk* in Sweden. Its iron—known by the brand name 'Hoop L' in Britain—was one of the most sought-after marks. Leufsta was a prime example of a manufacturing facility that, despite its position in a largely non-monetized rural location, was attuned to the pulse of a world market. A third vantage point is Stockholm, the focal point of Sweden's international trade. The city was a great commercial centre. It was also the seat of government, home to state bureaux such as the *Bergscollegium* and the *Kommerscollegium* that determined Sweden's industrial and commercial direction. It was also the venue for the *Riksdag*, the assembly of estates that did much to determine state policy during Sweden's *Frihetstiden* (Age of Liberty) of 1720–1772. These institutions defined the environment in which international trade took place. The success of the export economy was rooted in particular political conditions and sustained by carefully designed state structures. From Stockholm's quayside the interaction of trade and state policy can be surveyed. Much the same can be said of St Petersburg, Russia's Baltic capital. The quays that Peter the Great had built along the banks of the Neva acted as a hinge between the new Siberian ironworks on Europe's eastern-most frontier and the markets of western Europe. They acted as the junction between the system of *dirigisme* and aristocratic privilege that underpinned industrial developments in the Urals and the unabashed capitalism that drew Siberian iron to the British market.

Birmingham offers a further telling vantage point. With its gunsmiths, 'toy' manufacturers, and metalware specialists, Birmingham was the destination for much of the high-grade Swedish iron sold by Prankard. It also served as a marketing centre for the smoke-shrouded industrial neighbourhoods to the west that would in time conjoin into the Black Country. As an industrial and commercial crossroads, Birmingham provides a point of entry into the myriad production flows that made the West Midlands so dynamic a manufacturing district in the eighteenth century, where much of humanity's industrial destiny was being foretold. Calabar, on the Bight of Biafra, offers an equatorial vantage point from which to scrutinise the Atlantic economy. The slave marts of Calabar saw Swedish iron and a host of Birmingham-made articles that embodied Swedish bar iron being bartered for the captives whose unremitting labour was called for on the plantations of the New World. Our final vantage point is Charleston, the colonial port that was the most regular destination for Prankard's ships. It was also the final port of call for many of the slave ships whose human cargo had been paid for, at least in part, with Swedish voyage iron. The nails, hoes, gunpowder and lead shot that were shipped into Charleston were instrumental in driving the plantation frontier westward. As they dug the irrigation channels upon which rice production depended, South Carolina's African slaves gave unwilling impetus to a quickening commercial circuit that stretched far across the northern hemisphere. The rasp of their hoes as they sliced into the floodplain of Goose Creek was answered by the thump of forge hammers in *Bergslagen* and the percussive crash of axes wielded by enserfed tree fellers in the Urals, half a world away.

The period with which we are dealing was defined by two revolutions in the international iron trade: the first was the sudden rise of Baltic iron on western markets in the mid-seventeenth century, the second was Baltic iron's still more abrupt retreat in the early nineteenth century. In this time, two peripheral areas of northern Europe were drawn into a close economic relationship with the most advanced regions of the continent. Specialised export zones were established in both Sweden and Russia, with profound effects on local society in *Bergslagen* and the Urals mining districts. The effects upon British society were not so immediately visible but they were profound nonetheless. The rise in imports, taken together with increased domestic production, allowed for a very substantial rise in the consumption of malleable iron in Britain: from approximately 26,000 tons in 1700, to 44,000 tons in 1750, then to nearly 62,000 tons in 1790.[13] In *per capita* terms, this implied a substantial rise in the first half of the century, from 10.9lb per head in England and Wales in 1700 to 16.1lb in 1750; in the second half of the eighteenth century *per capita* consumption was maintained at this historically high level, despite very rapid population growth.

Had supplies of Baltic iron not proved elastic the Georgian economy would have been hobbled. Yet the international iron trade is not well understood. British historians have, as we have seen, devoted little attention to the iron trade, let alone its Baltic dimension. By contrast, there is a considerable literature in Swedish about that nation's most significant industrial sector. It is a literature that acknowledges the importance of exports, but it has not concerned itself with the fate of Swedish iron once it entered overseas markets. This is a serious deficiency, for the different commercial circuits into which Swedish and Russian bar iron were introduced in Britain and the wider Atlantic economy exercised a determining influence over production networks in the Baltic.

Eighteenth-century Swedes were alert to this. Officials of the *Bergscollegium* were aware that Britain constituted the most expansive market for Swedish iron. For that reason many of them visited Britain in an attempt to understand its peculiarities and to assess the competitive position of Swedish iron.[14] Eric Odelstierna, who arrived in Britain in the early 1690s, was one of the first to do so. Odelstierna noted the range and extent of Britain's international commerce. This, he concluded, had acted as a spur to the metal trades that were so 'plentiful in England, and more omnipresent there than in any other place in Europe'. Yet the market for bar iron was complex, with bars of different qualities competing against one another on different regional markets. English iron, Odelstierna thought, was far inferior to Swedish or Spanish. Yet it survived because some of the principal British markets were closed to imported iron. Ironworks in the Forest of Dean, for example,

---

[13] Peter King, 'The production and consumption of bar iron in early modern England and Wales', *Economic History Review*, LVIII, 1 (2005), 1–33, especially table 2. These figures are for England and Wales alone. Customs records from the 1750s suggest that the Scottish and Irish consumers absorbed an additional quantity equivalent to 20 per cent of the English-Welsh market: see TNA: PRO, CUST 3, 14, and 15.

[14] Sven Rydberg, *Svenska studieresor till England under Frihetstiden* (Uppsala, 1951).

faced little challenge in the West Midlands because importers of iron could not, so Odelstierna reckoned, navigate the 'difficult currents of the Bristol Channel'. Other regions were more vulnerable to import penetration. Generally speaking, Spanish iron was to be found along the south coast, Swedish on the east. Odelstierna had less to say about specialised niche markets. He noticed that anchors and other items of naval hardware were usually manufactured from Swedish iron, but he paid little attention to the steel industry, an industrial sector in which Swedish iron was to attain a critical importance. Steel making in Britain was was as yet in its infancy.[15]

The steel industry came to be of far more interest to those who followed in Odelstierna's footsteps in the eighteenth century. Anders Swab, who toured Britain in the 1710s, informed his superiors at the *Bergscollegium* that a substantial quantity of Swedish iron was used in steel manufacture, and that English expertise in this area had greatly improved, a state of affairs he attributed to the advice of a 'worker who had run off from Wira bruk in Sweden'. Henric Kahlmeter, a decade after Swab, elaborated. The English steel industry had undergone a major expansion since the beginning of the century, and as it had done so it had become reliant on particular brands of Swedish iron. Bars from the *bruk* at Leufsta and Österby were the most keenly sought after.[16]

Kahlmeter accepted the basic analysis that Odelstierna had made a generation earlier. The English market was divided into distinct eastern and western sectors: the east was the domain of Swedish iron, but the west, equipped with more luxuriant woodlands and shielded from import penetration by higher transport costs, remained the preserve of native ironmasters. Yet Kahlmeter also detected the beginnings of change. The appearance of iron from 'the Russian places, Archangel, Petersburg and Narva', albeit in tiny quantities, was worthy of note. He also drew the attention of the *Bergscollegium* to the occasional presence of pig iron from the north American colonies on the British market. A more complex and diverse market was in the making.

---

[15] 'Om Bergwercken uti Engeland utdragit ur Afledne Assessoren i Kungl Bergscollegio Eric Odelstiernas Relation åhr 1692', Bergskollegiets arkiv, D VI: 13, RA.

[16] See the two letters from Anders Swab to the Bergscollegium in 1712, the report from Swab to minister Gyllenborg, Kahlmeter's letters to the Bergscollegium in 1721 and 1724, and his 'Berättelse om Bergwerkens Tillstånd uti Engeland hwarwid i början något anföres om Scotland. Ingifwen til Kungl. BergsCollegium den 26 aug. 1725, af Hindric Kalmeter', all of which are in Bergskollegiets arkiv, D VI: 13, RA.

Kahlmeter's insights were amplified by Samuel Schröder in 1749. Schröder's 'Remarks on the English iron trade' furnished his superiors in Stockholm with a vivid overview of developments in the second quarter of the eighteenth century. He reveals a far more fluid situation than that described by Odelstierna fifty years earlier. The regional distinctions that had prevailed at the start of the eighteenth century had been greatly softened. Moreover, a diversification of the market, which Kahlmeter had so presciently anticipated in the mid-1720s, had come to pass. Russian iron now featured heavily on the West Midland market, challenging English and Welsh-made iron. Bar iron forged from colonial pig iron also jostled for its place. Nationally, Swedish iron still retained the most important market share, but Basque iron had a foothold, and small parcels of German iron sometimes made an appearance. In short, the voracious British market now drew to itself iron from across the nothern hemisphere.

Tracing the tangled connections that made Britain the centre of the European iron market is confessedly difficult. And the connections that stretched across the North Sea, although the most important, are among the least tractable. There is a paucity of sources below the level of official trade statistics. Indeed, one historian has spoken of how 'tantalisingly little' is known about the British merchants who were active in the trade.[17] There is, however, one body of archival material that throws considerable, if not massive light on the operation of the bar iron market in the early eighteenth century. This is the business archive of Graffin Prankard.[18]

Prankard was the son of a Somerset maltster. His father was a Quaker, connected with the Alloway family, Quaker merchants of Minehead and Bridgwater who traded in fish, cider and cheese from the West Country to ports in Ireland, France and Spain. Through the Alloways, the younger Prankard found a bride and a calling. Graffin Prankard married Sarah Alloway in 1708 and used his wife's dowry

[17] H.S.K. Kent, *War and trade in northern seas: Anglo-Scandinavian economic relations in the mid-eighteenth century* (Cambridge, 1973), p. 68. But see J.G. Duncan, 'A Scottish trading house in eighteenth-century Gothenburg: Carnegy and Shepherd', *Northern Scotland*, XI (1991), 1–9, which looks at the relatively small-scale trade between Sweden and Scotland, for an exception to this rule.

[18] J.H. Bettey, 'Graffin Prankard: an eighteenth-century Bristol merchant', *Southern History*, XII (1990), 34–47; J.H. Bettey, 'From Quaker traders to Anglican gentry: the rise of a Somerset dynasty', *Proceedings of the Somerset Archaeology and Natural History Society*, CXXXV (1991), 1–9.

to establish himself as a merchant in Bristol. He soon featured in industrial partnerships in which other Quaker Bristolians played a major role. He was, for example, a partner of Abraham Darby's at Coalbrookdale in 1712. He was also a founding partner in a metalworking enterprise at Tern in Shropshire, comprising a 'mill for Rowling of Brass Plates and Iron Hoops and slitting of bar Iron into Rods for Making of nails'.[19] Prankard's involvement in these works was relatively short-lived, however. His main business in the 1710s appears to have been as an Atlantic merchant, shipping ironmongery and other 'dry' goods to the north American colonies. That changed in the 1720s as Prankard turned to the Baltic. He began to import timber, hemp, and bar iron from Sweden on a large scale. He did so through an agent in Stockholm, Francis Jennings.

Francis Jennings, like many of Stockholm's merchant community, was a foreigner. A native of Belfast, he settled in the Swedish capital in 1719 and soon displayed an enviable commercial acumen, becoming the city's leading iron exporter. Commercial success brought social prestige. Francis Jennings ended his life as a member of the Swedish nobility and an estate owner. These prizes were hard won, however. They depended upon a willingness to explore new markets, especially those in western Britain. London had long been the entrepôt for the Baltic trade, and in the early eighteenth century Swedish iron was landed at London and just a few other east coast ports, principally Hull. Jennings, together with Prankard, pioneered the direct shipment of Baltic iron to western Britain.

Swedish observers of the early eighteenth century were adamant that the British bar iron market fell into two distinct segments: the east and the west.[20] Graffin Prankard's accounts clearly demonstrate that this ceased to be the case during the 1720s. In 1721 Prankard imported a mere 4 tons of foreign bar iron into Bristol. Thereafter the figures vault upwards: to 198 tons in 1723, then 395 tons in 1726, then to 933 tons in 1728.[21] At first, Prankard's sales seem to have been in the immediate

[19] Barbara Coulton, 'Tern Hall and the Hill family: 1700–75', *Shropshire History and Archaeology*, LXVI (1989), pp. 99–100.

[20] This perception is also to be found in modern scholarship. See Sven-Erik Åström, 'Swedish iron and the English iron industry about 1700: some neglected aspects', *Scandinavian Economic History Review*, XXX (1982), 129–41.

[21] Data supplied by Åsa Eklund from the Bristol port books in the National Archives (PRO, E190 series). By way of comparison, the Maister family, at the head of one of Hull's leading Baltic houses, distributed just 304 tons to customers in the North and

hinterland of Bristol, but from 1725 Swedish iron began to pour into the West Midlands, the heartland of British metalware manufacturing.[22] By the end of the decade Prankard was also dealing in Russian iron. Initially, he bought up supplies on the Rotterdam market, but in 1730 he established direct trading links with Messrs Vigor & Davenport in St Petersburg. Thereafter Russian iron assumed a major importance in Prankard's business, making up between a quarter and third of his total bar iron sales in the late 1730s.

Just as Francis Jennings won pre-eminence in Stockholm's trade, so Graffin Prankard came to dominate Bristol's trade with the Baltic, accounting for 54 per cent of the Swedish iron entering the port in 1730.[23] He was the most important and respected iron merchant in western Britain. 'The Bristoll Chester & Leverpool Traders are but Slippery', one Stockholm factor told a Hull merchant, 'except one Prankard of Bristoll.'[24] He was an important innovator, not just in opening up the market for bar iron in the Severn valley, but in yoking together the Baltic and Atlantic trades into a single commercial loop.

The Baltic trade was a highly specialised field of enterprise in the seventeenth century. Its practitioners did not much concern themselves with other aspects of commerce. They remained focused on Europe and on northern Europe in particular.[25] The Marescoes, for example, an Anglo-Netherlandish family active in London's Baltic trade in the 1660s and 1670s, imported Swedish iron, copper, pitch, and tar on a large scale. Charles Marescoe sent sugar, spices and tropical dyestuffs to Hamburg and Amsterdam in return; but these he bought on the London market, he did not engage directly in the Atlantic trades.[26] Bilateral trade was typical of the late seventeenth century; Baltic merchants

---

the Midlands in 1714, and 350 tons in 1715: Brynmor Jones Library, University of Hull, DP/82. See Gordon Jackson, *Hull in the eighteenth century: a study in economic and social history* (Oxford, 1972), pp. 122–23, for the Maisters' importance.

[22] Hussey, *Coastal and river trade*, p. 79.

[23] Åsa Eklund, 'Iron production, iron trade and iron markets. Swedish iron on the British market in the first half of the eighteenth century', (Licenciate thesis, Department of Economic History, University of Uppsala, 2001), p. 121.

[24] Hull City Archives, DFB/78, William Maister to Thomas Broadley, 25 August 1729.

[25] And for their part, Atlantic traders were highly specialised. See Nuala Zahediah, 'Making mercantilism work: London merchants and Atlantic trade in the seventeenth century', *Transactions of the Royal Historical Society*, 6th series, IX (1999), 143–58.

[26] Henry Roseveare (ed.), *Markets and merchants of the late seventeenth century: the Marescoe-David letters 1668–1680* (Oxford, 1987).

did not insert themselves into a multilateral system of exchanges that might harness the Baltic to the Atlantic world. The activities of Adam Montgomerie, a Scottish factor who arrived in Stockholm in 1699, were probably typical. Montgomerie shipped iron, brass wire, tar, and timber to correspondents in Scotland and the north of Ireland. He balanced his accounts by importing bulk products such as Clyde herring.[27]

Insofar as British merchants did organise multilateral exchanges, these extended no further than western Europe or the Mediterranean. There were, three Baltic merchants explained in 1725, 'annually imported from Sweden into this Kingdom very great quantityes of Iron, Timber, Pitch, Copper & other Commodityes, for which end it has been very usuall to Send many Brittish Ships... Laden with Salt or Wines from France, Portugall, La Matte, or Isle of May, to deliver their Cargoes in Sweden & there lade back for England'. Alternatively, ships from France, Portugal, or the Mediterranean could be despatched 'to the Ports of Denmark, Pomerania, Prussia, Dantzig, or the Russian Ports in the Baltick & thence after delivering their first Cargoes, relade them at their delivering ports with Corn, Hemp, or other Commodityes for Sweden whence they are reladen for Britain with Iron or other Merchandize usefull here'.[28] Despite the occasional foray to the Cape Verde islands, this was a trade that was restricted to European waters.

After the peace of Utrecht in 1713 the separation of the Baltic and Atlantic trades became less pronounced. Britain became the hinge that joined the Baltic and Atlantic basins together in an interlinked trading circuit. The process can be seen in the activities of Josias Wordsworth, one of the partners in Crowley Hallett & Co. Wordsworth was a major importer of Swedish iron and naval stores, using his kinsman Samuel Wordsworth as his Stockholm-based agent. He was no mere trader, though. Josias Wordsworth was a manufacturer in his own right. He was *inter alia* part-proprietor of anchor shops on the Thames and the Tyne,

---

[27] See Montgomerie's letterbook for 1699–1702: Mitchell Library, Glasgow, SR352.

[28] TNA: PRO, CO 388/25, Board of Trade correspondence, a memorandum on the Swedish shipping ordinances from Henry Norris, Josias Wordsworth and Richard Thomlinson, 13 April 1725. The Glasgow merchant Henry Smith, for example, shipped malmsey wine to Königsberg and used the proceeds to buy brass wire in Stockholm 'for ye use of his pin manufactory' in Scotland: Mitchell Library, Glasgow, SR352, Adam Montgomerie to William Gray, 6 June 1700.

as well as a rolling mill and a suite of nailing shops in Northumberland.[29] As a partner in Crowley Hallett & Co. Wordsworth was also a wholesale ironmonger, buying hardware from Midland manufacturers such as the Finches, the Homfrays, and the Molineuxs. Enormous quantities of this hardware were lodged in a complex of warehouses in London and Deptford, bound for the West Indies and the North American colonies. Shipments of hoes, axes, machetes, nails, and chains were despatched to the sugar plantations; remittance was made in raw sugar.[30]

The same procedure—involving the importation of Baltic raw materials, the fabrication of manufactured goods in Britain, and their export to the Americas in exchange for colonial produce—was followed by Graffin Prankard. Prankard not only imported Baltic iron to Britain, he became implicated in metalware manufacturing in Bristol's hinterland. He employed many of his customers as subcontractors. The Homfray family of Stourbridge, for example, put out nail rods on Prankard's behalf to domestic nailers in their neighbourhood. They returned hundreds of bags of nails to Prankard that he marketed in Charleston. Similarly, Prankard used the Shallard family, proprietors of a cementation furnace outside Bristol, to convert Swedish iron into steel on his account.

In the 1720s and 1730s Prankard fashioned a multifaceted production and marketing chain that stretched from *Bergslagen*, via the manufacturing districts of western England, to the lowcountry of South Carolina. In doing so, he was responding to changes in the Atlantic economy that held out the possibility of employing his ships on a year-long circuit between the Baltic and the newest British colonies in north America. The carrying of rice from South Carolina was a key factor here. Rice cultivation took off in the Lower South after 1700 to become South Carolina's leading export. These exports, which had amounted to only a few hundred thousand pounds per annum at the close of the seventeenth century, reached 1.5 million pounds by 1710, 6 million pounds in 1720, and nearly 20 million pounds in 1730.[31] This development was intimately connected to events in the Baltic. The onset of the

[29] TNA: PRO, C 11/822/3, Remnant and Legas versus Wordsworth, Hallett and Smith; Chris Evans, 'Manufacturing iron in the North-East in the eighteenth century: the case of Bedlington', *Northern History*, XXVIII (1992), 178–96.

[30] BL, Oriental and India Office Collections, MSS Eur F 218/113–15.

[31] John J. McCusker and Russell R. Menard, *The economy of British America, 1607–1789* (Chapel Hill and London, 1985), p. 176.

Great Northern War in 1699 had disrupted the movement of cereals from the southern Baltic to western Europe. As supplies of Polish grain slackened, it was Carolina rice that filled the gap.[32]

Swedish iron and Carolina rice had complementary production cycles, of which Prankard took full advantage. He despatched one of his ships, the *Parham* or the *Baltick Merchant*, to Charleston in the autumn, just as ice was closing the more northerly Baltic ports to shipping. While the *Parham* crossed a rough, wintry Atlantic, the Carolina rice crop, planted in April–May and harvested in September–October, was being prepared for shipment. When the *Parham* arrived at the year's end, Prankard's agent in Charleston would dispose of the nails, iron pots, steel, and gunpowder with which she was loaded. Having taken on hundreds of barrels of rice, a quantity of logwood, and perhaps some indigo, the *Parham* would sail for Europe in February or March.[33]

As Prankard's ship cleared Charleston, bar iron was already on the move from forges in *Bergslagen* to the staple towns from which it would be exported. In the depths of winter sledges carrying bar iron were being dragged over the frozen lakes and snowy roads of the Swedish midlands. Most deliveries were to be completed by the spring thaw that re-opened the Baltic ports to international trade. The successful completion of the transatlantic circuit Prankard had initiated the previous autumn now required careful synchronization amongst his agents across northern Europe. Rice, as an enumerated commodity under the Navigation Acts, had to be landed at a British port before it could be shipped on to a foreign market. The *Parham* would accordingly make the briefest of stops at Cowes or Poole, then on to Hamburg in May,

---

[32] Peter A. Coclanis, *The shadow of a dream: economic life and death in the South Carolina Low Country, 1670–1920* (New York, 1989); R.C. Nash, 'South Carolina and the Atlantic economy in the late seventeenth and eighteenth centuries', *Economic History Review*, XLV (1992), 677–702; and Max S. Edelson, *Plantation enterprise in colonial South Carolina* (Cambridge MA, 2006). See also: Kenneth Morgan, 'The organization of the colonial American rice trade', *William and Mary Quarterly*, 3rd series, LII (1995), 433–552; Stephen G. Hardy, 'Colonial South Carolina's rice industry and the Atlantic economy: patterns of trade, shipping, and growth', in Jack P. Greene, Rosemary Brana-Shute, and Randy J. Sparks (eds), *Money, trade, and power: the evolution of colonial South Carolina's plantation society* (Columbia SC, 2001), pp. 108–40; and Robert M. Weir, '"Shaftesbury's Darling": British settlement in the Carolinas at the close of the seventeenth century', in Nicholas Canny (ed.), *The Oxford history of the British Empire. Volume 1. The origins of empire: British overseas enterprise to the close of the seventeenth century* (Oxford, 1998), pp. 375–97.

[33] In 1731–32 over 54 per cent of sailings from Charleston for Britain and Ireland occurred in the months of February, March and April. Ian K. Steele, *The English Atlantic 1675–1740: an exploration of communication and community* (Oxford, 1986), p. 289.

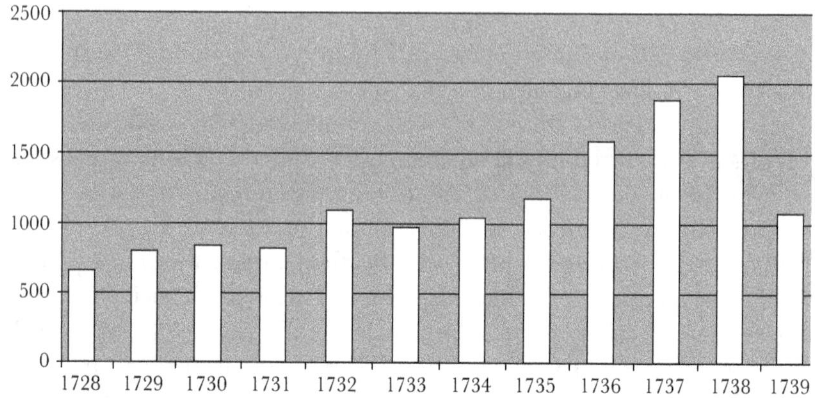

Source: SA, DD/DN 433, 438 and 439

Figure 2.1. Bar iron sales by Graffin Prankard, 1728–1739.

where David Skinner & Co would dispose of the rice for Prankard. The *Parham* would usually pass the Sound in ballast, tying up at Stockholm in June. The cargo of iron, timber and naval stores that Francis Jennings had assembled would be loaded without delay, for the ship was to return to Bristol in time for St James's fair in July, the highpoint of the city's commercial calendar. St James's fair attracted buyers and sellers from across the West Country, the West Midlands, and Wales. Accounts were settled and orders placed. It was the forum in which Graffin Prankard met the ironmongers and manufacturers who bought his iron, and who supplied him with the iron wares that he exported to the Americas, allowing the transatlantic cycle to begin anew.

If we follow the iron imported by Prankard on its journey into the manufacturing zones of western Britain we will be able to understand better the connections that brought the Baltic and Atlantic worlds together. Figure 2.1 presents the overall trajectory of Graffin Prankard's sales of bar iron between 1728 and 1739.

His sales breached the 1000-ton barrier for the first time in 1732. After a slight relapse in 1733, his sales moved upwards once more, exceeding 2000 tons in 1738. The slump in sales shown in 1739 is partly real—reflecting the curtailment of iron exports that followed the outbreak of the War of Jenkins' Ear with Spain—but partly artificial—reflecting the closure of Prankard's extant accounts in September of that year. Data for the period 1728–32 have been drawn from a ledger that provides basic information on the identity of the customer and the volume of iron that he or she bought. But from March 1732 onwards data have been drawn from a rather richer source, Prankard's waste books.[34] These supply far more valuable information, including: the location of the customer; the 'nationality' of iron that was purchased (Swedish or Russian); and the physical form of the commodity (¾ squares, 2½ inch broads, narrow flats, or any other of the gauges into which bar iron was hammered). Very often, the waste books specify the brand of iron concerned, recording the trademark with which the bars were stamped: the sable (or 'rat' as Prankard had it) that was the mark of the Demidov family; the imperial double eagle that was carried on bars of 'Government Siberia'; or the 'double bullets' (two touching circles) that distinguished iron from the forge at Österby. Taken together, these data allow an analysis of Prankard's market that discriminates between different regions, between different types of iron, and between different sorts of customer.

Graffin Prankard sold iron over a wide area of western Britain. He sent iron up the Severn as far north as Shrewsbury. The river port of Bewdley provided access to the West Midland plateau, whilst the southern Midland counties could be reached via the Warwickshire Avon. The Midlands formed a competitive frontier along which iron from Bristol confronted Baltic iron brought in from Hull via the Humber and Trent. John Huddesford, the Coventry ironmonger, was Prankard's most easterly customer, William Butler of Stafford his most northerly. To the west, Prankard sold iron to customers all around the Bristol Channel. On the northern shores of the Channel he had customers in the industrial towns of Neath and Swansea, as well as clusters of demand in west Walian ports such as Pembroke, Carmarthen, and Haverfordwest. To the south, Prankard supplied a variety of customers in Somerset and north Devon but had no commercial presence to speak of beyond Bideford.

The failure to penetrate the mining zone of Cornwall or the textile districts of south Devon may reflect the residual strength of Spanish iron in the region. The entry of iron imports from Bilbao to southwestern ports had been a feature of the regional economy since the Middle Ages.[35] By the eighteenth century the great days of the Basque iron

---

[34] SA, DD/DN 433 (ledger 1728–32) and 438–439 (waste books 1732–39).

[35] W.R. Childs, 'England's iron trade in the fifteenth century', *Economic History Review*, XXIV (1981), 25–47; Rafael Uriarte Ayo, 'Anglo-Spanish trade through the port of Bilbao during the second half of the eighteenth century: some preliminary findings',

Illustration 2.3. Swedish brands of iron in the eighteenth century.

Courtesy of Riksarkivet.
Caption: This excerpt from the 'stamp book' maintained by the *Bergscollegium* shows the stamps of some of the most renowned forges in Uppland. The clerk has copied the stamps as they appeared to him; that is, as the mirror image of the mark as it would appear on the bars of iron. Heading the list is the 'Hoop L' mark of Leufsta. Österby ('bullets') follows, then Gimo, and Åkerby ('P.L. & Crown' to its English users). Source: "Relation om Bergwärken uti Upland och Roslags samt Giästrike och WästerNorlands BergMästaredöme Åhr 1737", Bergskollegiums arkiv, Bergverksrelationer, m.m. vol E Iif: 4, RA.

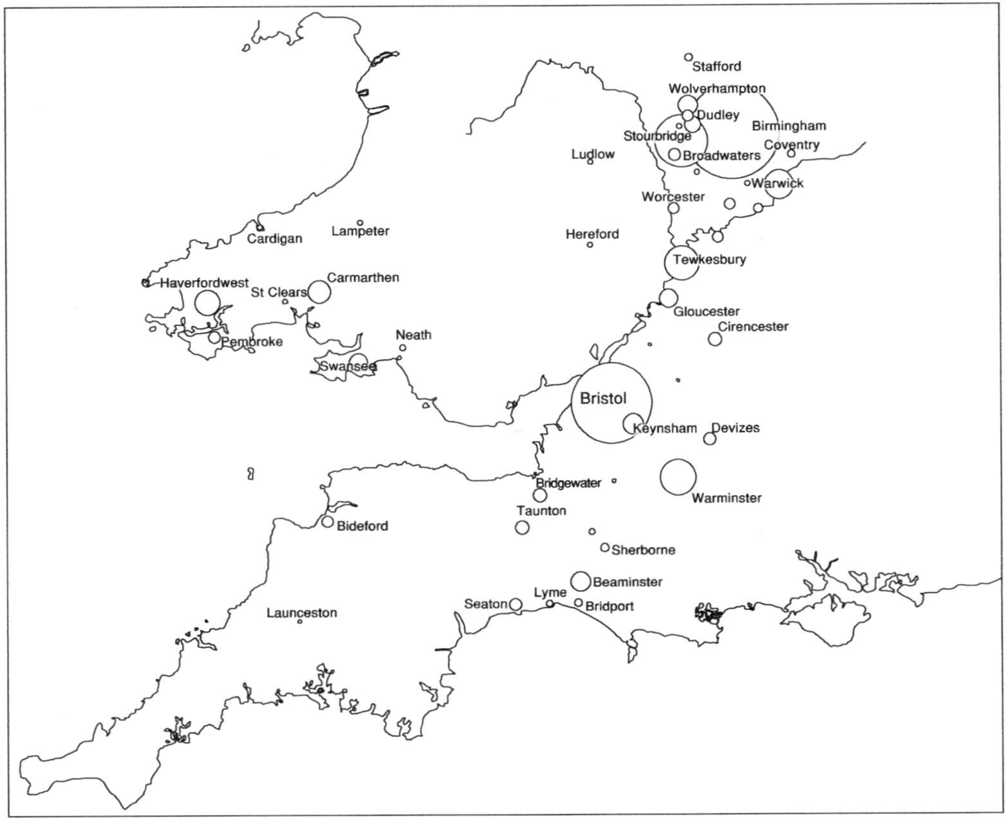

Source: SA, DD/DN 438 and 439

Map 2.1. Graffin Prankard's bar iron sales, 1732–1739.

industry had passed, but in Prankard's time imports to Britain persisted, organised by merchants in Plymouth, Exeter, or Weymouth. Another peculiarity of the southwest that must have held back Bristolian domination was the presence of the Royal Dockyard at Plymouth. Supplies of iron for Plymouth Dock, most of it Swedish, were furnished centrally by Navy Board contractors in London. As a result, it was claimed, the dockyard officers 'sometimes get more than they know how to make use of'.[36] The availability of cheap, navy-surplus iron evidently kept civilian wholesalers at bay in Plymouth's hinterland. Prankard did have a customer base further to the east, however, in Dorset. West Dorset towns such as Sherborne, Bridport, or Lyme were supplied from Bristol.

Using the data from Prankard's waste books the regional distribution of his iron sales between March 1732 and September 1739 can be mapped out. Five regional markets are distinguished: (i) the West Midlands (Worcestershire, Herefordshire, Staffordshire, Warwickshire, and Shropshire); (ii) the West Country (Gloucestershire, Wiltshire, Somerset, Dorset, Devon, and Cornwall); (iii) the city of Bristol itself; (iv) south Wales (Monmouthshire, Glamorgan, Carmarthenshire, Pembrokeshire, and Cardiganshire); and (v) the re-export market. Trifling amounts of bar iron were also sold in London and Ireland, whilst 3.9 per cent of sales (by volume) cannot be located.

The West Midlands was consistently Prankard's largest market, accounting for 37.2 per cent of sales across the period. The region had been home to a variety of specialised metalware producing districts since the sixteenth century.[37] Prankard supplied iron both to general purpose ironmongers-cum-manufacturers such as John Finch of Dudley, and to specialists such as John Podmore, the saw manufacturer of Broadwaters in Worcestershire, or Joseph Farmer the Birmingham gun-maker. The re-export trade—slavers trading to Africa—ranked as Prankard's second largest market, with 24.1 per cent of sales overall. Bristol customers were also significant, with 18.4 per cent overall, and in odd years (1734 and 1739) they bought more than the slave merchants. The city was an important centre of consumption in its own right, and it had a

International Journal of Maritime History, IV (1992), 193–217; *idem*, 'El hierro vasco y los mercados europeo y colonial durante el Antiguo Régimen', *Revista de Estudios Marítimos del País Vasco*, IV (2003), 313–26.

[36] *Angerstein*, p. 86.

[37] Marie B. Rowlands, *Masters and men in the West Midland metalware trades before the Industrial Revolution* (Manchester, 1975).

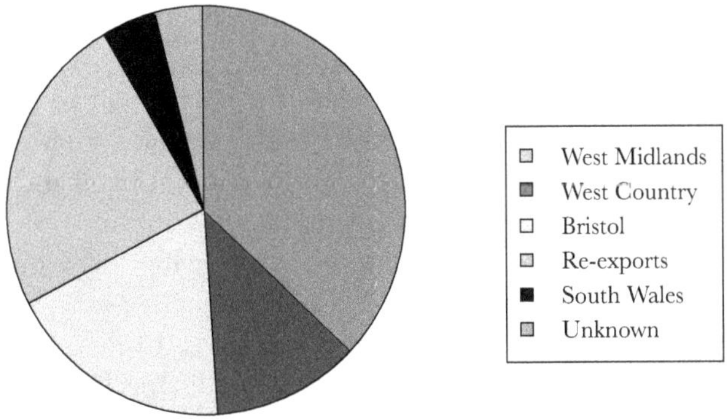

Source: SA, DD/DN 438 and 439

Figure 2.2. Regional distribution of Graffin Prankard's bar iron sales, 1732–1739.

busy industrial hinterland. The West Country and south Wales were comparatively small markets, taking 11.7 per cent and 4.5 per cent of sales respectively. The position is summarised in Figure 2.2.

The dynamics of these different markets can be better understood if account is taken of the types of iron that Prankard sold. He dealt in a wide variety: Swedish, Russian, some English, a very small amount of Spanish, even a little German. Swedish iron accounted for most of his sales, yet Swedish iron was itself a plural phenomenon, coming in a range of qualities, and intended for specialised markets. For example, nearly one-third of Prankard's sales of Swedish iron were of 'voyage iron'. This was a distinctive type of bar, made to very precise specifications, that was traded on the African coast for slaves.[38] Prankard's customers were all to be found in Bristol, Europe's greatest slave port in the 1730s. 'Orground' iron, on the other hand, was used for quite different purposes and found an entirely different market. Made at fewer than twenty *bruk* in the eastern county of Uppland, 'Orground' (so called because most of it was shipped to Stockholm through the Baltic port of Öregrund) commanded a premium price on international markets. It was coveted by a few specialised users for its exceptional toughness. The Navy Board deemed it essential for the manufacture of anchors in

[38] David Richardson, 'West African consumption patterns and their influence on the eighteenth-century English slave trade', in H.A. Gemery and J.S. Hogendorn (eds), *The uncommon market: essays in the economic history of the Atlantic slave trade* (New York, 1979), pp. 303–30.

the King's Dockyards, whilst English steel-makers considered nothing else worthy of conversion to blister steel.

That 'Orground' bars were destined for conversion to steel is confirmed by the regional distribution of Prankard's sales. Figure 2.3 shows the distribution of sales of Swedish 'common sorts', the standard varieties. Major sales were rather evenly divided between the West Midlands, the West Country, and Bristol. Figure 2.4, showing the distribution of 'Orground' sales, reveals a very different pattern. Some 74 per cent of 'Orground' iron went to the West Midlands. In fact, almost all of this iron went to just two customers, John Kettle and Francis Homfray, steel manufacturers of Birmingham and Stourbridge respectively. Much of the 'Orground' iron that was sold in Bristol, the only other market of any significance, was probably used in steel making as well, since the Shallards, who converted iron into steel on Prankard's behalf at their Keynsham furnace, did the same for a number of Bristol ironmongers.[39]

The markets for both 'Orground' and voyage iron were closely defined. Each involved a rather small number of specialist users who bought from Graffin Prankard on a regular basis and did so in bulk. The market for Swedish 'common sorts' was quite different. The number of customers was far larger; they were more evenly distributed through the Severn valley and the West Country; and their purchases were rather more spontaneous and *ad hoc*. This can be seen in the seasonality of Prankard's sales. As is shown in Figure 2.5, sales of 'Orground' iron were bunched in the summer and early autumn, the period in which Prankard's ships ordinarily arrived in Bristol with cargoes from the Baltic.

As 'Orground' iron had been ordered in advance by Prankard's clientele of steel-makers, it could be transferred immediately onto river craft for shipment up-Severn. The sales of 'common sorts', as shown in Figure 2.6, were distributed very differently. There were two distinct peaks: one in January, the other in July. The buyers of 'common sorts', in other words, tended to make their purchases at the Bristol's two major fairs, St Paul's in the New Year, St James's in high summer.

Russian iron displayed quite another pattern. Like 'Orground' iron, it found its main market in the Midlands. The regionality of Prankard's

---

[39] See below pp. 127–51.

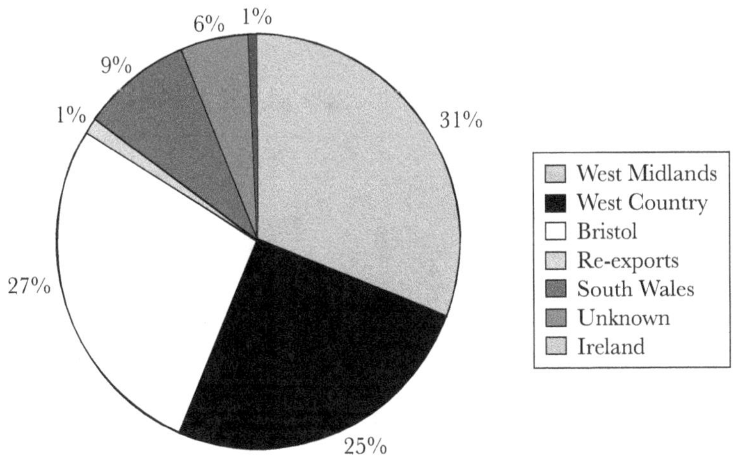

Source: SA, DD/DN 438 and 439

Figure 2.3. Regional distribution of Swedish 'common sorts' sold by Graffin Prankard, 1732–1739.

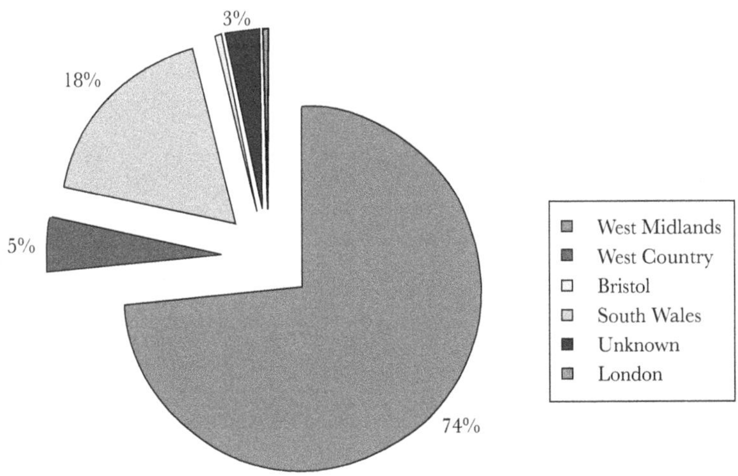

Source: SA, DD/DN 438 and 439

Figure 2.4. Regional distribution of Graffin Prankard's sales of 'Orground' iron, 1732–1739.

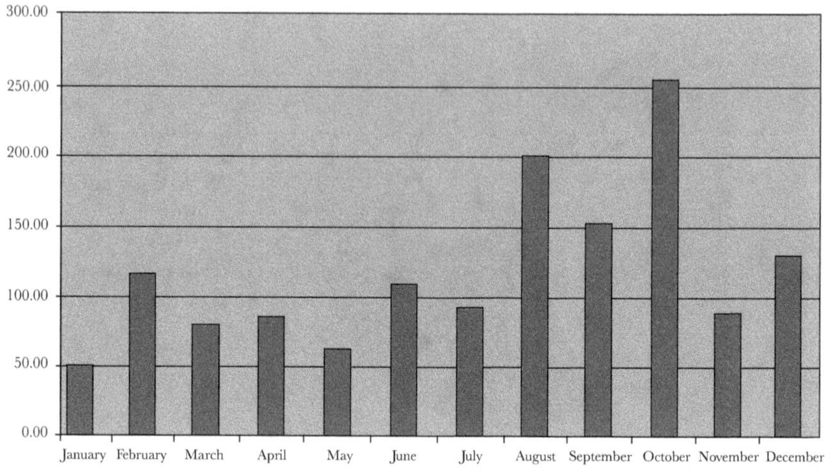

Source: SA, DD/DN 438 and 439

Figure 2.5. Graffin Prankard's monthly sales of 'Orground' iron, 1732–1739.

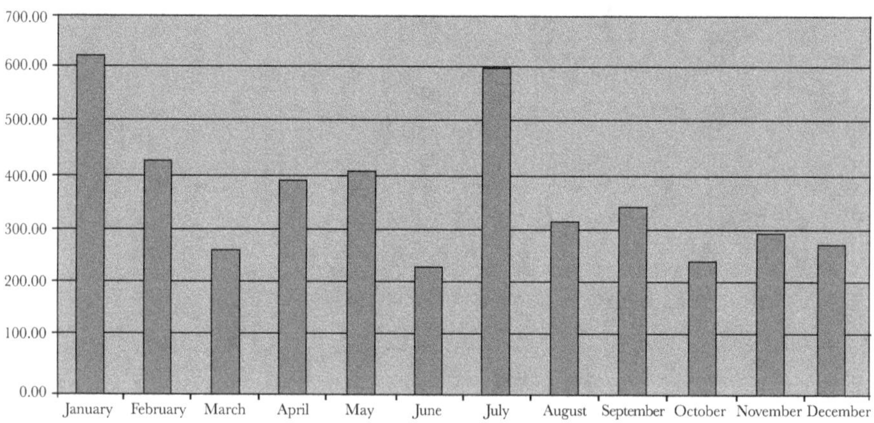

Source: SA, DD/DN 438 and 439

Figure 2.6. Graffin Prankard's monthly sales of Swedish 'common sorts', 1732–1739.

sales is striking in this respect (Fig. 2.7). The West Midlands accounted for 66 per cent of the 2,495 tons sold. But Russian iron, a rather brittle, 'coldshort' metal, was slit into nail rods, not converted to steel. Indeed, no less than 42 per cent of *all* the Russian iron Prankard sold went to a single customer, Sampson Lloyd, owner of the slitting mill at Birmingham.[40] A further 536 tons was sold in Bristol, but much of this was to disappear into the maw of the Midlands nail trade as well, having been processed at the slitting mill run by the Bristol merchant William Donne at Congresbury, a dozen miles to the west of the city.[41]

That Russian iron was destined for the nail trade is confirmed by the seasonal distribution of Prankard's sales, as shown in Figure 2.8. The seasonal pattern is stark. Almost nothing was moved up the Severn valley in the spring and early summer months. Deliveries began in earnest in the late summer and accelerated during the autumn, reaching a crescendo in December. This distribution conforms to what might be expected of an industry that was water-dependent. In dry summers a slitting mill would only work intermittently; it was in the autumn and winter, with adequate water supplies, that the rolls could turn continuously.

Different sorts of iron followed different paths when they entered the British market. They were used by consumers who had very varied requirements of what they bought. This had consequences for the ways in which production was shaped and commodities traded east of the Sound. The backward linkages from the British market to *Bergslagen* and beyond must be traced if the impact of the Atlantic world on the Baltic and its vast hinterland is to be understood. The first of these backward linkages is that which led from John Kettle's cementation furnace on Steelhouse Lane in Birmingham to the De Geer family's estate at Leufsta, the source of much of the 'Orground' iron that Kettle converted into steel.

40 H. Lloyd, *The Quaker Lloyds and the Industrial Revolution* (1975).

41 G. Bedingfield, 'Congresbury's mills and the iron industry', *Bristol Industrial Archaeology Society Journal*, 31 (1998), 28–30.

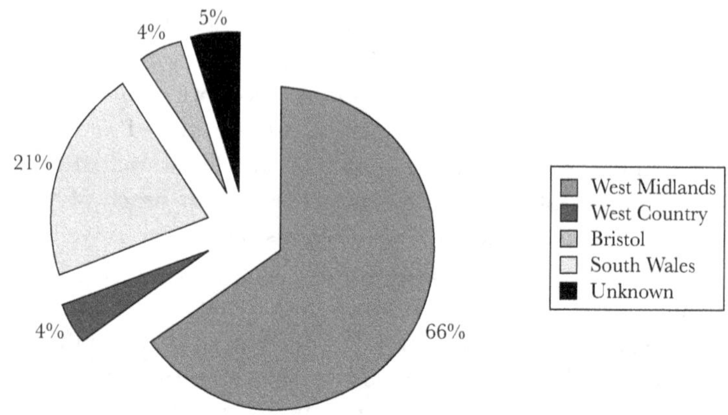

Source: SA, DD/DN 438 and 439

Figure 2.7. Regional distribution of Graffin Prankard's sales of Russian iron, 1732–1739.

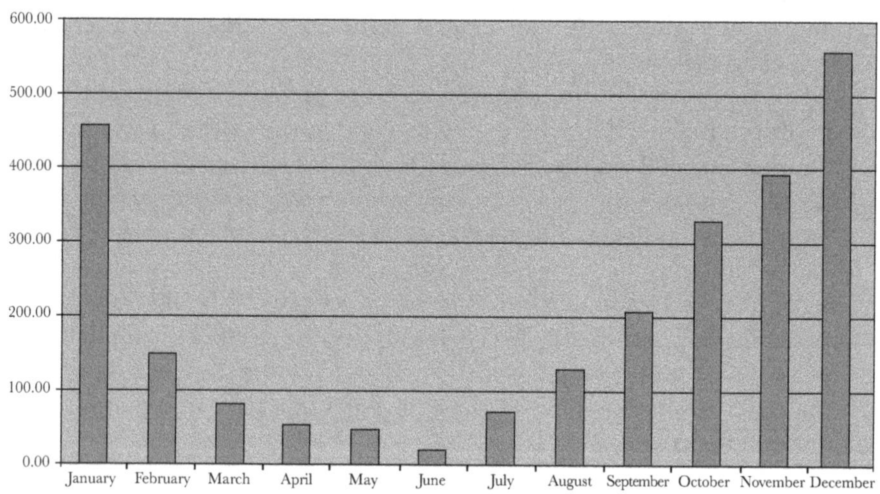

Source: SA, DD/DN 438 and 439

Figure 2.8. Monthly distribution of Graffin Prankard's sales of Russian bar iron, 1732–1739.

Leufsta Bruk

There had been a forge at Leufsta since the sixteenth century, yet the settlement that welcomed visitors in the 1730s was almost entirely new. The *bruk* had been remodelled after the devastation wreaked by Russian raids along the Uppland coast in the closing stages of the Great Northern War. Russian troops marched into Leufsta on 25 July 1719. Within a few hours they had destroyed the manor house, the church, the workers' housing, and most of the industrial facilities. The destruction was so complete as to leave the *bruk* authorities a *tabula rasa* on which to work. The map of the new *bruk*, drawn up in 1735, reveals their response.

The most striking feature of the settlement was the long, tree-lined avenue (*bruksgatan*), guarded at each end by ornamental gates, that ran north-south. 'On one side of the street', wrote Christer Berch, a young intellectual who visited Leufsta in 1753, 'live the workers (*bruksfolket*) in well-built and uniform houses.'[42] Each cottage housed two families in separate apartments, each consisting of a large room (*stuga*) with a fireplace and an oven, and a smaller room (*kammare*). There were perhaps 80 such apartments. Running parallel with the rear of *bruksgatan* was the 'cattle street' where barns, coldstores, and cattle sheds for the use of the workers were ranged.

At the centre of *bruksgatan* were two stone-built structures that loomed above the wooden cottages. Here was the seat of day-to-day secular and sacred authority in the *bruk*: the *bruk* office and the church. At one, the works *directeur* and his clerks monitored the work performed at the furnace and the forges. At the other, the pastor expatiated on the duty that the *bruksfolk* owed to their Creator. Just to the east, set back among formal gardens, was a further and still more imposing seat of authority: the manor house (*herrgård*) of the De Geers. This was a residence of considerable magnificence. Indeed, it was a miniature palace of the late baroque age.

The industrial plant was distributed along a stream that flowed roughly south-north through the community. This was dammed at four points to provide power for the forges and blast furnace. *Opphammaren* (the upper forge) lay in the forest southwest of the village (in the top

[42] 'Dagbok öfver en Resa till Roslags Bergslagen hållen 1753 af Christer Berch. Stipend Wred.' KB, M 172, folios 44ff. and 61.

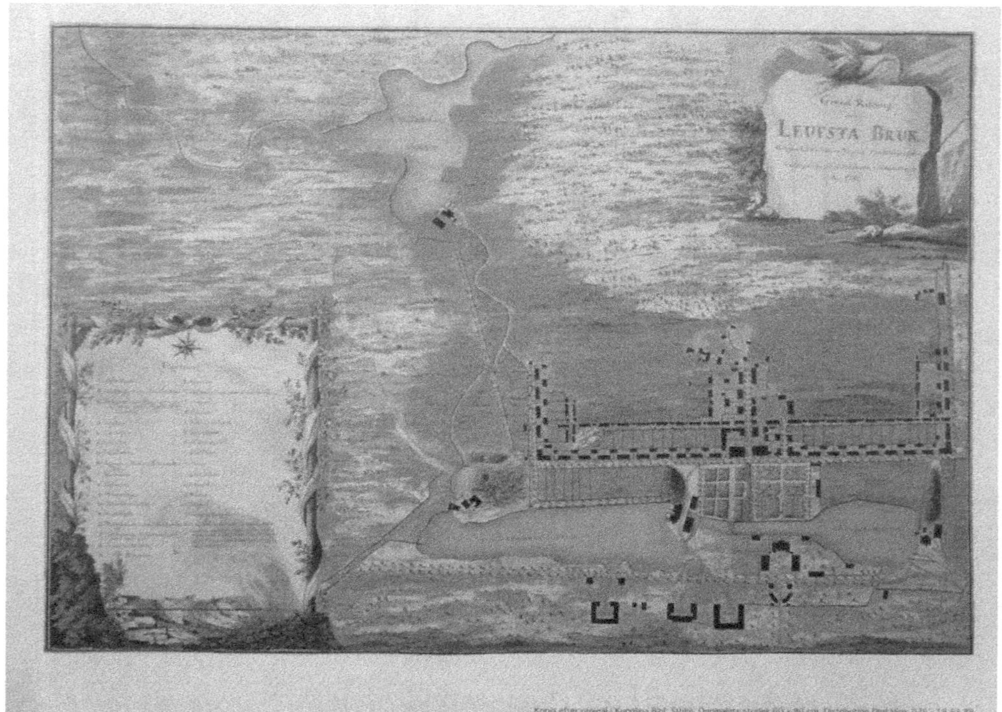

Illustration 2.4. Leufsta *bruk* in 1735.

Courtesy of Kungliga biblioteket, Stockholm.
Reference: KB, KoB, Tilas IV:5 nr: 55.

Illustration 2.5. Leufsta *herrgård.*

Courtesy of Jernkontoret.
Caption: The *herrgård* in the mid-eighteenth century, viewed from the north, from the direction of the lower forge. To the right is the *bruksgatan*, leading to the church and, just beyond it, the *bruk* office.

left-hand corner of Illustration 2.4). The blast furnace stood at the southern entrance to the village. Downstream, sitting side-by-side on top of the dam that penned back the main forge pond, were *Storhammaren* and *Spikhammaren* (the big forge and the nail forge). *Nederhammaren* (the lower forge) lay a little to the north, at the end of the smaller forge pond.

Because Christer Berch arrived at Leufsta on Saturday afternoon the *bruk* was unusually quiet. 'The forges were completely silent, as the forgemen were busy on Saturdays, as is their custom, weighing and measuring the production they had made during the week'. The blast furnace also stood idle, adding to the unnatural stillness. But on other days the thump of forge hammers, the roar of water, and the creaking of wooden machinery would have contributed to a distinctive soundscape, announcing all too clearly that here was a major industrial site. Leufsta, with its blast furnace and four forges, was among the largest *bruk* in Sweden.

Surviving tax ledgers indicate that between 400 and 450 people lived at Leufsta in the 1740s. These numbers do not include children below the age of fifteen. Studies of other *bruk* indicate that children made up between one third and one quarter of the total population, so the number of *bruksfolk* at Leufsta probably amounted to around 600. Many of the male household-heads in these tax ledgers were ironworkers—around 50 can be characterised as such—but other households were headed by artisanal workers whose presence is confirmed in an inventory of 1741 that describes workshops for joiners, carpenters, wheelwrights, farriers and blacksmiths, as well as a 'knitting house' and a corn mill. The largest group of Leufsta workers were day workers. Even with a population of 600 persons, Leufsta was a large community by Swedish standards. It was, Christer Berch noted, more like 'a neat town in the Dutch fashion than a *bruk*: and we hope that this place will be given the privileges of a town'. Indeed, Leufsta was sometimes styled 'Leufstad' (Leuf-stad) to denote its urban credentials.[43]

---

[43] Leufsta Arkivet, vol. 168 and 202, RA; Leufsta bruksarkiv, vol. 270; Hans Norman, 'Befolkningsförhållandena vid två uppländska vallonbruk. En studie av Forsmark och Söderfors 1775–1855', in Anders Florén & Gunnar Ternhag (eds), *Valloner—järnets människor* (Hedemora, 2002), pp. 177ff. Berch, 'Dagbok... 1753', folio 50. Two complete lists of inhabitants do exist, for the years 1749 and 1762, but they are not confined to Leufsta *bruk*; they include various people connected to the *bruk* but living outside the core community such as those who worked at the coastal warehouses at Löten and Ängskär. These lists indicate a total population of 1200 people.

Leufsta's Dutch appearance was quite appropriate, for the *bruk* embodied Dutch capital, and many of its inhabitants could trace their ancestry to the Low Countries. About 1,200 Walloons had migrated to Sweden between 1620 and 1655, following the flow of Dutch investment organized by Louis De Geer, Willem de Besche, and other Amsterdammers.[44] It was to Uppland, to the group of forges that ringed the mine at Dannemora, that most of the migrants from the southern Netherlands gravitated, using the forging methods of Wallonia to produce 'Orground' iron. De Geer owned three of the largest 'Walloon ironworks' (*Vallonbruk*)—Leufsta, Österby, and Gimo—and controlled perhaps one third of 'Orground' production.[45]

The links between the De Geer family and Amsterdam remained strong throughout the seventeenth century. Although the Dutch share of Sweden's iron export fell rapidly after 1650 in the face of British competition, iron from the *Vallonbruk* continued to be marketed via Amsterdam. 'Orground' iron, it was said in 1701, was exported nowhere 'save to Holland...so yt it's only to be had through Holl[and] factors'.[46] Even in the 1720s, when British merchants dominated the export of 'common sorts' from Stockholm, iron from Leufsta, Österby and Gimo was still shipped to Amsterdam by the Grills, a merchant dynasty of Dutch origins.[47] It would require a new source of demand to wrench 'Orground' iron from Dutch hands. The British steel industry was to provide that demand, but the keenness of its appetite did not become apparent before 1730.

The Uppland forges were unusual in their fidelity to the commercial patterns of the seventeenth century, but their conservatism reflected a wider lassitude. The investments made between the 1620s and 1650s had not been sustained, so that many *bruk* were visibly run-down by the early eighteenth century. Problems at Dannemora exacerbated the situation. As the mine was driven to ever deeper levels the cost of

---

[44] Bernt Douhan, *Arbete, kapital och migration. Valloninvandringen till Sverige under 1600-talet* (Uppsala, 1985), pp. 194ff.

[45] The standard biography of Louis De Geer is Erik W. Dahlgren, *Louis De Geer 1587–1652. Hans lif och verk* (Uppsala, 1923). His industrial undertakings are discussed at pp. 375ff.

[46] Mitchell Library, Glasgow, SR 352, Adam Montgomerie to John Crosse Senior & Co, 27 April 1701.

[47] Leos Müller, *The merchant houses of Stockholm, c. 1640–1800: a comparative study of early-modern entrepreneurial behaviour* (Uppsala, 1998), pp. 84ff, and GS to JJDG 5 April, 1731, Leufsta Arkivet, RA.

drainage mounted and with it the cost of ore, and when most of the pumping equipment was destroyed in a massive cave-in in 1693 production was halted for several years.[48] Thus, the Russian fury of 1719 capped a long period of decay. Leufsta was the most grievous sufferer, with damage estimated at 350,000 *daler silvermynt*, half the total losses incurred in that traumatic summer, but Leufsta was not alone.[49] The *bruk* at Harg and Forsmark had already been razed, and after leaving Leufsta the Russians went on to level Åkerby and Wesslands.[50]

The catastrophe of 1719 sparked a renewal, one led by the De Geer family. They, after all, had been the principal losers. Attacks had been made not just on the family *bruk* at Leufsta and Åkerby, but on the coastal warehouses of Gimo and Österby as well. Stung by these setbacks, they reverted to the expansionist policies of the great Louis De Geer. Indeed, they revived an ambition that had been unfulfilled at the time of Louis De Geer's death in 1652, that of monopolising 'Orground' iron. The architects of the new strategy were Carl De Geer, the proprietor of Leufsta and Åkerby, his brother Jean Jacques De Geer, and the latter's three sons, Louis, Charles and Antoine.

The wrecked *bruk* were rebuilt and reorganised, taking advantage of the eight-year tax holiday granted by *Bergscollegium* to victims of the Russian raids. Only one of the old Leufsta's two blast furnaces was rebuilt; the other was relocated to a new *bruk*, Carlholm, authorised by *Bergscollegium* in 1728. Hargs *bruk*, one of the largest in the region, was purchased at the same time, as were several landed estates. The acquisition of new estates was of crucial significance, for the leaseholders who cultivated the land paid their rent in charcoal, thereby sustaining industrial production at the furnaces and forges. The foundations were being laid for an increase in production. The effects were soon felt: Georg Swebilius, the manager at Leufsta from 1722 until his death in

---

[48] Svante Lindqvist, *Technology on trial: the introduction of steam power technology into Sweden, 1715–1736* (Uppsala, 1984), pp. 229–31; J. Wahlund, *Dannemora grufvor. Historiska skildring* (Stockholm, 1879), pp. 72ff. See also 'Relation om Bergwärken uti Upland och Roslags samt Giästrike och WästerNorlands BergMästaredöme Åhr 1737', Bergskollegiums arkiv, Bergverksrelationer, m.m. vol. E Iif: 4, RA.

[49] 'Relation om Bergwärken uti Upland och Roslags samt Giästrike och WästerNorlands BergMästaredöme Åhr 1737', Bergskollegiums arkiv, Bergverksrelationer, m.m. vol. E Iif: 4, RA, and Magnus Mörner, 'Vår östersjökust nedbränd 1719–1721. Terror avpassad för att framtvinga fred', *Militärhistorisk tidskrift* (2004), 178 and 186.

[50] For a description of events at Leufsta, and other ironworks in Uppland, see Bergskollegiets Arkiv, Bergverksrelationer Uppland och Västernorrland, vol. E II f:1, folio 929–991.

1735, noted that bar iron output at Leufsta and Åkerby rose from 675 tons in 1722 to 1,200 tons at the end of the decade.[51]

When Carl De Geer died childless in 1730 the best part of his industrial empire was bequeathed to his nephew Charles. The ten-year-old inherited not just Leufsta *bruk*, with its blast furnace and four forges, but Åkerby *bruk* (a furnace and a single forge), and the furnaces at Toboborg and Carlholm. During the young heir's minority, which would run until 1741, this formidable assemblage, known as *Leufstawerken*, was to be administered by his brother Louis De Geer.[52]

Carl De Geer's programme of refurbishment was copied by his brother Jean Jacques, who had superintendence of Österby and Gimo. Both *bruk* were dilapidated. Georg Swebilius thought Gimo in need of 'total reformation and reconstruction'; the workers were 'in their nature spoiled', requiring 'correction' by an experienced manager.[53] When Jean Jacques acquired Gimo outright in the early 1730s the necessary steps were taken. New charcoal-yielding estates were bought, as was the small *bruk* at Wellnora. The forge at Wellnora was immediately closed, but its furnace was enlarged. At Gimo the reverse took place: the furnace was downgraded while forge capacity was enhanced. The outcome was a rationalization of plant and forest resources. Bar iron output at Gimo duly increased.[54]

By 1732 the holdings of the De Geer family in Uppland had been substantially extended and consolidated. *Leufstawerken*, Österby and Gimo were united under a single management. Since Jean Jacques, the head of the family, remained on his estate at Rhijnhuizen near Utrecht, active management was delegated to his son Louis and Georg Swebilius. Young Louis directed operations from *De Geriske Stenhuset*, the family's Stockholm mansion; Swebilius ran the office at Leufsta. The

[51] 'Relation om Bergwärken uti Upland och Roslags samt Giästrike och WästerNorlands BergMästaredöme Åhr 1737', Bergskollegiums arkiv, Bergverksrelationer, m.m. vol. E Iif: 4, and GS to JJDG 5 April 1731, Leufsta Arkivet, vol. 106, RA.

[52] In his will Carl De Geer had stated that Baron Eric Oxenstierna, his nephew and the owner of Hargs *bruk*, and the *bruks directeur* Georg Swebilius should act as guardians to his heir Charles De Geer until his he came of age. Neither of them was willing to shoulder this responsibility, so Charles's father Jean Jacques became his guardian, and then, after the death of Jean Jacques in 1738, his brother Louis De Geer. See Leufsta Arkivet, vol 5 and 106, GS to JJDG, 21 December 1730, RA.

[53] GS to JJDG, 24 April 1732, Leufsta Arkivet, vol. 106, RA.

[54] 'Relation om Bergwärken uti Upland och Roslags samt Giästrike och WästerNorlands BergMästaredöme Åhr 1737', Bergskollegiums arkiv, Bergverksrelationer, m.m. vol. E Iif: 4, and GS to JJDG, 1730–1734, Leufsta Arkivet, vol. 106, RA.

De Geers now controlled about half the output of 'Orground' iron. In 1733 additional plant was acquired when Ullfors *bruk* was bought on behalf of young Charles. The following year three more *bruk* were added to his patrimony: Wessland, Hillebola and Strömsberg. Hillebola was absorbed into *Leufstawerken*, while Ullfors, Wesslands and Strömsberg were united in a new entity, *Strömsbergwerken*. The new combine consisted of three blast furnaces and three forges, to be added to the five furnaces and six forges in *Leufstawerken*.[55] The expansion drive culminated in 1738 with the acquisition of the prestigious *bruk* at Forsmark, one of the estates that the great Louis De Geer had most coveted but which had always eluded him. His great-grandchildren, after much devious manoeuvring, succeeded where he had failed.[56]

By 1740 Jean Jacques De Geer's three sons controlled three-quarters of the make of 'Orground' iron: 4,000 of the 5,500 tons forged yearly. Charles owned *Leufstawerken* and *Strömsbergswerken*, Antoine had inherited Österby and Forsmark, while Louis was the master of Gimo and the lessee of Wattholma.[57]

The changes sweeping the *Vallonbruk* in the 1720s and 1730s coincided with reform at Dannemora. It was the quality of Dannemora's ore that underpinned 'Orground' iron's international reputation. Eric Touscher, who succeeded Swebilius as general manager of *Leufstawerken* in 1735, was adamant on this point. In 'En liten handbok angående Leufsta Bruk', a *vade mecum* prepared for Charles De Geer on the occasion of the young *brukspatron*'s visit to Leufsta in 1739, *Directeur* Touscher gave thanks for the mineral bounty that providence had bestowed on the Crown, the Swedish nation, and the De Geer family. The mine was a source of immense 'utility and subsistence'. Without Dannemora, he proclaimed, the *Vallonbruk* would cease to exist; but, he was careful to add, without De Geer capital the mine would long ago have languished.[58]

---

[55] For this and following paragraphs, see Bergmästarämbetet i Gävleborgs, Uppsala och Stockholms län, Bergmästarens tjänsteberättelser 1737, ULA.

[56] Österby bruksarkiv, vol. B 1: 8; Forsmarks bruksarkiv, vol. K 22–24, and F Ib:6; Vattholma bruksarkiv, F2:23, ULA; LDG to ET 19 1736–1738, Leufsta Arkivet, vol. 105, RA.

[57] Bergmästarämbetet i Gävleborgs, Uppsala och Stockholms län, Bergmästarens tjänsteberättelser 1737, ULA. See Göran Rydén, 'Vallonbruk, vallonsmeder och vallonsmide—en precisering av ett kunskapsläge,' in Florén and Ternhag, *Valloner*, pp. 107–35, for further discussion of the expansion in the production of 'Orground' iron. The influence of the De Geer family also extended to Harg *bruk*, making about 500 tons of 'Orground' annually, which was owned by Louis De Geer's brother-in-law.

[58] 'En liten handbok angående Leufsta Bruk &. Wälborne Herr Carl de Geer, wid ankomsten i Orten af En Des Tienare, öfwerlemnat 1739', in RA, Leufstaarkivet, kartong 152 (hereafter 'En liten handbok'), pp. 72–79.

Map 2.2. Uppland and its ironworks in 1742.

Courtesy of Uppsala Universitetsbibliotek.
Reference: UUB, Kart och bildenheten.
Caption: This detail from Georg Biurman's *Charta öfwer Upland och Södertörn* (1742) has the Dannemora mine (1) at its centre, with the university city of Uppsala at its foot. Between Dannemora and the Baltic coast was a ring of *Vallonbruk*: Österby (2), Gimo (3), Harg (4), Forsmark (5), Leufsta (6), Akerby (7), Wesslands (8), Carlholm (9), Strömsberg (10), and Ullfors (11). The port of Öregrund, which lent its name to 'Orground' iron, lay between Harg and Forsmark.

Louis De Geer and his successors had indeed played a crucial role in the development of the mine. Yet for all that, Dannemora remained in the possession of the Crown throughout the seventeenth century. The royal *Bergmästare* (Mine Inspector), as the presiding officer of the Mine Court, determined which *bruk* was to extract ore from what location within the mine. Over time, a variety of customary practices emerged. Certain *bruk* established an exclusive *de facto* right to take ore from particular sectors of Dannemora. More usually, however, parts of the mine were classified as *companigruvor*—mines that were to be exploited collaboratively by different *bruk*. A *companigruva* was worked by several *bruk* sequentially. At *Ödesgruvan*, for instance, Leufsta had the right to extract ore for four weeks. Then, proprietorial rights would pass to Österby for four weeks, then to Gimo for two further weeks before the cycle started again. Each *bruk* had its own mine bailiff who hired miners from the 350-strong *corps* that worked at Dannemora. These miners would work at their designated place for their allotted period and then cede the workings to another work-crew, hired by a bailiff from a different *bruk*.[59] It was a system that made for confusion, rancour and wastefulness.

The first steps towards reform came in 1723 when formal ownership of Dannemora was transferred from the Crown to 19 *brukspatroner*.[60] The state, in the person of the *Bergmästare*, retained overall oversight, but greater authority now passed to the *brukspatroner*. The De Geers took advantage of this, appointing in 1731 a single bailiff to coordinate ore extraction from the family's various mining concerns. These were growing rapidly in tandem with the family's tightening grip on iron making in Uppland. The gains in efficiency were clear for all to see, which encouraged *bruk* that were not owned by the De Geers to subscribe to the more centralised system. By 1737, the *Bergmästare* reported, only Elfkarleby, Schebo, Wattholma and Ljusne stood outside.[61] The organisational changes were accompanied by technological refurbishment as the mine workings were driven into deeper, harder strata. Fire-

[59] This picture of Dannemora is based on remarks by Touscher in his 'En liten handbok'.

[60] Wahlund, *Dannemora grufvor*, pp. 16ff. See also 'En liten handbok', p. 77, and Bergmästarämbetet i Gävleborgs, Uppsala och Stockholms län, Bergmästarens tjänsteberättelser 1737, ULA. Two of the nineteen *bruk* extracting ore at Dannemora were located outside the area, along the northern coast of the Baltic.

[61] Bergmästarämbetet i Gävleborgs, Uppsala och Stockholms län, Bergmästarens tjänsteberättelser 1737, ULA.

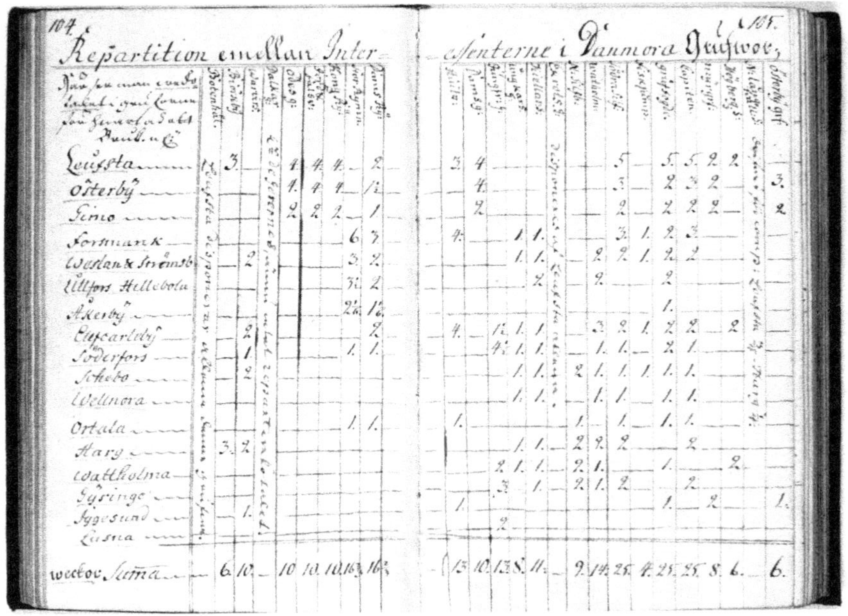

Illustration 2.6. The allotment of working weeks at the Dannemora mine, 1739.

Courtesy of Riksarkivet.
Reference: Leufstaarkivet, vol. 152, pp. 104–05.
Caption: This excerpt from Eric Touscher's 'En liten handbok' reveals the byzantine complexity of mining rights within Dannemora. The 17 *bruk* that worked the *companigruvor* are listed in the left-most column. Across the top are the names of the 25 *companigruvor*. *Ödesgruvan*, which was divided in a fairly simple manner between Leufsta, Österby and Gimo, is the fifth mine from the left. The division of other mines was usually more complicated. *Werviergruvan*, the third from the left, for example, was worked by Wessland and Strömsberg for 2 weeks, then Elfkarleby for two weeks, Söderfors for the following week, Schebo for the two weeks after that, Harg for the ensuing fortnight, and then Iggesund for one week.

setting, which had been the standard technique for breaking up rock in the seventeenth century, gave way to gunpowder in the 1730s. The same decade saw heavy investment in lifting equipment and drainage technology, including a Newcomen machine, Sweden's first.[62]

The 1730s were therefore years of radical change at Dannemora. The De Geers, by taking control of several *bruk* that took ore from the mine, concentrated shares in Dannemora in their hands. This, in turn, allowed them to enforce organisational changes and technological innovation. Touscher's boast in 1739 that the De Geers commanded 'the best mines' within the Dannemora complex was no empty one. And access to the best quality ores further facilitated the De Geers' programme of acquisition. When Louis De Geer leased Wattholma in 1736, he did so with the promise that he would use his own ore resources to enhance the quality of Wattholm iron, which had in recent years been in disrepute.[63]

Once ore had been hauled to the surface it was piled up in anticipation of the winter, for it was only after the snows had come that the movement of ore to the *bruk* could begin. Tellingly, the routes from Dannemora to the surrounding *bruk* were measured in 'winter distance'; that is, the distance along the icy tracks that the ore-laden sledges followed. Ore was measured in *lass*, equivalent to the load of a single sledge. In the first half of the eighteenth century 35,000 *lass* left Dannemora every winter. Of these, 4,000 sledge-loads went to Österby, 2,200 to Gimo, and 4,500 to Forsmark. The largest number of sledges, however, left for Charles De Geer's domains, with more than 6,000 going to *Leufstawerken* and 4,000 to *Strömsbergswerken*.

The deliveries were made to the different blast furnaces, where the ore was given a preliminary roasting in open pits before being tipped, together with a measure of charcoal, into the flaming throat of the furnace. The furnace was a massive stone structure, some six or seven metres high. Attached to it were ancillary buildings that sheltered the casting area and the bellows; adjacent was a timber-built charcoal shed, usually far larger than the furnace itself. Despite the size of the furnace complex, it was run by a very small workforce. A furnace keeper (*masmästare*) took charge, helped by an assistant keeper (*hyttdräng*), two chargers (*uppsättare*), two ore crushers (*bokare*), and an ore-carrier (*malm*

[62] Bergmästarämbetet i Gävleborgs, Uppsala och Stockholms län, Bergmästarens tjänsteberättelser 1737, ULA. See also Lindqvist, *Technology on trial*.

[63] 'En liten handbok'; Vattholma bruksarkiv, vol. F2: 23, ULA.

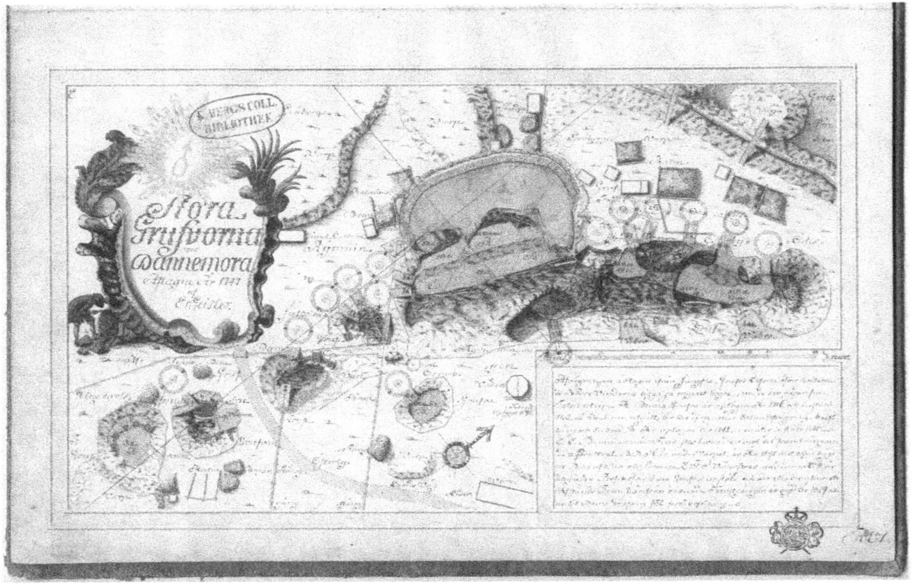

Illustration 2.7. Dannemora Mine.

Courtesy of Riksarkivet.
Reference: RA, Kommerskollegii Gruvkartor Uppsala, Dannemora Gruva, nr. 12.
Caption: 'We never thought to have a better notion of rocky Arabia then when we saw Dannemora', wrote Christer Berch on visiting the mine in 1753. Scattered about the surface were 'terrible piles of iron-stone and waste rock'; below, 'frightful abysses' yawned. The map reproduced here is the first sheet of an album prepared in 1747; each succeeding sheet descended further and further below ground, allowing the reader to build up a three-dimensional image of the workings. Mining had begun in the late fifteenth century; by the eighteenth century the excavations were immense, plunging deep into the earth. The circular features that fringe the ore pits were *Hästvandringar*, horse-drawn winding gins: horses plodded around the tracks, pulling tubs of ore to the surface. To the left of the cartouche a man is shown bent over, scrutinising a lump of the precious ore.

Table 2.1. Pig iron output at the blast furnaces of *Leufstawerken* in 1736.

Tobo	3,797 *skeppund*	738 tons
Carlholm	2,666 *skeppund*	518 tons
Strömsberg	2,447 *skeppund*	476 tons
Hillebola	2,302 *skeppund*	448 tons
Ullfors	2,195 *skeppund*	427 tons
Wessland	1,774 *skeppund*	345 tons

Source: Leufsta bruksarkiv, vol. 269. Data on pig iron production at Leufsta and Åkerby have not survived.

skutare). Once the furnace was ready to be charged, smelting continued day and night, with one half of the furnace crew alternating with the other every twelve hours.[64]

The furnace was tapped twice a day, allowing the liquid iron that had accumulated in its hearth to gush out, sparking and hissing, into a long depression in the floor of the casting house, before solidifying into a *geuse* that weighed about 1½ tons. The number of *geuses* made annually could vary substantially according to the availability of water, the supply of charcoal, the state of repair of the furnace, and the efficiency of the furnace crew. The output of *Leufstawerken*'s furnaces in 1736 demonstrates the point.

As *Directeur* Touscher maintained that a blast furnace should be able to turn out about 20 *skeppund* of *tackjärn* (cast iron) daily, it would appear that Tobo furnace was in blast for 24 weeks but Wessland for just 11 weeks.

The furnaces depended upon charcoal as an energy source. Most of this was supplied by the *bruk*'s tenant farmers in accordance with their leasehold agreements with the *brukspatron*. Indeed, ironworks estates were designed to ensure that charcoal production and iron making remained in balance, that industrial production did not press too hard on forest resources. The *bruk* tenantry delivered *Egna kol* ('own charcoal'). It was supplemented with *Köpekol* ('bought coal')—charcoal purchased from local freeholders. The market for charcoal was not, however, a free one; it was state-regulated. Freeholders who produced for the market had to sell to specified buyers at a fixed rate.

[64] Wahlund, *Dannemora grufvor*, pp. 131ff; 'En liten handbok', pp. 116ff. See also Sam Owen Jansson, *Måttordenboken* (Stockholm, 1995), p. 155.

Illustration 2.8. Österby *bruk* by Elias Martin.

Courtesy of Jernkontoret.
This detail of Martin's drawing, made towards the end of the eighteenth century, shows Österby's squat blast furnace. A ramp rising from the left allows charcoal and ore to be carried up to the charging platform. The furnace was tapped below, in the roofed-over area. Workmen are shown using a windlass to drag a *geuse* from the casting house.

Together, leaseholders and freeholders could realise prodigious quantities of charcoal: some 28,000 cubic metres were delivered to *Leufstabruk* in 1735 by 224 peasants.[65] Charcoal that was not used at the blast furnaces was destined for the forges. Indeed, the four forges at Leufsta were voracious consumers of fuel, for they processed the *tackjärn* not just from Leufsta's own blast furnace but from Tobo and Hillebola as well. Each forge had two hearths after the Walloon fashion: one, the finery, at which *tackjärn* was melted down and refined; the other, the chafery, at which the refined metal was reheated before being drawn out into bars under the forge hammer. The German forging technique employed at most Swedish ironworks used just one hearth for both refining and reheating, and a single forge crew was responsible for both the fining of the metal and the making of the bars. In Walloon forging the workforce was more differentiated, with finers and hammermen playing specialised roles.

The forge crew was made up of ten workmen, five of whom were always at work. One shift consisted of a master finer (*mästersmältare*) and his apprentice at the finery; a master hammerman (*mästerräckare*) and his helper at the chafery; and the *goujar*, the charcoal carrier, who served both hearths. The other five forgemen made up the workforce at the other shift. Sometimes an additional helper (*hielpekarl*) might be present. Although the work was divided among those who worked at the finery and those who attended the chafery, and between two different shifts, the ten-strong forge crew was considered to be a single unit. No disinction was made between the output of the different shifts. The forge crew was paid collectively for the entire week's make. The master finer and hammerman were each paid one-and-a-half *kopparmyn*t per *mille*, the traditional Walloon weight equivalent to 510 kilograms. The others were paid proportionately less, as the evidence from *Opphammaren*, presented in Table 2.2, suggests.

[65] Despite the efforts that were made to keep iron production in step with what the resource base of the *bruk* would allow, there are signs that the authorities at Leufsta permitted peasants to overstep the ecological limits. Leaseholders often delivered a surplus over and above what was stipulated in their contracts, for which they were paid in cash. Peasants were thereby encouraged to cut more timber than could be replenished within the usual growth cycle. The tendency to over-harvest may explain the gradual decline in charcoal deliveries across Leufstawerken from the 1750s onwards. See chapter 5 of Arnold Renting, *I Skuggan av Lövsta bruk. Järnbruksrörelsens inverkan på agrarsamhället in norra Uppland 1630–1930* (Stockholm, 1996), although the author is not familiar with the structure of Swedish iron industry and fails to distinguish between Leufsta *bruk* and *Leufstawerken*, which makes using his findings rather problematic.

Table 2.2. The forgemen at the upper forge at Leufsta *bruk* 1736.

Title	Name	Paid per Mille
Mästersmältare, Master finer	Jacob Tillman	1½ *kopparmynt*
Räckarmästare, Master hammerman	Michael Gilliam	1½
Mästersven, Finer's hand	Carl Bovie	1¼
Räckardräng, Hammerman's hand	Eric Gilliam	1–1¼
Räckardräng, Hammerman's hand	Philip Louison Bovie	1–1¼
Smältardräng, Apprentice finer	Anton Holm	1–1¼
Smältardräng, Apprentice finer	Per Gilliam	1–1¼
Goujar, Charcoal carrier	Jacob Bovie	½
Goujar, Charcoal carrier	Jean Claeson Martinell	½
Hielpekarl, Helper	Eric Jägare	½

Source: Leufstaarkivet, vol. 116

This was a stable workforce. In the second half of the 1730s many of the Leufsta master forgemen were more than fifty years old—two were over sixty—and most of them had spent their entire working life in the service of the De Geer family. There was a clear dynastic element at work, for most master forgemen had brought up their sons to follow them in that service. Eight of the forty men at work in 1738 bore the name of Tillman, and four of them were master finers or hammermen. Other prominent families included the Gilliams, Boives, Bonneviers, and Martinells. As these names suggest, all were of Walloon origin and had been present in *bruk* in Uppland since the first half of the seventeenth century.[66]

The finers and hammermen worked relentlessly. A pattern of four-hour shifts (*tourneijs*) allowed production to continue around the clock: 'when each have done their work', wrote Berch, 'in its time and Tourneij, he steps down, and new people come in and continue work. The forgeman goes home sweaty, first to eat then to sleep, until the time he is to return'. The working week began at six o'clock on Sunday evening when the master finer and his assistant, in 'their long white shirts, with their leather aprons' as Berch described them, arrived to kindle the charcoal at their hearth, adjusting the bellows so that the force and entry-angle of the air current would be at its optimum. After a while, when the finery hearth had reached a melting heat, the fining

[66] Leufstaarkivet, Bruksböcker from assorted years; Douhan, *Arbete, kapital och migration*.

could begin. One of the finery walls had an apperture through which the *geuse*, mounted on wooden rollers, could be introduced. Soon, the end of the *geuse* began to liquefy and droplets of iron trickled down through the charcoal bed, coagulating on the floor of the hearth. Once a sufficient volume of iron had accumulated, the finer had to lever up the mass of viscous metal and slag debris from the bottom of the hearth, exposing it to the air blast, and in this way oxidising most of the carbon impurities that had been present in the *geuse*. Once the iron had reached the requisite purity the spongy mass of iron (*smältan*) was hauled from the hearth and dragged across the stone-flagged floor to anvil of a large water-powered hammer where the cinder was pounded out and the iron shaped into a rough block called a *smältstycke*.[67]

The *smältstycke* was thrown into the middle of the floor for the hammermen to take up. The master hammerman and his assistant reheated the *smältstycke*—not to melting point but to a so-called welding heat that would allow the metal to be reshaped under the forge hammer. Gradually, in the course of repeated hammering, the squat *smältstycke* was drawn out into a thin bar, three to four metres in length, the form in which iron was traded internationally. With this, the transformation of brittle *tackjärn* into malleable bars was complete.

The week's labour, which had begun on Sunday night, concluded on Saturday morning after a sequence of 40 *tourneijs* was completed. The finers were expected to make seven *smältstycken* in each *tourneij* and the hammermen seven bars, making a notional total of 280 bars at each forge over the week. *Directeur* Touscher maintained that 'when it functions well [weekly production] will always be 40 skeppund and a little more'; that is, about six tons weekly or 300 tons over the year. Of course, under pre-industrial conditions it was unlikely that this ideal would be consistently achieved. Weekly output figures are not available for Leufsta, but the accounts from the lower forge at Gimo show output climbing and plunging in an erratic fashion. Even disregarding weeks when production was weighed off before Saturday, implying that far fewer than forty *tourneijs* were worked, the forgemen at Gimo sometimes managed to make no more than two tons in a week. At the other extreme, they might achieve an output in excess of 8 tons.[68] Much of this inconsistency no doubt reflects water shortages or plant

---

[67] Berch, 'Dagbok... 1753', folios 71ff.
[68] Leufstaarkivet, Gimo bruksarkiv, vol. G1a 2) Bruksbok 1738.

Illustration 2.9. A Walloon forge in Leufsta *bruk* *c.* 1790.

Courtesy of Jernkontorets bruksbildkatalog.
Caption: This famous painting by Pehr Hilleström is as imaginative as it is naturalistic. The artist appears to cramming a series of actions that would have happened sequentially into a single frame. Curiously, his forgemen are not dressed in the calf-length white shifts and clogs that contemporaries described and which were worn in some Swedish forges into the twentieth century. Nevertheless, Hilleström provides a powerful impression of the dusty and over-heated gloom in which the finers and hammermen laboured.

breakdowns that cut short the working week at various points in the year. Nevertheless, the very high output peaks recorded in the Gimo accounts hint at some elasticity in the organisation of production.

Despite the scope for wide variations in output from week to week, production at *Leufstawerken* was strikingly stable from year to year. Most of the forges were very close to or exceeded the operating capacity hoped for by Touscher. Only Carlholm, new-built in the mid 1730, was laggard.[69] All of this suggests that the forges at Leufsta and Åkerby operated close to a customary norm of about 300 tons but that an intensification of effort was possible. The accounting data from Gimo indicate that there was every possibility of boosting production levels in response to increased demand. The potential for extra production was there, but unlocking that potential was dependent upon social rather than technical factors. Additional output required the delivery of additional charcoal from leaseholders and freeholders in and around the *bruk*. More importantly, it required the acquiesence of finers and hammermen. And this was by no means assured.

Writing in 1739's 'En liten handbok', *Directeur* Touscher revealed that the relationship between the finers and the hammermen was a difficult one, and that cooperation within the forge crew could not be guaranteed. On the face of it there were few grounds for dissension—finers and hammermen were rewarded equally, paid a common piece-rate based on the output of bar iron over the week—but the work practices of the finers bred resentment among the hammermen. According to Touscher, the finers were apt to make very heavy *smältstycken*, as this was the quickest way of completing their work. The hammermen felt themselves to be disadvantaged by having to process these more ponderous *smältstycken* as they required prolonged reheating. Their preference was for a larger number of lighter *smältstycken* that could be processed more quickly, otherwise they were left at their hearths long after the finers had completed their work. Management had some sympathy with this. The production of over-heavy *smältstycken* imposed a burden on hammermen that threatened to disrupt the smooth changeover of workers at the end of each four-hour *tourneij*. Indeed, *Directeur* Touscher

[69] A problem for this new works, built right on the coast, was that sea water rose into the river mouth in the summer, causing the forge to flood. In 1737 an annual output of about 150 tons was thought to be the ceiling. Bergmästarämbetet i Gävleborgs, Uppsala och Stockholms län, Bergmästarens tjänsteberättelser 1737, vol. B II: 5, ULA.

sought to have a clerk present in the forge to ensure that the finers did not make *smältstycken* that were too heavy.[70]

The problem arose from the nature of Walloon forging. Christer Berch noted that the size of the bars made at Leufsta appeared to be decided upon in a quite arbitrary way: 'in Walloon forging one sort [of bar] is not drawn in sequence, but some become flat, some square, [some] broader or narrower according to whether the *smältstycke* is large or small'. This was in sharp contrast to German forging, where great care was taken to ensure that each bar matched its predecessor in weight and shape. Walloon forgemen laid enormous stress on the 'inner quality' of the iron but attached little importance to its physical form. After all, 'Orground' iron was an outstandingly obdurate material. It was, a Scottish factor reported, 'the very Choicest and Softest metall yt's made here yet it's so tough & hard in working that they will not ingage to strike it to any Thin or Certain Sizes'.[71]

Leufsta's forgemen were accustomed to making bar iron as they saw fit, without outside interference. They were especially averse to making 'Thin or Certain Sizes'. In the 1730s, however, their freedom to do as they wished was being eroded as pressures from the international market impinged upon the day-to-day conduct of work within the forges of Uppland. It was pressure emanating from the British market that proved particularly disruptive. The trade in 'Orground' iron, as we have seen, had long been the preserve of the Dutch, but in 1730 the Grill family were ousted as the De Geer family's export agents and their place taken by Robert Campbell (d. 1758), a Scotsman. For the Leufsta forgemen, looking back from the late 1730s, when relations with their employer had become discordant, this was a key moment. There had been no complaints about their work before the coming of 'an Englishman [sic] calling himself Mr Campbell'.[72] Since then, the forgemen alleged, there had been nothing but trouble, as British merchants tried to dictate how bars were to be made. The forgemen were not to carry on making large, heavy bars as was their wont. Smaller, lighter bars were required as well.

The fraught atmosphere at *Leufstawerken* at the end of the 1730s arose from the increased demand for 'Orground' iron in Britain. The

[70] 'En liten handbok', p. 150.

[71] Mitchell Library, Glasgow, SR 352, Adam Montgomerie to John Crosse Senior & Co, 27 April 1701.

[72] Leufstaarkivet vol. 43B.

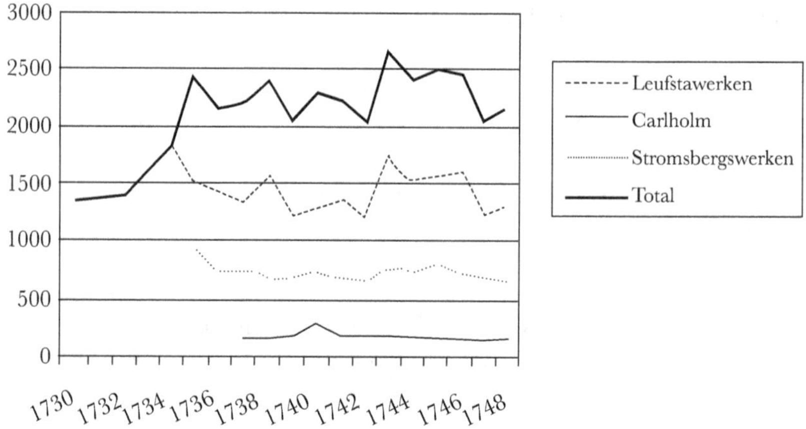

Source: Leufsta bruks arkiv.

Figure 2.9. Total bar iron production at Charles de Geer's works, 1720–1750.

changes made in Uppland in the 1720s and 1730s, largely at the behest of the De Geer family, must be understood in this changed international context. The De Geers did not just add to their collection of *bruk*, they restructured and streamlined production at every point. Dannemora was reorganised; new furnaces were erected; and forges were rebuilt. Production was to grow. It was also to be more responsive to signals from the British market. For that to happen, the movement of 'Orground' iron onto international markets had to be quickened. This was to be done by expediting communications between the *bruk* and Stockholm, the antechamber to the world market.

Stockholm

The heartland of the Swedish iron industry lay in heavily forested areas, rich in ore, in central Sweden: *Bergslagen*. Because the core of *Bergslagen* lay so far from the sea, when most Swedish iron was destined for the international market, transport was a crucial problem. Indeed, the early modern iron industry might more accurately be classified as a transport organisation than an 'industry'. The mercantilist policies of the Swedish state played a critical role here. The export of iron, as of other commodities, could only take place via specified towns. Those towns through which bar iron passed were authorised—indeed, compelled—to have a *Jernvåg* ('iron weigh') at which the bars were weighed and their quality monitored.

Of the exporting cities, Stockholm and Gothenberg were of special importance. Most iron passed through these ports, with the Baltic port of Gävle coming a poor third. In central *Bergslagen* iron was routed through inland ports such as Västerås or Arboga on the shores of Mälaren, the vast lake system that drained into the Baltic at Stockholm. Iron from the western parts of *Bergslagen*, in Värmland, on the other hand, was taken to Kristinehamn and shipped across lake Vänern and then down the Göta valley to the North Sea coast.

It has long been assumed that the overland transport of bar iron was restricted to the winter months, when sledges could run over frozen lakes and moors. The experience of the *bruk* in Uppland suggests otherwise. None of the *bruk* at which 'Orground' iron was made was particularly far from the sea; most were close to coastal depots from which the iron could be shipped quickly to Stockholm. The records of *Leufstabruk* indicate that bar iron was carried to the inlets at Ängskär and Löten right through the year, without a seasonal break. Eric Touscher took a keen interest in upgrading road links and took particular pride in the new route that stretched between Leufsta and Ängskär. Indeed, Ängskär began to replace Löten as the main point of shipment for Leufsta iron in the 1730s, a process that culminated in the 1750s with the building of the formidable stone warehouse that still stands at the head of the bay.[73]

The road-building programme of the 1730s showed the commitment of the *bruk* management to moving iron promptly into the hands of buyers. Summer transports were essential, otherwise iron made in the spring could not be brought to market for almost a year: the bars would have languished at Leufsta or Åkerby over the summer, then at Ängskär over the winter, when ice brought navigation in the Baltic to a standstill. Indeed, it was the freezing over of the Baltic that was the most serious hindrance to the transport of iron. Nothing could be done to counteract it; it simply had to be endured.

Ice-free navigation was usually possible by April, and as soon as the ice broke shiploads of iron were rushed southward. Roughly 20

[73] 'En liten handbok', pp. 48–63, and Leufstaarkivet, vol. 132 and 133.

shiploads of *Leufstawerken* iron, each of 100 tons, left Ängskär and Löten each summer in the 1730s and 1740s. The majority of sailings were during the first six to eight weeks of the shipping season, when the three ships owned directly by the Leufsta estate were augmented by chartered vessels. Only two or three sailings were made per month thereafter. As the autumn drew in, and the Baltic grew stormier, the managers at Leufsta grew ever more anxious. Shipping in October, as Georg Swebilius told Jean Jacques De Geer, made for 'much anxiety of the mind'.[74]

The ships from Ängskär and Löten crept down the coast and into the Stockholm archipelago. After paying a toll at *Lilla Sjötullen*, the internal customs station on Djurgården island, the ships hove into sight of Stockholm itself. The long quay of Stadsholmen (the Old Town of today) came into view, crowded with vessels of all sizes. The bar iron from Leufsta had arrived at the pivot around which Swedish commerce turned.

Stockholm was by far the most important urban settlement in Sweden. With 70,000 inhabitants, it stood at the head of the urban hierarchy, far ahead of second-placed Karlskrona, the southern naval base. The city was not merely a commercial centre; it had an important manufacturing sector as well. It could boast the biggest concentration of textiles production in Sweden. The woollen, worsted and silk industries were clustered on Södermalm, the large island to the south of Stadsholmen. Here, centralised factory establishments and a myriad of smaller workshops were brought together in a flexible production system that gave employment to thousands. The shipyards that ringed the harbour provided work for hundreds more.[75]

Stockholm was, of course, the national capital. To come to Stadsholmen was to come to the very core of the Swedish state.[76] At the north-east corner of the island was the Royal Palace, the work of the great court architect Nicodemus Tessin the younger. It stood on the site of the old royal castle that had burnt down in 1697. Progress on the new

---

[74] GS to JJDG, 2 and 23 October 1732, Leufsta Arkivet, vol. 106, RA. See also numerous letters from ET to LDG, Leufsta Arkivet, vol. 105, RA. Sure enough, in 1735 Leufsta lost a shipload of iron at sea: see Leufsta bruksarkiv, Leufsta, Bunt 32.

[75] For information on Swedish urbanisation see Sven Lilja, *Tjuvehål och stolta städer. Urbaniseringens kronologi och geografi i Sverige (med Finland) ca 1570-tal till 1810-tal* (Stockholm, 2000).

[76] Eva Eggeby and Klas Nyberg, 'Stad i Stagnation 1720–1850', in Lars Nilsson (ed.), *Staden på vattnet. Del I: 1252–1850* (Stockholm, 2002), pp. 187–276.

Illustration 2.10. Stockholm in 1697.

Courtesy of Kungliga Biblioteket
Reference: KB, KoB, Erik Dahlbergh's *Suecia Antiqua et hodierna*. 1:13.
Caption: The Swedish capital as seen from the east, the direction from which shipping entered the harbour. Stadsholmen occupies the centre, with Södermalm, the southern island to the left.

palace had been was slow, however, and in the 1730s it was still uninhabited. There was a symbolic aptness to the empty palace, for this was the Age of Liberty (*Frihetstiden*), when the authority of the crown was in abeyance. After the death of Charles XII in 1718 the nobility had succeeded in curtailing the prerogatives of the monarchy. Effective power had switched to *Riddarhuset* at the north-western corner of Stadsholmen, the assembly of the noble estate. The other estates (the clergy, the burghers, and the peasants) also met on Stadsholmen when the Diet was in session. This made for a lively, concentrated political culture, nourished in the taverns and coffee houses of the Old Town.[77]

Stadsholmen was the administrative as well as the political hub of the kingdom. It was home to the different royal offices (*Collegierna*) through which policies decided upon in the Diet were effected. Most importantly from our perspective, the Old Town housed a set of interlinked institutions that governed the iron trade. The Board of Mines (*Bergscollegium*), based on the north side of Stadsholmen, had overall direction of the mining and processing industries. Close by was the *Riksbank*, which played an important role in facilitating the iron export. *Riksbanken* underwrote 'assignations', the financial instruments that enabled large sums to be transferred from the merchant class in Stockholm to *brukspatroner* in *Bergslagen*, thereby allowing production to continue in the mining districts. In years to come the *Riksbank* would be supplemented by *Jernkontoret*, the ironmasters' association (literally the 'iron bureau'), founded with state approval in 1747, which was also to furnish credit to the iron industry on a large scale.[78]

Whilst the state supplied a mercantilist framework for the economy as a whole, it was largely up to individual economic actors to set commodity flows in motion. The key actors were to be found along the seaward quay of Stadsholmen: *Skeppsbron*. This was the point of departure for over 60 per cent of Sweden's iron and tar, the two major export commodities. Likewise, most of the grain shipped in from the southern Baltic came ashore on these wharves.[79] The quay was lined

---

[77] See Karin Sennefelt, *Den politiska sjukan. Dalaupproret 1743 och frihetstida politisk kultur* (Hedemora, 2001), for a discussion of political life in Stockholm in relation to an uprising of peasants from Dalarna.

[78] Bertil Boëthius and Åke Kromnow, *Jenrkontorets historia, Del I–III* (Stockholm, 1947–68).

[79] For figures of Swedish foreign trade see Eli Heckscher, *Sveriges ekonomiska historia från Gustav Vasa. Andra delen. Det moderna Sveriges grundläggning* (Stockholm, 1949), pp. 99–104.

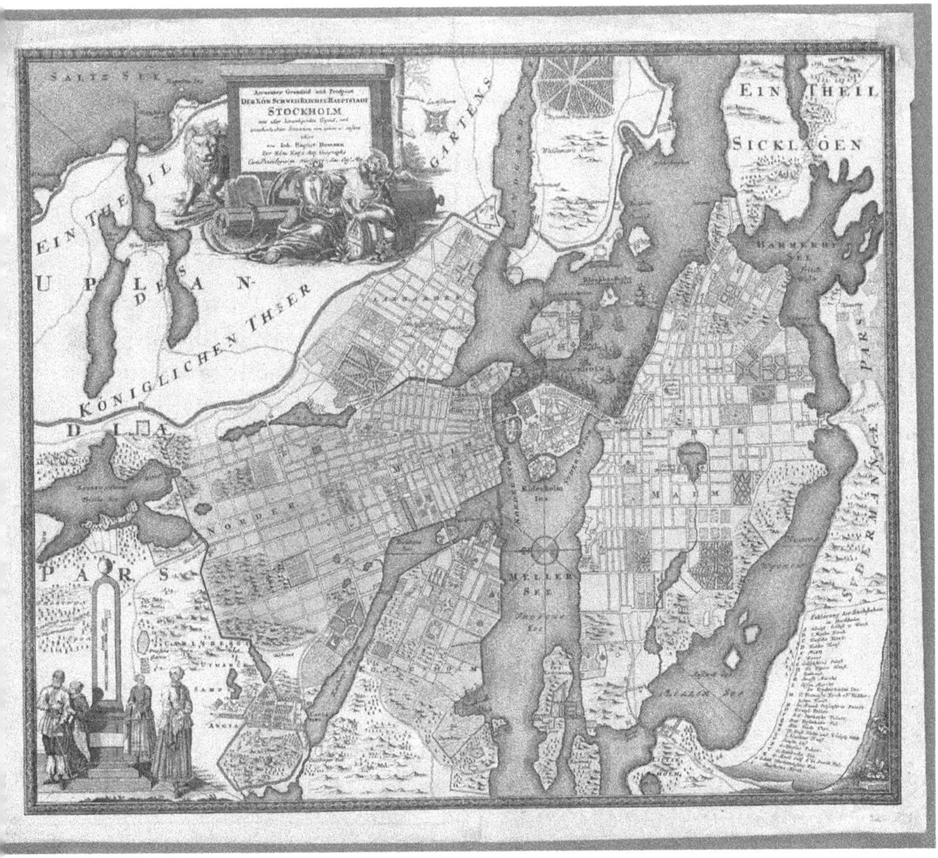

Illustration 2.11. Stockholm *c.* 1720.

Courtesy of Kungliga Biblioteket.
Reference: KB, KoB 3a.
Caption: This map of Stockholm and its area is orientated east-west, with Mälaren ('Meller See') at the foot, lapping against the freshwater quays of Stadsholmen. The cartographer Johann Baptist Homan shows sea-going shipping crowded against *Skeppsbron* on the other side of the island. The far larger island of Södermalm is to the right. At the northern end of the quay the old royal castle, visible in Illustration 2.10, has given way to Tessin's baroque palace with its rectilinear layout.

with the tall, imposing houses of the great merchants: the Plomgren brothers, the Hebbes, the Bedoires, the Grill family, Samuel Worster, and others. Those who did not live on the quay itself lived in close proximity, on one of the main north-south thoroughfares leading through Stadsholmen. Francis Jennings, who had a house on Västerlånggatan, was one such.[80]

The export trade was dominated by this clutch of powerful merchants, few in number and growing progressively fewer as the century wore on. At mid century the largest seven iron exporters handled between 40 and 50 per cent of the total. The process of concentration had gone furthest among those who exported to the British market. In 1730, for instance, 94 per cent of the bar iron export to Britain was conducted by just ten Stockholm merchants. Six of these were of British origin, reflecting the tendency of *Skeppsbron*'s merchants, many of whom were of foreign extraction, to export to their ancestral country.

At the southern end of *Skeppsbron* bridges crossed over the great sluice through which the waters of lake Mälaren emptied into the sea. On the other side of the sluice, on a tip of Södermalm, was *Jernvågen*, the 'iron weigh'. One side of the *Jernvåg* faced onto Mälaren, allowing lake craft from the interior to tie up. On the seaward side a flotilla of lighters stood ready to empty the holds of ships from Uppland. Special iron-carriers (*jernbärare*) shouldered the bars ashore for six *öre* per *skeppund*. It was at *Jernvågen*, once the bars had been checked by the master-weighman and the weigh fee paid (another six *öre* per *skeppund*), that iron from *Leufstawerken* passed into the hands of the exporting merchants.[81]

In 1737 the Leufsta 'Bar Iron Account' concluded with the entry: 'Weighed—to Samuel Worster'. This was overly terse, for the bars had in fact been bought by a trio of merchants acting in concert: Worster, Samuel Wordsworth and Francis Jennings.[82] The iron was already partially paid for. The merchant triumvirate had made the first

---

[80] Only the very grandest merchants could afford to be far from *Skeppsbron*. Jennings bought a sumptuous house on the neighbouring island of Riddarholmen in 1747, when he reigned supreme as Stockholm's biggest iron exporter, but it is likely that this move signalled his semi-retirement after twenty-eight years spent in the iron trade. Jonas Norrby, *Jennings* (Köping, 1991), pp. 7–14.

[81] For a general account of the procedures in *Jernvågen* see *Forsmark och Vallonjärnet* (Stockholm, 1987), pp. 114–16. See also Nilsson, *Staden på vattnet*, p. 235; Leufstaarkivet vols 132 and 268; E.W. Dahlgren, *Järnvräkeri och järnstämpling. Ett bidrag till den svenska järnhandelns historia* (Stockholm, 1930).

[82] See the accounts for several years between 1735 and 1740 in Leufstaarkivet vol. 268.

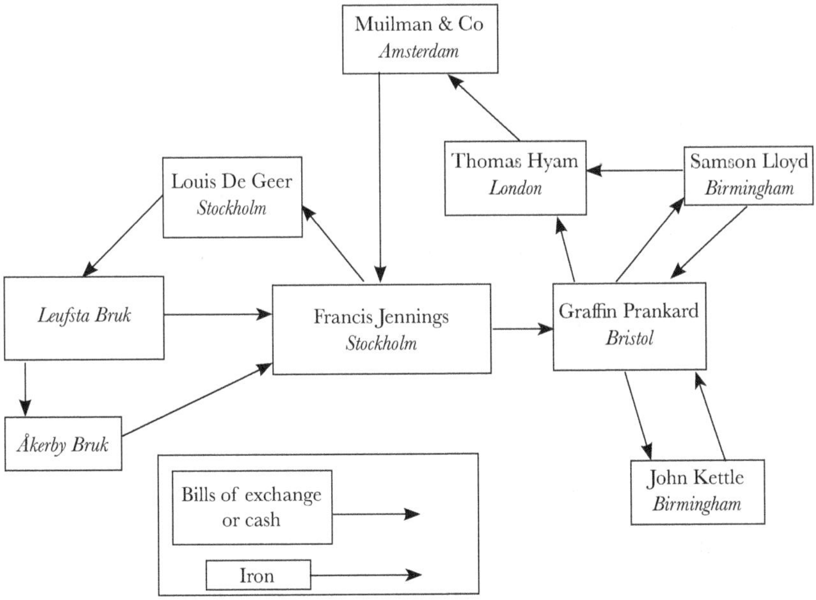

Figure 2.10. Financial links between Leufsta *bruk* and Graffin Prankard in the 1730s.

Caption: A schematic view of the payments made by Midland manufacturers that led back via Prankard, his bankers in London and Amsterdam, and Francis Jennings in Stockholm to the De Geer estates in Uppland.

of nine monthly payments in January, passing 'assignations' to Jacob Swedmark, the De Geer family's head clerk (*Cammererare*) in Stockholm. This Swedmark was responsible for paying the tolls and weigh fees that *Leufstawerken* iron incurred as it passed through Stockholm. It was also for him to procure and ship up-country essential supplies, such as salt, that could not be produced at the *bruk* themselves. Most importantly of all, he had to advance credit to the different De Geer *bruk*.[83] Without credit from the Stockholm merchants production in the interior would have atrophied and sailings from *Skeppsbron* dwindled.

The *Carolina Merchant* departed Stockholm on 31 June 1737 under the command of George Gibbs, loaded with 163 tons of bar iron and 2,280 timber deals. The cargo was destined for Graffin Prankard in

[83] For this see Leufstaarkivet vols. 109, 132 and 268.

Bristol and constituted a new link in the commodity chain that began at Leufsta. 1737 had already been a busy year for Master Gibbs and his crew. The *Carolina Merchant* had crossed the Atlantic at the start of the year with a lading of rice from Charleston. Having tied up briefly at Cowes on the Isle of Wight in mid-March, she sailed on to Bremen. After disposing of her cargo of rice there, she looped north around Denmark, passing the Sound on 10 June. The *Carolina Merchant* entered the harbour at Stockholm just two days later. Upon arrival, Gibbs reported to Francis Jennings. As usual, the Irishman had had advance notice of the sorts of iron required by Prankard, but despite the Quaker merchant's strident pleas and Jennings' best efforts the sought-after bars were evidently not in stock at the *Jernvåg*, for the *Carolina Merchant* remained at Stockholm for over five weeks. She did not pass the Sound westward until 31 July.[84]

The *Carolina Merchant* was not alone in leaving Stockholm with bar iron—far from it. About 350 vessels sailed from the Swedish capital every year in the early eighteenth century laden with bars. Some weighed anchor as soon as the ice broke. The first clearance of an iron-bearing vessel in 1737 was on 4 March. Sixteen more sailed before the end of March. These, however, were mostly small craft, operating within the Baltic. The larger, ocean-going vessels that were to pass west of the Sound did not thread their way through the Stockholm archipelago until May. The period between May and September was the peak season for shipping, when 50 or so iron-laden vessels cleared Stockholm every month. The export of iron slowed in October—just 35 clearances from *Skeppsbron* in 1737—and slumped in November to 22 clearances. The last parcel of iron to be shipped from Stockholm in 1737 left on 8 December. The shipments made by Francis Jennings correspond to this general pattern. Most of the 31 vessels loaded by him sailed during the May-to-September peak season. The *Carolina*

[84] In writing this section on the iron trade from Stockholm (and Sweden) we have received much help from Åsa Eklund, whose 'Iron production, iron trade, and iron markets', marks a considerable advance on earlier studies. We have also benefitted from having access to her data from three important sources: Tolagsjournalerna, 1720–1754, Statskamrerarens arkiv, Stadens verifikationer, Stockholms stadsarkiv, Stockholm; Manufakturkontorets arkiv, Handlanden Petrer Westmans arkiv, utskeppningsböcker 1729–1745, RA; TNA: PRO, E190 series, Bristol Port Books, 1720–1740. To this we have also added the 1737 Toll Accounts, Rigsarkivet, Köpenhamn, microfilms S.15.085 and S.15.086.

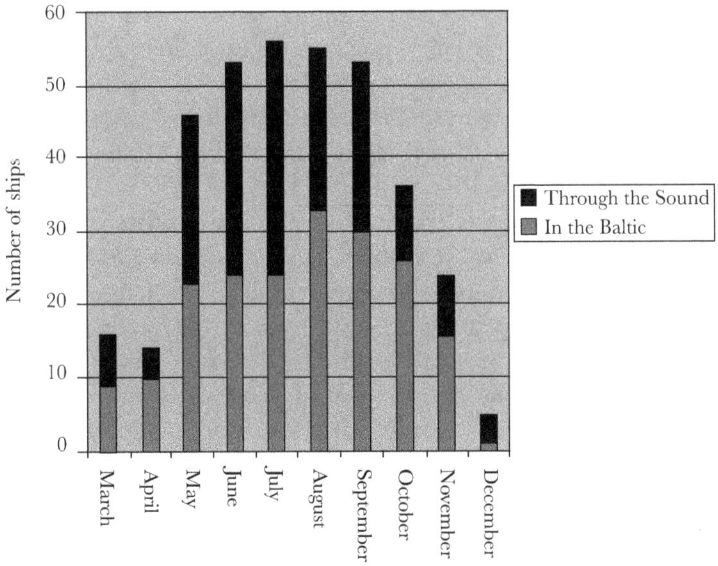

Source: Tolags journalerna 1737, SSA

Figure 2.11. Monthly departure of ships from Stockholm in 1737.

Merchant, in setting out on 31 June, conformed very closely to the norm for the 1730s.[85]

What was unusual about the *Carolina Merchant* was her destination: Bristol. When Britain supplanted the Netherlands as the principal market for Swedish bar iron in the last years of the seventeenth century London was the principal point of entry. Only small amounts of bar iron were shipped direct to the outports, and those ports that did receive Swedish iron were on the east coast: Hull, Lynn, and Newcastle upon Tyne. This changed in the early eighteenth century as the outports ate away at London's supremacy, and as Swedish iron began to penetrate the markets of western Britain for the first time. More than half of the iron export from Stockholm to reach England in 1700 was landed at London. By mid century the proportion had fallen to a third, with

[85] Manufakturkontorets arkiv, Peter Westmans utskeppningsböcker 1729–45, vol. 490, RA.

substantial volumes of iron now being directed to Hull and Bristol.[86] The picture is still clearer when it comes to 'Orground' iron. London's share of 'Orground' shipments to the British market fell from 66 per cent in 1737 to 43 per cent in 1748. It was Hull and Bristol that prospered at London's expense, increasing their share of the 'Orground' import to 23 and 15 per cent respectively.[87]

The Ulsterman Francis Jennings led the way in opening markets for Swedish iron in western Britain. It was in these markets that he specialised and these that he dominated. In 1737 Jennings shipped 2,993 tons of bar iron to Britain. Of this, 1,100 tons went to Bristol, 825 tons to Irish ports, 404 tons to Liverpool, and 100 tons to Scotland. Only 409 tons (13 per cent of the total) went to London.[88] By the mid-1730s Francis Jennings had established himself as one of Stockholm's premier iron exporters. Graffin Prankard, his Bristol correspondent, had likewise consolidated his position as the leading iron merchant in western Britain. Together, they exercised complete control over the supply of 'Orground' iron to Bristol and its hinterland.[89] The exchanges between the two during the 1737 shipping season therefore provide an instructive, micro-level insight into the organisation of the Anglo-Baltic iron trade.

It was Prankard who took the initiative. He told Jennings of his plans at the start of April. His own ships, the *Parham* and the *Baltick Merchant*, were to sail for St Peterburg in 1737, by-passing Stockholm altogether. In their stead, Prankard had 'chartered 4 ships for Stockholm viz ye King David yt is [already] gone [the] Carolina [Merchant], [the] Kingsweston and [the] Severn all wch I aprehend will fall succesivly to Stockholm and load on my accot 700 tons iron and about 6000 deal'.[90] In subsequent letters he was more specific, stipulating the composition

[86] Stockholm was by far the most important Swedish iron exporting port, with as much as 60 per cent of the total. See Eklund, 'Iron production', p. 52 and Staffan Högberg *Utrikeshandel och sjöfart på 1700-talet. Stapelvaror i svensk export och import 1738–1808* (Lund, 1969), pp. 62ff.

[87] Eklund, 'Iron production', pp. 55–63.

[88] Manufakturkontorets arkiv, Peter Westmans utskeppningsböcker 1729–45, vol. 490, RA.

[89] Eklund, 'Iron production', pp. 63–67.

[90] GP to FJ, 6 April 1737. Another chartered vessel, the *Seaflower*, was to accompany the *Parham* and the *Baltick Merchant* to Russia. Between them they carried 302 tons back to Bristol. 1737 Toll Accounts, and TNA: PRO, E190/1214/1, Bristol Port Books, 1737. SA, DD/DN 427.

of each cargo. The *Carolina Merchant*, for example, was to be loaded with 170 tons of bar iron and 1,600 deals.[91]

To ship me on the Carolina as undermentioned or near it

Tons	
60	of [Leufsta] and [Åkerby] flats one half or 40 tons of it 2½ wide including 3 or 4 tons of 2 inch sq and 5 or 6 tons of 4 inch wide thin and free from flaws or cracks
30	fine narrow flats about 64 to the ton and thin
10	¾ squares all the common orgrounds if possible
15	[Strömsberg] flats 5 tons of it 4 inch wide thin drawn
20	voyage iron
10	3 inch wide xx thick box iron
5	2 ½ wide ½ inch thick box iron
5	2 ½ wide and ¾ thick
10	2 ¼ and 2 ½ wide thin drawn not quite ⅜ thick
5	1 ⅛ sq

170 tons with about 16 × of deales at 120 to the hundred

When the *Carolina Merchant* finally cleared Stockholm she did so with 163 tons of bar iron in her hold, not an exact match for what had been ordered on 13 June, but 'near it' (as Prankard had put in his original instructions).[92]

Prankard had chartered four ships to sail to Stockholm, but it soon became apparent that the demand for Swedish iron was greater than he had anticipated. As midsummer approached Prankard told Jennings that he was to assemble an additional cargo: 'I am under an obligation to shipp 80 tons of iron on board the ship Elizabeth Thos Read master when she arrives at Stockholm so that I desire thee to make some provision for her also.' A sixth ship followed. By the early autumn Prankard's initial order of 700 tons for the 1737 season had grown to 900 tons, and when all the deliveries were completed nearly 1,000 tons had been landed at Bristol.

This was very much a demand-driven process, given force and urgency by Prankard's knowledge of the market for malleable iron and iron wares in Britain's Atlantic empire. Prankard issued detailed

[91] GP to FJ, 13 June 1737.

[92] Forty-eight tons of 'Orground' iron was stowed aboard, not the 60 tons Prankard had ordered, with 8 tons of squares, not the 10 tons that was required. As for the remainder of cargo, it is listed in the shipping register simply as *stångjern* without further elaboration.

instructions to Jennings as to the quantities and sorts of iron that were required. It was Prankard who owned or chartered the merchant fleet that carried the iron westward; he who arranged for insurance to be paid on the ships; and he whose property the iron became once it was lodged in the hold of the *Carolina Merchant* or the *Elizabeth*. Jennings sought payment as soon as the iron had left the *Jernvåg*, drawing bills of exchange on Prankard via bankers in Hamburg and Amsterdam.[93]

This was very different from the practices that governed the export of bar iron to Holland. The Grill family, the principal players in the iron trade between Stockholm and Amsterdam in the first half of the eighteenth century, sent out parcels of iron to merchants who acted as commission agents. The iron remained the property of the Grills throughout. It was the Grills who paid the freight charges, the insurance bill, the Sound dues and all other incidental costs associated with bringing the iron to market. The Dutch merchants to whom the iron was consigned took care of sales in Amsterdam or Rotterdam for which they charged a commission of two per cent, but the risk lay entirely with the Grills in Stockholm, since the commission agents would receive their percentage even if the iron was sold at a loss. The trade was directed from the supply-end of the commodity chain rather than by those who sold bar iron along the canal-sides of the United Provinces. Supply took priority over demand.[94]

Commission sales were common enough in England. The excess iron that was sent to Prankard in 1737, over and above the 900 tons he had ordered, was very probably shipped by Francis Jennings on a fee basis. Indeed, in 1740 and 1741, when Prankard was on the brink of bankruptcy and his affairs had been placed in the hands of his brother-

[93] Jennings drew bills on Messrs Smith & Lake in Hamburg or Muilman & Son in Amsterdam. They in turn drew upon Thomas Hyam, Prankard's London banker. Hyam was furnished with cash or bills by Prankard. Alternatively, Prankard would ask a major customer to make pay Hyam directly. Sampson Lloyd bought 310 tons of iron from Prankard in 1737–38 at a cost of £3,894. Of this, £1,565 was paid into the hands of Thomas Hyam. SA, DD/DN 435 and 442. See also Åsa Eklund, Chris Evans and Göran Rydén, 'Baltic iron and the organisation of the British iron market in the eighteenth century', in Patrick Salmon and Tony Barrow (eds), *Britain and the Baltic: studies in commercial, political and cultural relations* (Sunderland, 2003), p. 141.

[94] The relationship between the principal in Stockholm and the commission agent in Holland was not so one-sided as might at first appear. Commission agents were expected to advance credit (in the form of bills of exchange) to their principals equal to the value of the iron consignment when the parcel was issued to them. This gave the commission agent a dual role: he was at once an employee of the principal and a major creditor of the principal. See Müller, *The merchant houses of Stockholm*, pp. 151–56.

in-law John Galton and son-in-law Caleb Dickinson, the younger men opted to take iron from Jennings as commission agents to minimise the risks they took.[95] Yet when Graffin Prankard was in his mercantile pomp he preferred to exercise the tightest possible control over the commodities in which he dealt. Close control was very dear to him, for in the 1730s he had embarked on a campaign to monopolise the supply of bars from Leufsta and Åkerby, the most sought-after marks of the most sought-after type of iron: 'Orground'.

'Orground' was the most prized variety of iron on the international market. Its superlative reputation rested upon the high quality of the materials used in its manufacture and the special standards of workmanship exhibited by the forgemen who made it. The *bruk* that produced 'Orground' iron were clustered around the renowned Dannemora mine, the source of a non-phosphoric ore of exceptional purity. This set them apart from other sectors of the Swedish iron industry. So did the use of a forging technique that was distinct from the 'German' forging method that had been in use in Sweden since the sixteenth century. 'Orground' iron was a unique material.[96] It was also a scarce material, with no more than 5500 tons being forged annually in the 1730s. This made 'Orground' iron a much sought-after commodity. For consumers who demanded its unrivalled toughness or its superior purity, there was no alternative. British naval bureaucrats insisted on its use in the making of anchors, and steel makers would allow little else to be used in their cementation furnaces. Yet some of the 'Orground' brands were more coveted than others. English steel makers hungered for bars from the forges at Leufsta or, better still, bars from the neighbouring forge at Åkerby: 'no other marks will answer here for steel', as Graffin Prankard reminded Francis Jennings. Georg Swebilius could also testify to the the superiority of Åkerby iron: 'the Leufsta, Österby and Gimo brands are the best in the country', he told Jean Jacques De Geer, 'Åkerby apart'.[97]

---

[95] SA, DD/DN 442.

[96] For a discussion of Walloon iron from a metallurgical standpoint see Wilhelm Ekman, 'Vallonjärnet—en kvalitetsprodukt med världsrykte' in *Forsmark och vallonjärnet* (Stockholm, 1987), pp. 121–49.

[97] SA, DD/DN 425, GP to FJ, 16 August 1732. Consumer choice, so Swedish observers reckoned, oscillated between a limited selection of brands: Leufsta, Österby, Åkerby and Strömsberg. Barraclough, *Steelmaking*, pp. 184, 215, 218. GS to JJDG, 1118 June, 1733, Leufsta Arkivet, vol. 106. RA.

From a Swedish perspective *Leufstawerken* and the other *bruk* owned by the De Geer family constituted a formidable industrial complex, capable of producing a large quantity of high-grade iron, but from a British perspective this output was frustratingly small and inelastic. Given the very finite quantity of 'Orground' iron that came to market every year, there was a strong incentive for merchants to attempt to monopolise that supply. Accordingly, the major merchant houses in Stockholm contracted with the different *bruk* for exclusive rights to the iron produced over an annual period: 'ye Iron works wch make it are under contract to part[icu]lar Men who ship it for holland and England'.[98] In the late 1720s, for example, control over the output from *Leufstabruk* rested with the firm of Carlos & Claes Grill.[99] Some of this iron was exported to Holland by the Grills themselves, whilst the remainder was shipped to the English market, either directly by the Grills or through intermediaries in Stockholm. One such intermediary was Francis Jennings, who secured bars for Graffin Prankard. Åkerby iron was handled by a different Stockholm merchant, Johan Adam Pettersson, who seems to have sold the entire output directly to England.[100]

In this way the most desirable 'Orground' brands made their way to the English market. Such a system was pleasing to those Stockholm merchant houses that could secure contracts with the leading *brukspatroner*; it was less attractive to English merchants like Prankard who were forced to pay a considerable premium to guarantee access to the best brands.[101] In 1730 Prankard was asked to pay 49 *daler kopparmynt* per *skeppund* for Leufsta bars that had cost the Grills 45 *daler*, which, the Bristol man snapped, 'is too much profitt on it'.[102]

The benefits of contracting directly with the Leufsta estate were obvious, and by the end of the 1720s Graffin Prankard had decided to do so. This was an ambitious undertaking. Contracting on such a scale was usually the preserve of long-established members of Stockholm's

---

[98] Mitchell Library, Glasgow, SR352, Adam Montgomerie to John Corse senior, 28 November 1700.

[99] The firm was headed by Carlos Grill (1681–1736) and his nephew Claes (1705–1767).

[100] GS to JJDG, 9 September 1730, and 13 August 1733, Leufsta Arkivet, vol. 106, RA.

[101] Swebilius noted that the English consumers of Åkerby iron rather wanted to trade directly with the De Geers; GS to JJDG, 9 September 1730, Leufsta Arkivet, vol. 106, RA.

[102] GP to FJ, 19 December 1730.

merchant elite. Their associates were usually merchant houses of equal stature in London or Amsterdam. Graffin Prankard hoped to prise open this gilded circle in cooperation with another provincial outsider, Samuel Shore, the Sheffield steel maker who had hitherto depended upon Johan Adam Petterson.[103]

Prankard and Shore aimed at engrossing the entire import of Åkerby and Leufsta bars to Britain. It was their particular wish to extinguish an open market in London for these key steel-making brands. Their initial thought was to reach an accord with the Grills. Prankard and Shore proposed to join Claes Grill in negotiating with Carl De Geer. The two Englishmen would take sufficient Leufsta and Åkerby bars to supply the entire British steel trade, bringing the whole quantity through Hull and Bristol. Grill would be left in sole command of the Dutch market, provided that he would agree not to release any part of his share of *Leufstawerken* iron onto the London market.

The scheme came to naught. Carl De Geer died in the autumn of 1730, and his executors made a new contract with Robert Campbell, the Scottish-born Stockholm merchant, not with the Grills. This was a setback, for despite the misgivings Prankard and Shore had entertained about Grill & Co, they had at least enjoyed a settled relationship with the firm. The Grills had been orientated upon the Dutch market, allowing Prankard and his Sheffield-based associate some leeway in the English market. Campbell's commercial affinities were different and potentially threatening. He was the Stockholm correspondent of Henry Norris, one of London's premier Baltic merchants. And Norris, for his part, was the London agent of Abraham Spooner, the largest ironmonger in the West Midlands in the 1720s and 1730s. From Prankard's perspective, this Campbell-Norris-Spooner axis was the most dangerous of liaisons. Spooner was the bitter rival of John Kettle, the Birmingham steel maker who was Prankard's main customer for Åkerby and Leufsta iron. If Spooner could aggrandise supplies of iron from *Leufstawerken*, he would not merely shut Prankard out of a highly profitable commercial circuit for the duration of Campbell's contract, he could permanently impair Kettle's business.

---

[103] As will become clear, we disagree with the interpretation advanced in P.W. King, 'The cartel in Oregrounds iron: trading relationships in the raw material for steel', *Journal of Industrial History*, VI, 1 (2003), 25–48, where it is claimed that 'Orground' iron was from the beginning of the eighteenth century under the control of a cartel of Sheffield steel makers.

By the autumn of 1731, when negotiations began for the distribution of iron in the 1732 season, Prankard was becoming desperate. He ordered Jennings to offer Robert Campbell in excess of 50 *daler kopparmynt* for Åkerby 'to prevent its falling into Norris hands'.[104] It was to no avail. Prankard and Shore were excluded once more. The following year brought no relief. Prankard had to yield to Henry Norris yet again, whilst picking up small parcels of Åkerby iron on the Rotterdam market. He had the mortifying experience of watching a shipment of Åkerby iron from Norris being landed at Bristol *en route* to Abraham Spooner.[105]

In the autumn of 1732 negotiations began anew over the distribution of iron from *Leufstawerken*. Prankard and Shore hoped initially for a three-way split between Campbell, the Grills, and themselves. That possibility receded as Campbell made plain his determination to retain his exclusive grip, and as the Grills proved—in Prankard's eyes at least—pusillanimous. The best that might emerge was a strictly subaltern role in a cartel headed by Campbell.[106] Prankard and Shore resolved that in future they would bid for the output of the *Leufstawerken* forges by themselves, accepting the tutelage of neither Robert Campbell nor the Grills. Prankard summarised their preferred terms:

> Shore & Self to take 350 Tons each of us Yearly of [Åkerby] & [Leufsta] viz all that Shall be Struck yearly of the [Åkerby] allowing it to be 280 Yearly or thereabouts and 420 Tons of ye [Leufsta] so that ye Remainder of ye [Leufsta] to be Shipt for Holland...and by agreement betwixt us not any of said Marks is to be Shipt for London but the whole for Hull & Bristol.[107]

Prankard showed an almost reckless determination in 1733. The 350 tons that he pledged to buy was in fact more than he could dispose of 'in the Steele way', but such was the importance of the steel market

[104] GP to FJ, 4 August 1731.

[105] It was, he told Jennings, 'very hard on me to See it Pass by me here & up into ye Markett & Sold by a Person that wont Sell it on any reasonable terms or really not at all to my best Chapp [i.e. Kettle] but endeavour to thwart his Interest to the utmost of his Power...' GP to FJ, 16 August 1732.

[106] Prankard was unimpressed: 'we might by Capittulating come in for a part but then its probable be und[e]r such Restriction as to ye quantum that it wont answer our end[.] besides as Long as A Spooners Agent hath it[,] it will be I doubt [not] a means for my Friend J Kittle always to thward at Birmingham & kept und[e]r by AS'. GP to SS, 9 December 1732.

[107] GP to FJ, 27 July 1733.

to him that he was prepared to dump up to 50 tons of high-grade 'Orground' in the guise of Swedish 'common sorts' rather than risk being deprived of Åkerby and Leufsta iron for a further year.[108] Moreover, Jennings was authorised to offer an unprecedented 55 *daler kopparmynt* per *skeppund*. Such extravagance was no guarantee of success, for Georg Swebilius was notoriously partial to Robert Campbell—'a most reasonable, correct and steady man', as he described him—and cautioned his masters against accepting the highest bid.[109] Swebilius, however, was a declining force, having already entered upon the long illness that was to bring about his death in 1735.

Francis Jennings (acting for Prankard), together with Samuel Worster and Samuel Wordsworth (acting for Shore), concluded a two-year contract with the De Geers early in 1734. The output of *Leufstawerken*'s two principal *bruk* was secured for 'about 53 Including Presents'. 'Mr Prankard hath the Bristoll London Birmingham and Ireland marketts to himself', Shore crowed, 'I the Hull & Newcastle Marketts as we may not prejudice each other.'[110] Getting access to the Leufsta and Åkerby iron was a startling success for Prankard and Shore, but success very quickly brought its own problems. Not least, when Prankard and Shore were apprised of the projected production of the Leufsta and Åkerby forges they found it far in excess of their expectations: 'the Quantity... Struck Yearly is near About 1470 Tons... at least 270 Tons More than Wee realy had a Notion off.' The problem was not insurmountable, but it required careful management. The excess production could be directed partially into the provincial markets that Shore and Prankard had command of. Some of it could go to Holland. The rest would have to be absorbed by the London market.

Shore and Prankard had, of course, resolved to prevent Leufsta and Åkerby bars from circulating freely on the London market. That remained their aim, but they were happy for their Stockholm partners

[108] GP to FJ, 28 July 1733.

[109] GS to JJDG 29 October 1733, Leufsta Arkivet, vol. 106, RA.

[110] SS to Francis Bird, 15 August 1735. The east-west division can be seen very clearly in the subsequent trading patterns of Francis Jennings and Samuel Worster. In 1737 Jennings exported 444 tons of 'Orground' (of all sorts) to Britain: 343 tons went to Prankard, 7 tons to Ireland, and 94 tons to London. During the same season 22 ships left Stockholm at Worster's behest with 'Orground' iron on board. Apart from individual ventures to Amsterdam, Newcastle upon Tyne, and an unnamed Scottish port, all of them sailed to Hull or London. Worster landed 851 tons of 'Orground' in the capital and 525 tons at Hull. Manufakturkontorets arkiv, Peter Westmans utskeppningsböcker 1729–45, vol. 490, RA.

to service certain institutional buyers in the capital. The Navy Board and the East India Company were both significant consumers of Swedish iron, but they were unlikely to interfere in the markets that Shore and Prankard hoped to master. The Navy Board bought iron for consumption in the Royal Dockyards; there was little or no leakage of its iron into the civilian market.[111] The East India Company purchased bar iron for re-export to Madras or Calcutta, not for further sale in northern Europe.[112] Lacking high-level metropolitan contacts, Shore and Prankard ceded these specialised markets to their more worldly associates in Stockholm.

There was one other weighty actor on the London market to be considered: Theodosia Crowley. The Crowley family firm was by pre-industrial standards gargantuan. The creation of Sir Ambrose Crowley (1658–1713), the business included three massive metalware factories on Tyneside, a central depot at Greenwich, and a set of warehouses that supplied outworkers across the Midlands.[113] The Crowleys were by some distance the largest producers of metalwares in Britain. They were necessarily major consumers of Baltic iron and, having their own steel making facilities at Winlaton Mill and Swalwell, they were per-

[111] The Navy Board invited tenders for the supply of bar iron every year. The successful bid in 1731 came from Henry Norris (TNA: PRO, ADM 106/2545, 17 March 1731). Norris was, of course, splendidly placed to fulfil this contract. He was a trusted associate of Robert Campbell, who controlled the flow of iron from *Leufstawerken* in the early 1730s. Indeed, Norris's pre-eminence as a naval contractor—he was sole contractor between 1727 and 1732—coincided with Campbell's supremacy in Stockholm (Navy Board minutes in TNA: PRO, ADM 106/2544 and ADM 106/2456, and—where the former are deficient—correspondence from Chatham Dockyard to the Navy Board in the National Maritime Museum, CHA/L/19–20). Conversely, the failure of Campbell to retain his exclusive contract with *Leufstawerken* brought a new Navy Board contractor to the fore in London: Josias Wordsworth, an ally of the triumvirate of Francis Jennings, Samuel Worster, and Samuel Wordsworth that had wrested control of the Åkerby and Leufsta brands from Campbell.

[112] Annual exports for the period 1727–28 to 1732–33 averaged 20,836 bars. This figure is calculated from returns of goods exported by the East India Company in the Board of Trade papers (TNA: PRO, CO 388 series). See H.V. Bowen, 'Sinews of trade and empire: the supply of commodity exports to the East India Company during the late eighteenth century', *Economic History Review*, LV, 3 (2002), 466–86, for broadly comparable export figures from later decades. This did not amount to much when compared to the Company's bullion shipments—bullion usually accounted for 80 per cent or more of the value of exports to the east, and bar iron for not much more than 1 per cent—but from the perspective of a Baltic merchant the East India Company's contract for bar iron was worth a handy £5,000 or £6,000 annually.

[113] See the inventory taken at the death of John Crowley in 1728: Suffolk Record Office (Ipswich), HAI/GD/5/1–17. The firm's assets were reckoned at £157,928.

force regular buyers of 'Orground' iron. Indeed, by virtue of its four cementation furnaces in the north east, equal to at least 20 per cent of Britain's steel making capacity in the 1730s, the Crowley business was the country's single largest customer for 'Orground' iron, taking in excess of 300 tons annually. Theodosia Crowley, the head of the firm from 1728, had been supplied with 'Orground' iron via Henry Norris during the period of Campbell's contract with the Leufsta estate. Shore and Prankard were happy for this to continue. For as long as Norris's residual supplies of Åkerby and Leufsta were shipped to Winlaton and Swalwell they could not interfere in the markets that were of most concern to Shore and Prankard. Indeed, Prankard positively enouraged the cordial relationship between Norris and the Crowley. 'Contrive it', he told his London correspondents, 'so as for Norris to work of what he has to the Lady Crowley by which means London would be Clear.'[114] It was an indulgence designed to exhaust whatever reserves of 'Orground' iron Norris held on to and to put him to the expense and inconvenience of dealing through Amsterdam if he wanted to acquire fresh supplies.

Henry Norris had been expelled from most of the markets for 'Orground' iron in Britain. He was not inclined to accept his ejection, however. He did what he could to disrupt the division of the English market that Prankard and Shore had settled upon. He released 100 tons of cut-price Leufsta iron onto the Bristol market in June 1735, just before the arrival of Prankard's ships from Stockholm. A further 100 tons was sent north to discomfort Samuel Shore.[115] This 'Politicall & Revengefull Stroke' would not, Prankard vowed, 'Cause me to Sink the Price without Reason'. Excluded from the contract with the De Geers, Norris's ability to thwart Prankard and Shore in the sale of Åkerby and Leufsta was limited and diminishing. Soon, one of Prankard's Midland customers was assured, 'Affairs may be Ordered So as to have the Sale of it Contracted into a Narrer Compass'.[116] This was to prove overly optimistic, for Norris had other weapons in his armoury.

Henry Norris began to broadcast the merits of some of the so-called 'second Orground' brands, arguing that they were of comparable quality to Åkerby and Leufsta. Particular attention was paid to iron from the

[114] GP to Pat & Robert Mackey, 20 September 1735.
[115] GP to Pat & Robert Mackey, 20 September 1735.
[116] GP to William Bowyer, 18 September 1735.

forge at Ullfors: 'Norris has used all Possible means to represent it of equall goodness... & Still Continues his Endeavours for So doing by Prevailing on Sundry Noted Steel Converters for make Assay & Tyralls of it.'[117] If steel manufacturers could be persuaded that iron from Ullfors or Strömsberg was an adequate substitute for Åkerby bars then Shore and Prankard's hard-won monopoly would be capsized. Norris was assisted by the fact that the manager at *Strömsbergwerken* was indeed trying improve the quality of the bar iron made at Ullfors, Wessland and Strömsberg. This was much to Prankard's bafflement since these three *bruk* had recently been purchased by the trustees of Charles De Geer. It could not, he reasoned, be of any value to the De Geers to boost the reputation of their '2d Orgrounds', for any increment in the quality of Ullfors bars would only reduce the price of the more established Åkerby and Leufsta marks.

Prankard was right. The De Geers, by extending their hold over the 'Orground'-making *bruk* of Uppland, had created something of a dilemma for themselves. By the mid-1730s they were marketing not one or two, but half a dozen 'Orground' brands. Once, it had been simplest to dispose of all their iron through a single Dutch merchant house, but as the volume of iron to be sold increased, and as a growing number of British merchants became intent on dealing directly with Stockholm, that possibility receded.[118] De Geer iron started to flow through several, competing channels. It was this that allowed Robert Campbell, deprived of the contract for Åkerby and Leufsta, to switch his allegiance to Ullfors, thereby supplying Henry Norris with the ammunition to harass Prankard and Shore.[119] The De Geers introduced a further element of instability into the once ordered market for 'Orground' iron by revamping the run-down *Strömsbergswerken*. In doing so, they were upsetting the established hierarchy in Uppland. At first, Louis De Geer relished the fact that improvements in *Strömsbergswerken* iron had discomforted Samuel Worster and his associates. 'It is fun', he told Eric Touscher in October 1735, 'that Worster is complaining that the material from Strömsberg, Hille &c works is too good, in all cases as good as Leufsta, which he says should have precedence, and wishes

[117] SS & Son to Worster, Wordsworth & Jennings, 15 August 1735.

[118] Swebilius urged a return to the habits of the 1720s in a letter to Jean Jacques De Geer, presumably hoping for the restoration of exclusive dealing with the Grills. He was to be disappointed. GS to JJDG 13 June 1733, Leufsta Arkivet, vol. 106, RA.

[119] GS to JJDG 22 April and 26 August 1734, Leufsta Arkivet, vol. 106, RA.

us to make it slightly worse. I smiled at him.'[120] Such self-satisfaction was not to last. Before long, De Geer and his management team were assailed by a barrage of complaint from their British customers.

Prankard and Shore were incensed. Not only did the greater care taken with the iron at Ullfors and Strömsberg tend to undermine the pre-eminence of Åkerby and Leufsta, but the quality of Åkerby iron seemed to deteriorate in tandem. The priority that Åkerby enjoyed on international markets was attributable not just to the superior materials used in its production, but to the peerless workmanship with which it was finished. The Åkerby hammermen had an unmatched reputation, even amongst their fellow Walloons. They managed, Prankard noted, 'to have less raw Ends in it than Either [Leufsta] or bullets [i.e. Österby]'; that is, they were supremely skilled in expelling slag inclusions from the ends of the bars.[121] But no sooner had Shore and Prankard achieved monopoly rights over Åkerby iron than they detected a lapse in standards. Samuel Shore was quick to complain: 'the Proprietor of Said Works is very Defficient in keeping it to Its usual Goodness So that Instead of making it Sound good & Free from Flaws & Cracks it dont Prove So good in that respect as the best Common Iron'.[122] Swebilius, it seems, had experimented with a different ore mix in smelting at *Leufstawerken*. The outcome was a poorer quality pig iron, with effects that resonated through the subsequent refining operations. The iron was not 'realy Clean from ye drossy part...which causes it to be so rotten...[that it is] not fit for Conversion into Steel'.[123]

Shore and Prankard faced a crisis. They had contracted to take a large amount of iron from the forges at Åkerby and Leufsta, but iron that was of increasingly uncertain quality. At the same time a high-quality product from Ullfors and Strömsberg was being offered to their customers at a bargain rate. So, when the contract with the Leufsta estate came up for renewal in 1736, Shore and Prankard demanded a rebate on the price they were paying for Åkerby and Leufsta bars.[124] More audaciously, they urged their Stockholm agents to contract for the output of Ullfors as well as that from *Leufstawerken*, as 'it very

[120] LDG to ET 6 October 1735, Leufsta Arkivet, vol. 105, RA.

[121] GP to FJ, 13 October 1731.

[122] SS & Son to Worster, Wordsworth & Jennings, 15 August 1735.

[123] SS & Son to Worster, Wordsworth & Jennings, 7 August 1738. See also GP to FJ, 13 December 1735.

[124] GP and SS to Worster, Wordsworth & Jennings, 17 March 1736.

much prejudice us in the sale of the [Åkerby] and [Leufsta]'.[125] In the event, Prankard and Shore agreed to take a whole range of 'common Orgrounds' rather than allow them to fall into the hands of Campbell and Norris. Prankard took 600 *skeppund* from Ullfors, 300 *skeppund* from Wattholma, 300 *skeppund* from Strömsberg, and 200 *skeppund* from Harg, in addition to his existing shipments from Åkerby and Leufsta.[126] This amounted to some 560 tons.[127]

Shore and Prankard had stifled the threat of competition but at the cost of taking far more 'Orground' iron than could possibly be absorbed by the markets they regularly supplied. Alternative uses had to be found for the glut of steel-making iron they had on their hands. Prankard pressed Jennings to have the 'common Orgrounds' struck in a different form. They should be hammered into square bars of a fine gauge, rather than the broad flat bars that steel makers preferred for their furnaces. Better still, they should be struck 'without any Stamp on it yt I might Sell it under ye Determination of English Iron'.[128]

Formerly, Louis De Geer had taken a high-handed attitude to such requests—let the British 'whine about the sorts' was his response to earlier complaints—but Prankard and Shore's willingness to handle all the 'Orground' iron destined for the British market persuaded him to be more obliging.[129] New instructions were issued. The forgemen at *Leufstawerken* were to concentrate on making bar iron of superlative 'inner quality', but their counterparts at the *Strömsbergswerken* were to attend to the outward form of the iron, producing finely finished bars in preset dimensions, just as Prankard and Shore demanded. *Leufstawerken* bars were to be sold at a premium rate to denote their superior qual-

[125] SS to Samuel Worster & Samuel Wordsworth, 30 July 1736.

[126] GP to FJ, 16 February 1737.

[127] In October 1736 Louis De Geer informed Touscher that he had concluded the new contracts for both iron from *Leufstawerken* and *Strömsbergswerken*, LDG to ET 28 October 1736, Leufsta Arkivet, vol. 105, RA. Leufsta and Åkerby iron was contracted to Worster and Company for three years at 49 *daler kopparmynt* per *skeppund*. The contract for *Strömsbergswerken*, was concluded with the Grills, and was for two years at 43 *daler kopparmynt* per *skeppund*. As Prankard was able to obtain Ullfors iron from Jennings we can assume a collaboration between 'Worster and Comp' and the Grills. This is hinted at by Louis De Geer: 'now the Englishmen Worster & Co and the Grills are interested in each other, so that all Roslags [the area around Dannemora] irons are in one hand, together with Forsmark and Harg, I think they can force the iron price up again.' LDG to ET 18 November 1736, Leufsta Arkivet, vol. 105. RA.

[128] GP to FJ, 16 February 1737.

[129] LDG to ET 2 June 1735, Leufsta Arkivet, vol. 105, RA.

ity; *Strömsbergswerken* iron was to be marketed on the strength of 'its assortment and beauty'.[130]

The concessions that were made to Graffin Prankard and Samuel Shore had important repercussions at the Uppland *bruk*. Walloon forgemen were not accustomed to having their work criticised, still less to following precise instructions as to the form that the bars should take. This became apparent when Prankard and Shore issued further complaints about the declining quality of Leufsta and Åkerby bars in the summer of 1738. Samuel Shore reviewed the situation: 'in order to Support ye Creditt of those marks [i.e. Leufsta and Åkerby] we have Joyn'd in Contracting for the other 2d orgrounds & Subjected our Selves to have it Struck in to Such Sorts for ye most part as yt it may be Sold for Common uses'. This was done so that 'ye market might not be overburthened with orgrounds Iron', but it was a loss-making strategy that was only justifiable if the premium quality of Leufsta and Åkerby iron led to compensating gains. Alas, the *Leufstawerken* brands had not been kept to their 'wonted goodness', despite 'fair words and Promises' from Leufsta. Indeed, they were so poorly wrought as to be unfit for conversion to steel.[131] Graffin Prankard developed the theme in a shrill letter of his own: 'My Cheafest dealer seems resolved to lay down his Trade if what Comes now ys year dont prove better wch if he should I must forbear Importing of it.'[132]

This prospect was taken very seriously at Leufsta, and a special meeting was convened in the *bruk* office on 18 August 1738 to consider the allegations made by the English contractors. The master furnace keeper at Leufsta (Noe Dandanell) was present, as were five forgemen: the master finer from Åkerby (Noe Tillman), and his counterparts at the four Leufsta forges (Jacob Tillman, Mårten Martinell, Jan Tillman, and Raphael Pouset). Eight clerks from Leufsta were in attendance, as was Magnus Kindel, *Leufstabruk*'s pastor, who was to witness the proceedings. *Directeur* Touscher presided.[133]

---

[130] Louis De Geer even wrote to Touscher saying that if the manager of *Strömsbergswerken*, Kiörning, was unhappy at not being able to fulfill his ambition of improving his works, De Geer was willing to compensate him. LDG to ET 28 October, 4 and 18 November 1736, Leufsta Arkivet, vol. 105, RA.

[131] Samuel Shore & Son to Worster, Wordsworth & Jennings, 7 August 1738.

[132] GP to FJ, 28 June 1738. Prankard's 'Cheafest dealer' was presumably John Kettle of Birmingham.

[133] What follows is based upon the record of this hearing in Leufstaarkivet, vol. 43B.

The situation was a delicate one for Eric Touscher. He was not long in his post. Moreover, as the forgemen well knew, he had no background in the iron industry. He was a lawyer who had been recruited by Georg Swebilius as 'a quick and honest man' to oversee the complex transactions involved in the reorganisation of the De Geers' affairs in Uppland in the mid-1730s.[134] But with Swebilius's death Touscher underwent an unexpected (and perhaps unwanted) elevation to the position of *directeur* at *Leufstawarken*. This promotion, daunting in itself, came at a time of unthinkable crisis, when the quality of the premium brands from Leufsta and Åkerby had been called into question.

It was Touscher's task to read out a letter, 'an austere and earnest letter', from Louis De Geer, requiring a full account of recent work at the blast furnaces and forges of *Leufstawerken*. Enclosed with De Geer's missive were copies of three letters from British customers. Although the copies were unsigned, they were plainly translations of letters from Samuel Shore and Graffin Prankard. The assembled clerks and workers, Touscher went on, were to respond to the points made: to 'answer in plain and confess'.

The forgemen were in no mood to confess to any failings on their part. They rejected indignantly the suggestion that their iron was of an inferior nature. They were more than willing for such iron as was left in the Stockholm *Jernvåg* to be closely examined, confident that the bars were superior to anything made at other *bruk* in Uppland. The forgemen saw no merit in Prankard's suggestion that there had been a deterioration in the quality of the pig iron they melted. There had been a short-lived attempt under *Directeur* Swebilius to alter the ore mix at the *Leufstawerken* furnaces, but that was now far in the past. The finers had no serious complaints about the pig iron delivered to them. The odd defective *geuse* would simply be laid to one side.[135]

Touscher himself was certain that the quality of *Leufstawerken* iron had not suffered. Smarting, no doubt, from the 'curses' that the forgemen directed at him, he made a strident defence of his management.[136] Louis De Geer sympathised: 'That Mr Directeur is very upset by the English-

[134] GS to JJDG 18 June 1733, Leufsta Arkivet, vol. 106, RA.

[135] The forgemen maintained that substandard *geuses* were used to make the iron railings that were being put up around the *herrgård* and its park in the 1730s.

[136] For this and much else on events in the summer of 1738 see Touscher's letter of 19 August 1738 to Anders von Drake, the governor of Stockholm: Leufsta Arkivet, vol. 167, RA.

men's complaints, I wonder not... I have myself been so provoked that I have felt an urge to hang them'.[137] Yet the forgemen themselves were aware that their product had been subject to criticism for some time, even though, in their eyes, its essential goodness remained unsullied. Complaints about Leufsta and Åkerby iron had first been heard when *Directeur* Swebilius had fallen in with 'an Englishman calling himself Mr Campbell'. Campbell was a Scot, not an Englishman, but the forgemen were not wrong in thinking that closer ties to the English market had brought a new, harsher tone to working life at Leufsta and Åkerby. Once, they lamented, the forge had been their own domain; they had governed the pace of work themselves. The criterion by which their work was judged was 'the goodness of the iron'. Little attention was paid to 'the fineness of the sorts'; that is, the exactness with which the bars were finished. Indeed, it was a matter of notoriety that Walloon forgemen concerned themselves with the inner quality of the material, not its external form. The composition of the meeting on 18 August testified to that: five finers were in attendance, but no hammermen. In the 1730s, however, the management at *Leufstawerken* began to insist that bars of very particular dimensions were made. Sometimes, the finers complained, 'so much of that sort is commanded, then of others'. When the forgemen were unable to comply with their instructions 'the clerks throw the iron back into the hearths, as it is too long, then too short and too thick, although this has never been asked for before'.

Such precision could be asked of forgemen who used the German forging technique. They devoted less care to the melting of pig iron than Walloon finers, but they were far more attentive to the size and shape of the bars, employing so-called 'cold-drawing' to give their bars a smoother finish. This was possible because the pace of work was far more leisurely in German forges than in the high tempo Walloon enclave. Yet conditions on the British market now dictated that more and more 'Orground' iron was finished to a precise standard. Because Prankard and Shore were taking far more iron than could be absorbed by the steel industry in Britain, it was necessary that part of the annual make was drawn into the slender square bars required by the generality of smiths rather than the broad bars used by cementation steel furnace

[137] LDG to ET 17 August 1738, Leufsta Arkivet, vol. 105, RA.

operators.[138] Hence the instruction given to the *Strömsbergswerken* forgemen in 1736 that their iron to was to be 'Struck in to Such Sorts for ye most part as yt it may be Sold for Common uses'.

The *Leufstawerken* forgemen had been entrusted with making the heavier bars that were destined for conversion to steel, but at the hearing convened by Touscher on 18 August 1738 the Leufsta and Åkerby finers alleged that they too were being asked to make a selection of smaller but more numerous *smältstycken*.[139] This implied a speed up in the pace of work and a threat to the 'limbs and health' of the forgemen 'until their dying days'. It also signified, they said, a slackening of standards. This was a damaging admission for the forgemen to make, coming on the heels of their heated denials that *Leufstawerken* iron had in any way deteriorated. The finers then made another *volte face*, suggesting that high-quality output and an accelerated rhythm of work were, in fact, reconcilable objectives—provided, that is, that they got higher wages. If, the forgemen announced, they 'were given more rewards' Leufsta iron 'could be made with an outer adornment that shall exceed all German forgings in the country, and be made into the finest rod or hoop iron that was ever made'.

The hearing of August 1738 ended with the workmen being enjoined to make better iron. Whether they did so to the satisfaction of their British customers is doubtful. The price paid for Leufsta and Åkerby iron in the years that followed suggests that they did not. The price had risen sharply in the early 1730s as Samuel Shore and Graffin Prankard battled to gain control of the *Leufstawerken* brands, from 46 *daler kopparmynt* per skeppund in 1730 to 52 in 1733. It remained at that level until 1736, the point at which Prankard and Shore demanded a rebate to compensate them for the declining quality of the iron. The price fell

[138] The 'right sort' of bars for steelmaking, one Sheffield merchant noted, 'must be from 2 to 3 Inches broad'. John Rylands Library, Manchester, B 5/4/1, Richard Dalton to Samuel Mould, 25 October 1735.

[139] In fact, no clear trend towards making lighter bars is visible in the works accounts. On average, 30 per cent of output at Leufsta consisted of squares in the period 1730–37. The proportion of squares fluctuated, with a high of 48 per cent in 1731 and a low of 21 per cent in 1733. In 1738, the year in which the Leufsta finers railed against the practice of making lighter bars, squares made up just 17 per cent of their output. At Åkerby no squares whatsoever were made in 1738; the forge was given over entirely to the making of steel iron. The manufacture of squares was actually concentrated at Carlholm, the newest part of Leufstawärken, where 62 per cent of output in 1738 was in the form of square bars. See Leufstaarkivet, vols 43B and 268. Such was the actuality; the forgemen's perception of developments was very different.

accordingly to 49 *daler kopparmynt* in 1737; then, in 1740, to 48. The fall continued through the 1740s, reaching 44 *daler kopparmynt* in 1744. This decline took place at a time when the steel industry in Britain was expanding and with it the demand for 'Orground' iron. That the *Leufstawerken* marks were not forced up in price suggests that problems with quality persisted, or perhaps that iron from *Strömsbergswerken* and other sorts of '2d Orgrounds' were increasingly accepted by steel makers in Britain, thereby wiping away the premium once commanded by Leufsta and Åkerby bars. That Leufsta and Åkerby iron no longer enjoyed the priority it once had is hinted at in the accounts of Carlholm forge. In 1740 a payment of 10 *daler kopparmynt* was made to Mårten Douhan, the master finer at Carlholm, for going to Åkerby to assist the master finer there, Noe Tillman, in introducing a better 'procedure in his finery'.[140] Formerly, no iron had been held in higher esteem than the 'P.L. & Crown' brand from Åkerby. That a forgemen had to be summoned from the subaltern works at Carlholm to advise the supremely experienced Tillman speaks of a shift in power.

Production at Carlholm remained at a far lower level than at Åkerby or any of the Leufsta forges. Quality was privileged over quantity. In this, Carlholm set the pattern for the future.[141] Production at all the *Leufstawerken* forges was cut back in the late 1740s. Although Leufsta remained in production until the early years of the twentieth century, 1743 remained forever the peak year for output. Charles De Geer commented upon this in later years. In 1774, as he prepared for retirement, De Geer wrote a memorandum on the running of his industrial empire: 'Information to my successor at Leufsta, founded in some experiences, which he can follow if that pleases him'. This short text reprised many of the themes that Eric Touscher had addressed in his 'En liten handbok' of thirty-five years earlier. There were sections on ore extraction, forestry and charcoal making, smelting and bar iron manufacture, as well as advice on dealing with the clerks at the works, the working people, and the local peasants. De Geer's remarks on bar iron making began with some obvious generalities: 'The bar iron should be well melted and sorted in the finery... smoothly hammered... without any cracks

---

[140] Landsarkivet i Uppsala, Karlholms Bruk, vol. A1:12, Bruksbok 1740.

[141] When distinguishing between *Leufstawerken* and *Strömsbergswerken* in 1736, Louis De Geer had suggested that Carlholm should occupy an intermediate position, making iron of high quality but drawing the bars into precise forms. LDG to ET 4 November 1736, Leufsta Arkivet, vol. 105, RA.

or flakes'. Then De Geer's tone became more emphatic: 'Too high a weekly production cannot be combined with a well-made iron; it is better to make less of a well-forged iron, otherwise the iron will end in disrepute, something that will not easily be cured'. The old *brukspatron* warned: 'Once the iron loses credit with the foreigner, it is difficult, or even impossible, to get it back'. No doubt de Geer was thinking back to the 1730s and 1740s when it had proved so troublesome to combine a high output with satisfying the demands of foreign buyers.[142]

In the 1720s and 1730s the market for 'Orground' iron was transformed. At the start of the eighteenth century the marketing of the elite Uppland brands had been centred on the United Provinces, as it had been for decades. The uses to which 'Orground' iron was put were evidently varied: it had no special affinity with Britain, nor with steelmaking.[143] After 1720 that changed as the uses to which 'Orground' iron was put narrowed and it became feedstock for the British steel industry. This shift in the form and focus of international demand prompted wholesale change. In Britain, Samuel Shore and Graffin Prankard strove to monopolise the import of the key *Leufstawerken* brands; in Uppland, the De Geers sought to tighten their grip over the 'Orground'-making *bruk*. The two initiatives were not entirely compatible. The De Geers, scenting an opportunity, were interested in expanding production. Shore and Prankard, on the other hand, did not want the volume of iron made at *Leufstawerken* to exceed what they could comfortably handle. Nor did the two Englishmen want the range of irons suitable for conversion to steel to be extended. The outcome was a period of strife and confusion as the English buyers sought to dictate production practices in Uppland. Two worlds came into collision: that of British merchant capitalism, driven by the pulsing Atlantic economy, and that of artisanal production, stable and orderly, in the *Vallonbruk* of Uppland. The tyro

[142] This document was kept at Leufsta Arkivet, RA, but is, however, sadly lost. The quotation is from Folke Thörnwall, *Leufsta. Ett gammalt upplandsbruk* (Tierp, 1968), pp. 183–86.

[143] A Swedish report of the 1670s makes no mention of 'Orground' as a material suitable for steelmaking and dismisses Britain as entirely marginal to European steelmaking, other than as a market for Spanish and central European steels: The Historical Metallurgy Group of the Swedish Ironmasters' Association, *Iron and steel on the European market in the 17th century: a contemporary Swedish account of production forms and marketing* (Stockholm, 1982), pp. 168, 174, 185.

manager of *Leufstabruken*, Eric Touscher, was left with the unenviable task of reconciling the two.

Birmingham

Reinhold Angerstein, the roving investigator of the *Bergscollegium*, left Bristol on 24 June 1754 to head up the Severn valley into the heartland of English metal working. After a detour to inspect the ironworks of south-east Wales and the Forest of Dean, he followed the river through Gloucester and Worcester before arriving at the river port of Bewdley, the gateway to the West Midlands.

> Bewdley is a small place, but business there is quite good, due to the harbour, which serves the manufacturing towns Birmingham, Wolverhampton, Stourbridge, Dudley, Wednesbury, etc, all located in Staffordshire. In this country there are many manufacturers of nails and other articles of steel and iron as well as of copper and brass, such as boxes and similar fine work. A great deal of this is shipped down the Severn to Bristol. Large quantities of Swedish and Russian iron and other goods are carried as return cargo to be worked up to steel, or in the slitting mills to rods for the nailers and also for other purposes, for which the iron from these countries is particularly suitable.[144]

Angerstein estimated that 2000 tons of Baltic iron passed through Bewdley every year, most of it imported by the great ironmongers of the region: Abraham Spooner and Sampson Lloyd in Birmingham, and John Finch in Dudley.

From Bewdley, Angerstein took the Birmingham road, tracking the route taken by Baltic iron to the manufacturing towns of south Staffordshire. He stopped to view the cementation furnaces at Broadwaters, where Leufsta iron was being converted to blister steel, and he lingered at Stourbridge, famous for its nail trade, its glassworks, and the quality of the local fireclay. Angerstein reckoned that 9000 tons of bar iron was slit annually into rods within a nine-mile radius of the town. Turning north towards Dudley, the Swede entered what was to become the Black Country, a landscape already marked out by the winding gear and engine houses of coal mines.[145] Dudley was the

---

[144] *Angerstein*, p. 174f.

[145] For definitions of the Black Country see Richard H. Trainor, *Black Country elites: the exercise of authority in an industrialized area 1830–1900* (Oxford, 1993), pp. 1–4.

main organising centre for the nail trade, although it was home to a variety of other manufacturing processes as well. Angerstein noted the making of 'malt and coffee-mills, hinges for doors and caskets, axes, and other large edge tools, screw vices, [and] horse-locks'.[146] This entire region was, in fact, host to an intricate network of specialisms. Wolverhampton was renowned for locks, as well as buckles and small chains, Bilston specialised in brass wares and enamel box making, while Wednesbury was famous for its gun-locks. Lorimers were concentrated in and around Walsall.

To the east, just beyond the coal measures, was Birmingham, 'the head of all manufacturing towns in iron, steel, [and] brass'. In the mid seventeenth century the town had been no more than a large village with 1500 inhabitants. Thereafter, a phase of rapid development began as the old staples of woollens and leather working were outstripped by metal manufacturing. Scythe making, cutlery, and sword grinding proved so successful that the town's population was propelled to 15,000 by 1700. New trades in the eighteenth century such as gun making, brass working, and 'toy' manufacture ensured that population growth continued on an upward trajectory, reaching 23,000 in 1750.

Birmingham was a working town, pulsing with business. William Hutton, the town's first historian, recalled his own first impressions as a young migrant in the 1740s:

> I was surprised at the place, but more so at the people: They possessed a vivacity I had never beheld: I had been among dreamers, but now I saw men awake: Their very step along the street shewed alacrity: Every man seemed to know and prosecute his own affairs: The town was large, and full of inhabitants, and those inhabitants full of industry.[147]

The industrial structure of the town was analysed by Samuel Schröder, another *Bergscollegium* investigator. He arrived in Birmingham in 1749 and was so struck by the place that he remained for two months, resulting in a very full report for his superiors.

> The main traffic and trade of this place is manufacturing in iron, steel and other metals, in particular the manufacturing of a wide variety of buckles and buttons. Glass buttons of all sorts, Snuff boxes, painted and lacquered iron sheets.... Tea trays, tea caddies etc from iron sheets,

[146] *Angerstein*, pp. 178–79.

[147] William Hutton, *An history of Birmingham, to the end of the year 1780* (Birmingham, 1781), p. 63.

> lacquered with black, gold and coloured, a wide variety of guns, coarser and finer muskets, pistols, swords and sabre blades.[148]

The more basic forms of metalware manufacture in which Birmingham had excelled in the seventeenth century tended to migrate to the coal districts to the west in the eighteenth century, leaving Birmingham to specialise in the production of higher status goods and semi-luxury novelties of the sort Schröder mentions. Birmingham was though, as Schröder made clear, a marketing centre for the entire West Midlands. Goods made by independent artisans were bought up by merchants for distribution to distant markets. Angerstein described how on market days workmen would congregate at inns where the major merchants had gathered:

> I had hardly entered my room at the inn, before scores of smiths came in to offer their wares for sale. Included in these were nails, tools, locks, hinges, key-rings, buckles, corkscrews, watch-chains, flat-irons, crimping-irons, sugar axes, snuffers and other similar goods in iron and steel which fetch a good price.[149]

Birmingham's merchant class connected the shoals of small artisan producers to national and international markets. There was a clear hierarchy within this merchant community. Some smaller dealers traded in a narrow range of goods, but at the pinnacle of Birmingham's commerce was a small group of great merchants who handled the full range of local wares, sending catalogues and samples to correspondents throughout the country. It was they who inhabited the prestigious new houses in The Square or worshipped at the modish parish church of St Philip, built between 1709 and 1715, that Samuel Schröder so admired when he attended divine service there with the family of Thomas Hadley, the leading gunmaker. Birmingham was not without elegance, despite the fuliginous coating that soot from thousands of industrial and domestic hearths gave to the town. (Schröder reported that between 50 and 60 waggon loads of coal rumbled into Birmingham from Wednesbury every day.) Even so, the counterpart to mercantile wealth was to be found in the densely packed streets of districts such as Digbeth, where the clang of hammers rang out and brass founders spent so long over their crucibles that their hair grew green from cupric contamination.

---

[148] Schröder I includes a very thorough description of Birmingham.

[149] *Angerstein*, p. 43. The description is actually of Wolverhampton, but can stand for Birmingham.

This was a town of small producers, often highly specialised, embracing new materials, new tools, and new products with alacrity. Birmingham had never been a borough and therefore lacked regulatory structures. In the absence of corporate regulation, Birmingham's tradespeople responded quickly to changed market conditions and introduced innovations with relative ease. Yet, Schröder thought, there was also a tendency for artisans to undersell one another in an unfettered market. Enrichment came to the merchant class before it came to the direct producers.[150]

Birmingham, with its varied and protean industrial hinterland, was central to the development of British metalware production. At the time of the visits of both Samuel Schröder and Reinhold Angerstein it was possible to inspect most, if not all, the links in a ferrous production chain from ore extraction to the making of very complex metal mechanisms within a short distance of the town. Pig and bar iron (albeit low-grade) was made in the area—Aston furnace and Bromford forge lay just to the east—and rod iron was slit at Sampson Lloyd's town mill. Steel was made at a number of cementation furnaces, both in Birmingham and its hinterland. These processing operations supported a broad spectrum of metalware manufacturing trades. Nail making was by far the most important and widely spread of the region's metal trades. Local specialisms abounded, as we have seen. In the case of Birmingham, the most important 'traffic', to use an expression of Schröder's, was toy making. ('Toy' was a generic term denoting any kind of decorative consumer good: shoe buckles, snuff boxes, steel buttons, and the like.) It was this, the surveyor Samuel Bradford announced, when presenting his 1750 *Plan of Birmingham* to the public, 'which has gain'd the Place a name & great esteem all over Europe'.

---

[150] The development of Birmingham during the first half of the eighteenth century is treated in Michael J. Wise, 'Birmingham and its trade relations in the early eighteenth century', *University of Birmingham Historical Journal*, II, 1 (1949), 53–79; M.J. Wise and B.L.C. Johnson, 'The changing regional pattern during the eighteenth century', in *Birmingham and its regional setting: a scientific survey* (Birmingham, 1950), pp. 161–186; Marie B. Rowlands, 'Continuity and change in an industrial society: the case of the West Midlands industries', in Pat Hudson (ed.), *Regions and industries: a perspective on the Industrial Revolution in Britain* (Cambridge, 1989), pp. 103–31. See also Gordon E. Cherry, *Birmingham: a study in geography, history and planning* (Chichester, 1993), chap. 2; Eric Hopkins, *Birmingham: the first manufacturing town in the* world (1989); Maxine Berg, 'New consumer industries in eighteenth-century England: products, markets and metal goods in Birmingham and Sheffield', in René Leboutte (ed.), *Proto-industrialization: recent research and new perspectives* (Geneva, 1996), pp. 211–36.

Illustration 2.12. Westley's *The East Prospect of Birmingham* (1731).

Courtesy of Birmingham Library Services.
Caption: The baroque church of St Philip, shown here looming exaggeratedly large, commands the horizon. The spire of St Martin's, a more ancient church, emerges from the densely packed courts of the lower town. The river Rea runs through the foreground of this improbably smoke-free vista.

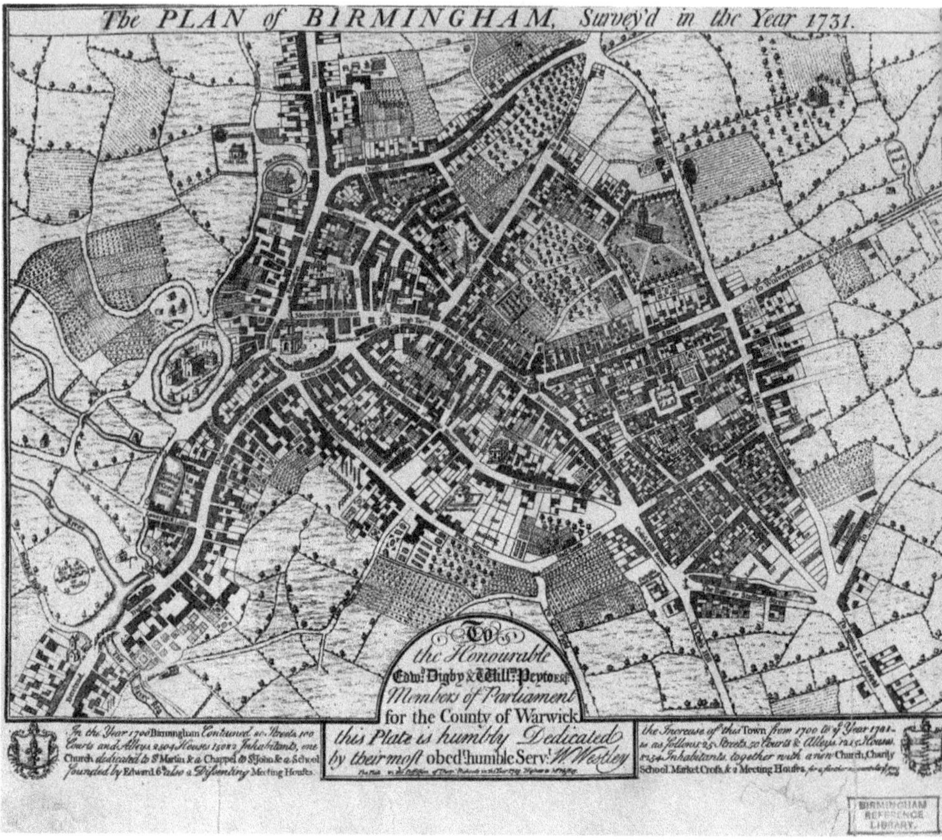

Illustration 2.13. Westley's *The plan of Birmingham, survey'd in the year 1731.*

Courtesy of Birmingham Library Services.
Caption: Westley's map shows a rapidly expanding town. 'Land for Building' is marked out in several places. The more salubrious parts are to the right of the map on higher ground, where the imposing church of St Philip stood. The metal ware trades were bunched on lower ground to the left. Sampson Lloyd's slitting mill was here, along the Digbeth road—the legend 'Lloyd's Slitting and Corn Mill' appears across the mill pond.

Such was the diversity of trades practised in Birmingham and its region that they defy comprehensive analysis. Yet there were two highly strategic Birmingham trades that demand a fuller examination: steel making and gun manufacture. They exemplify the interplay between imported iron and the domestic product in the English Midlands, and they illuminate the region's linkages, backward to the Baltic and forward into the Atlantic basin. As manufacturing trades, they were highly specialised and highly localised within the town. A third trade, that of nailing, was more far-flung, flourishing best in the semi-rural industrial villages to the west of Birmingham. Nevertheless, it shared with steel making and gun manufacture the characteristic of being part of a commodity chain whose links stretched back and forth, east through the Sound to the Baltic and west through the widening Bristol Channel to the Atlantic world at large.

John Kettle's steel works was on the northern edge of Birmingham, backing onto open fields. There were two furnaces or 'steel houses' that are clearly visible on Westley's 1731 map of the town. A cementation furnace was an imposing piece of industrial plant, made visible to Kettle's neighbours by the conical flue that rose to a height of eight metres or so. At the base of the flue was a brick-lined vault in which the conversion process took place. Swedish bar iron was loaded into the vault; English steel emerged many days later when the process was complete.

The bar iron that Kettle bought from Graffin Prankard was an iron almost completely free of carbon. This gave bar iron a ductile quality, allowing it to be readily forged into different shapes when brought to a red heat. What bar iron lacked was hardness and tensile strength. It was easily deformed, and when ground to a sharp edge it lost that edge very rapidly. Bar iron was therefore of little use to edge tool makers or to manufacturers of instruments in which components had to maintain their precise physical form. Such people required a super-hard material, an alloy of iron and carbon: steel.

The cementation furnace was a means of infusing carbon into bar iron, transforming it into steel. The vault at the base of the conical flue contained two, occasionally three, stone chests in which the bars of iron were placed. The bars were first examined for defects or dirt: 'all raw ends cutt off, all flawed or cracky ends layd by or cutt off', as one steelmaker specified. 'If any pitch be upon the iron designed for steele, it must be burnt off; if any clay it must be washed or beat off;

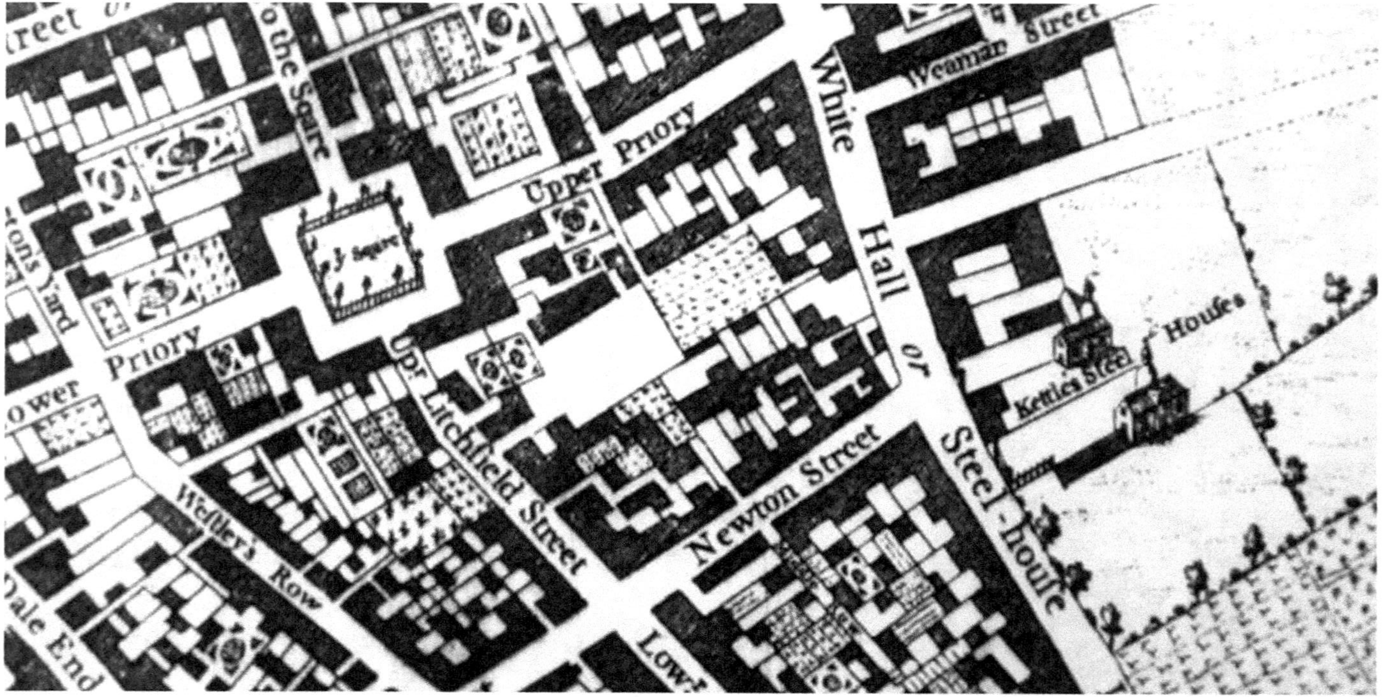

Illustration 2.14. 'Kettle's Steel Houses'.

Courtesy of Birmingham Library Services.

Caption: Just north of The Square, eighteenth-century Birmingham's most prestigious address, home to wealthy ironmongers like the Lloyds and the Pembertons, stood John Kettle's cementation furnaces. They were considered sufficiently notable to be rendered in perspective view, with smoke issuing from their flues, in Westley's *Plan of Birmingham*.

if any rust it must all be beat off.'[151] Those bars that passed muster were now laid, one by one, in the bottom of the chest. Each bar was packed about with charcoal dust, preferably from beechwood or juniper, separating it from its neighbour and from those bars that were to be placed on top. Once the chests had been filled, a layer of fine sand was packed down on top to provide an air-tight seal. The mineral coal that occupied the grate beneath the vaulted chamber could now be fired, enveloping the stone chests in flame and hot gases.[152]

Once the furnace had achieved the proper temperature, which took about fifteen hours, the workmen had to ensure that an even heat was maintained for the next five to six days, allowing carbon from the charcoal dust to penetrate the bars, imparting a steely hardness. When the conversion was judged to be complete, the fire was raked out and the furnace left to cool. This would take six days at least, much more in summer. (Hence the desirability of operating furnaces in tandem, as John Kettle did: as one was left to cool, the other could be fired.) When the heat had subsided to bearable levels the workmen could crawl into the vault, break up the sandy crust that sealed the chests, and prise the bars from their charcoal bed. When brushed down the bars were found to have surface blemishes that gave the product of the cementation furnace its distinctive name: 'blister steel'.

The cementation technique had been introduced to England from the Netherlands in the early seventeenth century, but steel making in Britain was slow to mature. The first furnace for which there is convincing documentation was built at Coalbrookdale *c.* 1620, but two generations later there were still only a handful of cementation furnaces in operation.[153] At the start of the 1690s blister steel was being made at Stourbridge and Abbots Bromley in the west Midlands, at Blackhall Mill in the North East, and at some imperfectly documented locations in south Yorkshire.[154] The capacity of these early furnaces was small—that

[151] Instructions for steelmaking at Winlaton, 1701, quoted in K.C. Barraclough, *Steelmaking before Bessemer. Volume 1. Blister steel: the birth of an industry* (1984), p. 201.

[152] David Cranstone, *Derwentcote steel furnace: an industrial monument in County Durham* (Lancaster, 1997); M.W. Flinn, 'Industry and technology in the Derwent valley of Durham and Northumberland in the eighteenth century', *Transactions of the Newcomen Society*, XXIX (1953–55), 255–62.

[153] The Coalbrookdale site was closed down, as a steel making facility at least, *c.* 1680. *Ex inf* Paul Belford, Senior Archaeologist for the Ironbridge Gorge Museum Trust.

[154] Charles Tooker of Rotherham, gentleman, appears to have experimented with steel making in the 1660s, but to have abandoned the trade by time of his death in 1680; while a 'Steel furnish' was listed by tax assessors in the township of Kimberworth,

Illustration 2.15. The steel furnace at Derwentcote.

Courtesy of English Heritage.
Caption: Derwentcote furnace in County Durham, built in the 1730s, is the only surviving eighteenth-century cementation furnace in the British Isles.

at Stourbridge converted just one ton at a time—so national output cannot have amounted to more than a few dozen tons annually.[155]

That was to change with a burst of furnace construction at the close of the seventeenth century. In 1697 Ambrose Crowley, who was a veteran of steelmaking at Stourbridge, began building a furnace at Winlaton Mill in the Derwent valley, just west of Newcastle upon Tyne. A decade later Crowley added extra plant at nearby Swalwell. By 1710 that formidable figure had command of four furnaces in the Derwent valley, making the North East the most important centre of blister steel production in England.[156]

Other regions also saw significant developments. In the west Midlands, the two steel furnaces operated in Birmingham by John Kettle 'were established at the end of the seventeenth century' (according to a nineteenth-century authority); a 'furnace for converting of Iron into Steel' was built at Tern in Shropshire in 1712–13; while Westley's 1731 *Plan of Birmingham* shows a third furnace on the north side of the town.[157] In south Yorkshire, the chronology of furnace construction cannot be established with any certainty, but the accounts of the Fell partnership, which sold steel made locally, indicate that cementation was being carried on at four sites between Sheffield and Rotherham in the first decade of the eighteenth century.[158] These were soon joined by two furnaces in the town of Sheffield itself, one of them the property of Samuel Shore: the existence of the first is documented by 1716, the second by 1720.[159]

Nationally, the stock of cementation furnaces certainly doubled and most probably tripled in the twenty-five years bounded by the Glorious Revolution and the Peace of Utrecht. This was a dramatic and unheralded transformation. Previously, the British Isles had been on the periphery of European steel making. Eric Odelstierna, the Swedish

---

neighbouring Sheffield, in 1672. It is not clear whether these sites were continually used. David Hey, *The fiery blades of Hallamshire: Sheffield and its neighbourhood, 1660–1747* (Leicester, 1991), pp. 183–95.

155 The description of the Stourbridge furnace that appeared in Robert Plot's *The natural history of Staffordshire* (Oxford, 1686) is reproduced in Barraclough, *Blister steel*, p. 154.

156 M.W. Flinn, *Men of iron: the Crowleys in the early iron industry* (Edinburgh, 1962).

157 Samuel Timmins (ed.), *The resources, products, and industrial history of Birmingham and the Midland hardware district* (London, 1866), p. 212; Coulton, 'Tern Hall', pp. 99–100. The third Birmingham furnace was identified as belong to a 'Mr Carlesse'.

158 Barraclough, *Blister steel*, pp. 69–80.

159 Hey, *The fiery blades of Hallamshire*, p. 191; Barraclough, *Blister steel*, p. 77.

official who visited England in the early 1690s, when this great change was just beginning, reported that some steel was made from Swedish iron, but only 'to a small extent'. Most of what was consumed in English workshops originated in the German-speaking lands, so Odelstierna reckoned, in Styria and Carinthia, the leading centres of European steel making since the middle ages.[160] German steel was brought down the Rhine to Holland and shipped thence to London. Such shipments ranged between 50 and 150 tons annually at the turn of the eighteenth century.[161]

Imports from Rotterdam nosed upwards in the wake of the peace of Utrecht, reaching a peak of 210 tons in 1737, but that was no more than a fraction of what was now made at British furnaces.[162] In 1737 the House of Commons was told that 1000 tons of Swedish iron was converted into steel annually in England. If true, that figure represented a startling rate of increase since the 1690s. In fact, the rate of increase was probably higher still. The British steel industry had a stock of about twenty cementation furnaces in the 1730s: a solitary outlier at Keynsham in the southwest; five, possibly six, furnaces in the West Midlands; six in the Sheffield district; and six in the North East. Just ten 'heats' per year, each of five tons, at each of these sites would have been sufficient to make 1000 tons. A still higher output is entirely plausible, for there was unquestionably a sharp upturn in furnace capacity in the first half of the eighteenth century. The single ton that was converted in a 'heat' in the 1680s had become a charge of as much as 10 tons by the 1750s. Bengt Andersson Qvist, when visiting Britain in the 1760s, reckoned that furnaces in the Sheffield area held eight tons, those in the Midlands nine tons, and those in the North East, ten tons. Moreover, a good deal of new capacity was coming on stream in the mid eighteenth century—at least eight new furnaces in the 1740s and 1750s.[163] Given

---

[160] 'Om Bergwercken uti Engeland utdragit ur Afledne Assessoren i Kungl Bergscollegio Eric Odelstiernas Relation åhr 1692', Bergskollegiets arkiv, D VI: 13, RA.

[161] TNA: PRO, CUST 3 series.

[162] TNA: PRO, CUST 3/37, ledger of imports and exports, Christmas 1736 to Christmas 1737.

[163] The Quaker ironmaster Sampson Lloyd and the Birmingham gunmaker John Willet built a new furnace at Tetbury in Gloucestershire in 1739: Religious Society of Friends Library, Lloyd MSS (TEMP MSS 210), 2/61, Sampson Lloyd II to Thomas Kirton, 24 October 1739. Samuel Walker and John Booth did the same at Masborough, near Rotherham, in 1748, whilst Samuel Galton and Joseph Farmer, two of Birmingham's greatest gunmakers, set up a steel furnace at Belbroughton in north Worcestershire *c.* 1750: A.H. John (ed.), *Minutes relating to Messrs Samuel Walker & Co.,*

all of this, the output of blister steel must have approached 2,500 tons nationally by the end of the 1750s. From being marginal to European steel making, Britain had become central.

This explains the determination of Graffin Prankard and Samuel Shore to engross the supply of steel-making iron from Leufsta and Åkerby. It explains how an appetite for the 'first oreground' brands was so quickly translated into a desire to appropriate the 'second oregrounds' as well. It also explains the willingness of steelmakers to experiment with raw materials that did not originate in the *Vallonbruk* of Uppland. Spanish iron had long been used to make steel that was serviceable, if not of the highest quality: 'for the sake of your Reputation', Ambrose Crowley advised in 1712, 'be carefull never to sell Spanish for Orgroon, but make & keep it Separate'.[164] But the supply of Spanish iron was no more elastic than that of 'Orground', so steelmakers began to look farther afield. Appearing before the House of Commons in 1737, the gunmaker Joseph Farmer claimed to have had bar iron from Maryland successfully converted to steel.[165] This was unusual. Most members of the iron trade thought colonial iron too coldshort to ever make good steel, but they had high hopes for the 'tough' sorts of Russian iron. The Crowleys were converting the 'Sable' iron of the Demidovs in the 1750s, and if Graffin Prankard's sales of 'Spread Eagle' steel can be interpreted as steel made from 'Government Siberia' bars that bore the imperial double eagle stamp, then Russian iron was being made into steel as early as the 1730s.[166]

The sales that Graffin Prankard made of 'Orground' iron shed a good deal of light on the organisation of the steel industry. On the face of it, furnace proprietors were independent industrialists, owning

Rotherham, iron founders and steel refiners, 1741–1829 (1951), p. 3. A new furnace was erected in Gateshead on Tyneside in the early 1750s, and two cementation furnaces at Broadwaters, between Kidderminster and Stourbridge in Worcestershire, were reported upon by the Swedish traveller Angerstein in 1754: Cranstone, *Derwentcote steel furnace*, p. 20; *Angerstein*, p. 175; Flinn, 'Industry and technology'. The 1750s also saw continued growth in the Sheffield region, with new furnaces set up by Roebuck & Sons and by the Cutlers Company: Neville Flavell, 'The economic development of Sheffield and the growth of the town, *c.* 1740–1820', (Ph.D thesis, University of Sheffield, 1996), p. 53.

[164] Religious Society of Friends Library, Lloyd MSS (TEMP MSS 210), 1/64, Ambrose Crowley to 'Bro James', 21 February 1712.

[165] SML, Weale MSS, 371/1, fo. 130.

[166] *Angerstein*, pp. 258–59; SA, DD/DN 438 for sales of 'spread eagle' steel to John Cook of Bristol and John Ellis of Gloucester in April, May and June 1732.

Table 2.3. Graffin Prankard's sales of bar iron to the Shallard family, 1732–39 (in tons).

	John & William Shallard	William Shallard	Christopher Shallard	Total
1732	10.94			10.94
1733	14.21			14.21
1734	20.39			20.39
1735	1.90	16.15		18.05
1736		10.64	7.22	17.86
1737			17.28	17.28
1738			11.49	11.49
1739			5.00	5.00

Source: SA, DD/DN 438, 439

their own plant, employing their own workforces, buying in their raw materials from specialist suppliers, and disposing of the end-product themselves. But appearances were misleading. Although steel makers purchased large quantities of 'Orground' iron for themselves, they also spent a good deal of time converting bar iron to steel for local clients on a fee basis, just as mill owners ground corn for farmers and dealers. The case is very clear for one of Graffin Prankard's 'Orground' customers, the Shallard family of Keynsham. The capacity of the Shallards' cementation furnace cannot have been much less than 50 tons per annum, but Prankard's sales to them fell far short of that.

For the best part of the year, it would seem, the Shallards were engaged in converting steel for Bristol-based clients rather than making steel on their own account. Angerstein remarked upon this when visiting the Keynsham furnace in 1754—'The proprietor who owns this works converts iron for the merchants in Bristol against payment per ton'—and the phenomenon is clearly visible in Prankard's books.[167] Prankard would sometimes send iron purchased by Reynolds & Daniel, the Bristol ironmongers, straight to Keynsham: 75 bars from Forsmark in September 1735, 19 bars from Leufsta a year later, and 19 bars from Ullfors in May 1738. The Shallards delivered the converted bars to Reynolds & Daniel back in Bristol. Prankard made plentiful use of the Shallards' services himself. He sent them over 72 tons of iron to

[167] *Angerstein*, p. 140.

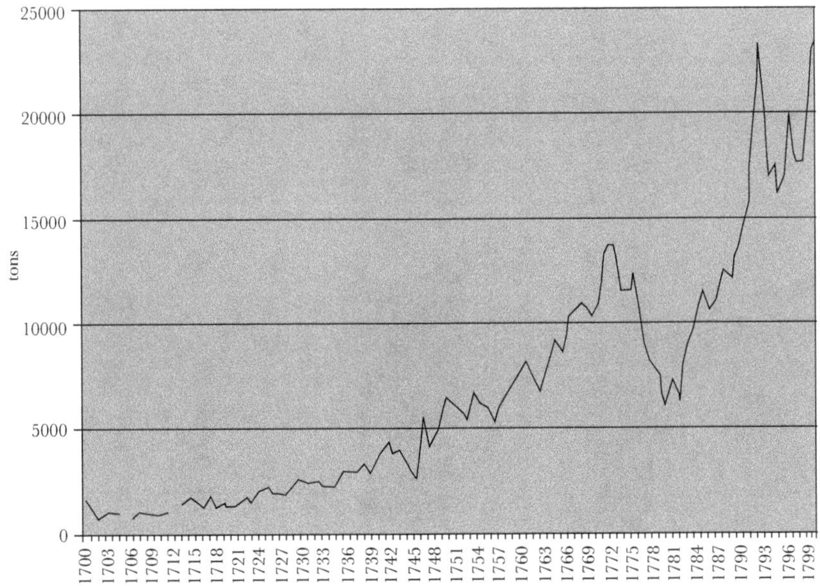

Source: B.R. Mitchell with P. Deane, *Abstract of British historical statistics* (1962). 'Wrought iron' was defined in a parliamentary act of 1672 as 'Axes, adzes, hoes, armour, bits, knives, locks, fowling pieces, muskets, pistols, scissors, stirrups, all carpenters' and gravers' tools, jackwork, clockwork and all ironmongery ware perfectly manufactured'.

Figure 2.12. Wrought iron exports from England and Wales, 1700–1799 (Great Britain from 1792).

be converted into steel on his account between 1732 and 1739, paying a fee of £2.50 per ton. The Keynsham furnace was the source of the blistered bars that were marketed as 'PRANKARD' steel in Charleston and Philadelphia in the late 1730s.

The demand for steel intensified in step with the rising demand for ironwares within Britain's Atlantic empire. Few articles that had a cutting function were made from iron alone; most had a steel cutting edge grafted on to an iron body. The manufacture of agricultural implements almost always involved 'steeling' of this sort. As the production of sugar, tobacco, and other exotic groceries vaulted upwards in Britain's New World plantations, so the demand for hoes, axes, spades, mattocks, and machetes boomed. Exports of 'wrought iron' goods from England and Wales began to accelerate from the 1720s onwards, beginning an upward surge that continued until the onset of the American Revolution.

Most agricultural tools could be manufactured with blister steel: 'for such works as Sythes & Syckles they use noe Steel in Engl^{d} but y^{t} is

Illustration 2.16. Birmingham-made tools of the mid-eighteenth century.

Courtesy of The Colonial Williamsburg Foundation.
Caption: All the tools shown here were exported to North America, and all were made by members of the Freeth family, Birmingham's foremost tool manufacturers. The turning chisel on the left carries the stamp of Sampson Freeth. A maker of that name was a regular customer of Graffin Prankard in the 1730s, buying between four and seven tons of Baltic iron annually.

made [there]'.[168] Blister steel was, however, a very imperfect material. The carbon content of blistered bars, and so their hardness, varied markedly. The surface of a bar had absorbed the most carbon, its core the least. Such variability could not be tolerated by more discriminating users. To satisfy their needs, blister steel required further treatment.

The usual method was for bars to be broken into shorter lengths, nine or ten of which would be bundled together, heated, and then subjected to a forge hammer. The hammer would compact the different parts of the parcel into a single mass, and intermix high-carbon and low-carbon portions of the blister steel. The outcome was a bar with a more uniform distribution of carbon.[169] If required, the operation could be repeated several times, yielding a material that became steadily more homogeneous in its internal structure and so more predictable in its properties. The outcome was sometimes called 'shear steel', sometimes 'Hayford steel' (after the pioneer of the technique, Denis Hayford), and sometimes 'German steel' because the process had been introduced from Germany in the late seventeenth century.[170] In the early eighteenth century the production of this material was centred on the North East of England, in the Derwent valley, where five different grades of shear steel were recognised.

> The softest of this kind of steel is called Sheerblade, & used for the large cloth sheers—The next, rather harder, marked with a sheer blade & star, may be employed for the same use—The third in hardness, called spur steel, makes pen knives at Sheffield, & the best razors—The next, double spur—The hardest of all, double spur & star: this is used by [en]gravers: razors are also made of it, & fine scissars...[171]

Steel mounted abruptly in price as these different refinements were made. Graffin Prankard would usually offer Leufsta bars for sale at between £17 and £18 per ton in the 1730s, but blister steel made from

[168] Mitchell Library, Glasgow, SR352, Adam Montgomerie to John Corse senior, 28 November 1700.

[169] Alternatively, bars of blister steel might be drawn out into smaller sizes and put back in the furnace for further conversion. Angerstein watched this being done at the Crowleys' Teams works. 'Twice-converted' steel might then be drawn down to a still finer gauge and converted a third time. *Angerstein*, p. 259.

[170] Brian G. Awty, 'Hayford, Denis (*c.* 1635–1733)', *Oxford Dictionary of National Biography*, Oxford University Press, 2004 [accessed 23 Sept 2004: http://www.oxforddnb.com/view/article/47478].

[171] CCL, MS 3.250, 'Mineral and Chemical History of Iron' by William Lewis, fo. 225, citing information provided by the Yorkshire ironmaster John Cockshutt.

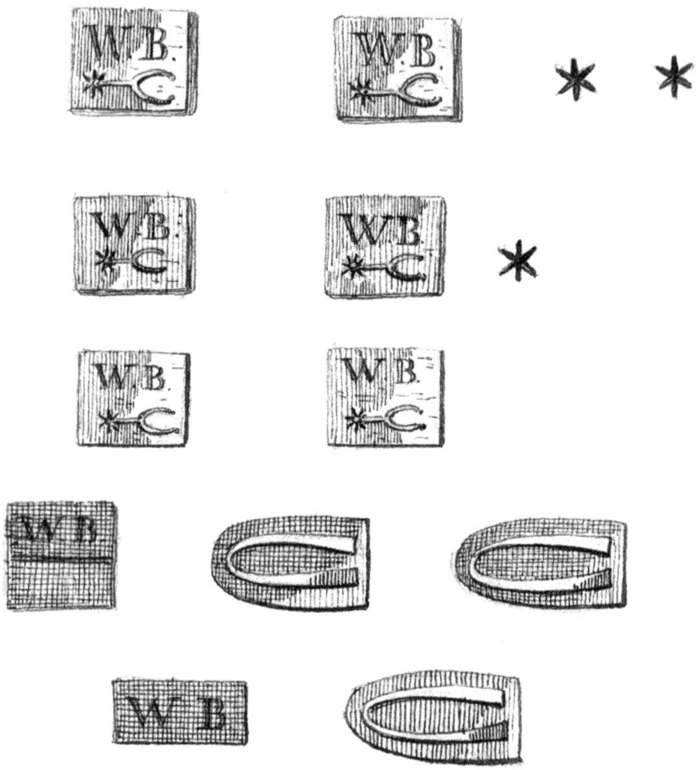

Illustration 2.17. Shear steel marks at Blackhall Mill in the Derwent valley.

Courtesy of Jernkontoret.
Caption: Reinhold Angerstein sketched these marks in 1754. The 'WB' stands for William Bertram, the native of Remscheid who had reputedly introduced the technique in the late seventeenth century. The top-most mark, 'Double Spur, Double Star', was the hardest. (Shear steel, aleady blessed with names enough, was sometimes called 'spur steel'.)

them at £24 per ton. The drawing out of blister steel added further costs. Blister steel was routinely drawn out into slimmer bars, so-called 'faggot steel', suitable for artisans to work up. Blister steel from Blackhall Mill in the North East sold for £26 per ton when Angerstein visited in 1754, but when 'drawn down to bars of the ordinary grade [i.e. faggot steel] for the East India Company, the price is £30 to £32 per ton.'[172] Shear steel was considerably more precious. It was the product of an extended process of hammering and reheating. It was not just the physical form of the bar that was being altered (as was the case with faggot steel); the internal structure of the material was being changed. The cost reflected this. It sold, as both Angerstein and Qvist agreed, for over £50 per ton.[173]

To complicate the picture further, foreign steel still circulated on the British market. Graffin Prankard had a particular preference for *Herz und klebatt* steel from Westphalia, or 'heart and clubb' as he described it. This was much in demand as it did 'not lose its temper by heating, as the Heyford [shear] steel does, hence best for such tools as require to be often heated'.[174] Steel from the Alpine provinces of the Habsburg monarchy also continued to enter the British market, to be used 'for Razors, Chirurgion's Instruments, Gravers, &c Because it will come to a fine and thin Edge'.[175] It was also employed in wire-drawing machinery and other uses where an exceptionally hard material was required.

This aspect of industrial development in eighteenth-century Britain is easily overlooked, but the making of tools was of critical importance. Without a profusion of files, stamps, drill bits, or dies—all of which were made of steel or case-hardened iron—the manufacturing trades that flourished in Georgian towns and villages would have perished. Machinery was almost always framed in wood, but the working parts—be they blades, wires, stamps, or needles—were usually ferrous. The level of specialisation in toolmaking, even in the early years of the eighteenth century, is worthy of note. Saws, for example, came in huge variety. Edmund Hoppus's trade guide of the 1730s, *Practical measuring made easy*, distinguished between '*Compass Steel* Saws, *Grafting*-Saws, *Hand*

---

172 *Angerstein*, p. 271.

173 *Angerstein*, p. 21; Qvist quoted in Barraclough, *Blister steel*, p. 196.

174 CCL, MS 3.250, fo. 227. Prankard imported 'Heart & Clubb' from Rotterdam in two grades, 'Razor' at 16 guilders per cwt, and 'Sorted Ax' at 15¾ guilders per cwt: Bristol Central Library, B17368/20.

175 Moxon, *Mechanick exercises*, p. 58.

and *Pannell*-Saws, *Tenant*-Saws, *Two-Hand-Peg-Tooth* Saws, *Whip Steel* Saws, [and] *Rib-Steel*-Saws'. These were just the main headings. Hoppus then listed the sub-varieties of each: 'Of these there are 13 different Sorts, which are sold from 6*s* 6*d*. to 30*s*. *a piece; viz*....'[176]

Precision tools required steel of very precise properties and of consistent quality, and it was the quest for consistent quality that stimulated technological change within the steel trade. Shear steel, despite the laborious procedures involved in its production, could never be wholly reliable. The forging of broken bars of blister steel into a single mass diminished but did not eliminate the uneven distribution of carbon through the metal, and it did nothing to eradicate the slag inclusions that had been present in the original bar iron. Because of these imperfections, even shear steel might behave in a fashion too erratic for those working at the top end of the market for tools. White, the Clerkenwell saw maker, was one who stood pre-eminent in his trade in the mid-eighteenth century, so much so that his products were ordered specifically by colonial joiners.[177] When Samuel Schröder visited his premises in 1749 he discovered that White subjected shear steel ('here called Newcastle Steel as it comes from that place and from Crowlis [sic] works') to a number of further refining processes. The saw maker's smiths told Schröder that shear steel was 'melted all to one lump' before being forged and cut into the appropriate shape. The saws were then hardened and tempered amid great secrecy. White performed the work himself at 'a large hearth in his cellar, during which no-one else can be present'.[178]

Schröder's reference to shear steel being 'melted all to one lump' is somewhat opaque. Quite what was happening in White's workshop remains unclear, but the intention was clear enough: it was to render the steel as uniform as possible in structure. White was not alone in this pursuit in the 1740s. It was a pressing concern for many artisans—not least clock and watchmakers who required pinions and springs made from the finest steel available. Only a spring drawn out from steel of consistent quality would behave with the unerring regularity expected

[176] Edmund Hoppus, *Practical measuring made easy* (7th edn., 1765), p. 194.

[177] Jane Rees and Mark Rees, *Christopher Gabriel and the tool trade in eighteenth-century London* (Ipswich, 1997), p. 45. A surviving tenon saw of White's making that was exported to Connecticut is illustrated in James M. Gaynor and Nancy L. Hagedorn, *Tools: working wood in eighteenth-century America* (Williamsburg VA, 1993), p. 84.

[178] Schröder II, entry for 24 July 1749.

of it. It was this that prompted Benjamin Huntsman to experiment with alternative methods of refining steel, including the melting down of blister steel.[179]

Huntsman was well aware of the inconsistent quality of English steel, even steel that had been subjected to faggotting and forge-welding. He was, after all, a consumer of shear steel. He was a clockmaker from Doncaster who had become frustrated with the indifferent quality of the springs and other steel-made elements he incorporated into his clock mechanisms. Huntsman, inspired, so it seems, by brass founding, turned to the possibility of melting steel. If blister steel was reduced to a liquid state, Huntsman reasoned, it would become a homogeneous product. Moreover, once the steel was molten any residual slag would float to the surface where it could be skimmed off. Only two obstacles stood in the way. One was the difficulty of achieving a temperature capable of melting steel; the other was the problem of manufacturing a crucible capable of enduring such a high temperature without cracking.

In 1742 Huntsman moved to Sheffield to begin his experiments. The solution to the problem of raising a sufficiently high temperature lay with coke. Charcoal would not do. It could achieve a temperature no greater than 1,425°C; high enough for brass founding, but not enough to melt steel. Coke could achieve the requisite temperature (1,600°C), yet different coals produced different cokes, and it took some time for Huntsman to isolate the locally available coals that would burn with the correct heat for a suitable period of time. Huntsman had also to devise a furnace structure that would amplify the heat of burning coke. His solution was to provide a very strong natural draught that would intensify the combustion of the coke to the utmost. Pieces of broken blister steel were packed into clay crucibles, each of which was lodged in a hole in the floor of the furnace building, surrounded by a bed of incandescent coke. The coke was maintained at a high temperature by the passage of air that was drawn from the cellar of the furnace building and through the bed of coke by the draught from a range of tall chimneys, one for each hole.

The crucible remained in the furnace hole for four or five hours. It had to be capable of withstanding intense heat without fracturing. This was the other major technical problem confronting Huntsman.

---

[179] The following discussion is based upon K.C. Barraclough, *Steelmaking before Bessemer. Volume 2. Crucible steel: the growth of technology* (1984), chap. 1.

Illustration 2.18. Crucible steel making plant as seen by the Swedish traveller Gustaf Broling.

Courtesy of Jernkontoret.
Caption: This comes from an unpublished set of prints at Jernkontorets Bibliotek, Stockholm, prepared to accompany Broling's report on his English journey ('Presentd inför Bergslagernes Deputerade i Jerncontoret den 5 Maij 1817, jämte 3ie Delen af Hr Bergsrådet och Riddn Brolings Resa'). The version that accompanied Broling's *Anteckningar under en resa i England åren 1797, 1798, och 1799*, 3 volumes (Stockholm, 1811–1817) showed the crucible plant in a verdant, romantic landscape.

He resolved it by experimenting with a variety of clays and mineral additives until he achieved a mixture that would soften slightly in the furnace but would not break. It was this yielding quality that allowed the workman to grip the crucible with a pair of tongs, lift it from its hole, and set it down on the floor of the workshop. Another workman, the 'teemer', then grasped the crucible with tongs of his own. Resting the tongs across his knee for steadiness, the teemer poured the liquid metal into a mould, forming an ingot of exceptionally pure steel.

The crucible process made Sheffield the world centre of steel making in the nineteenth century, but its impact in the eighteenth century was somewhat restricted. In the late 1780s, forty years after Huntsman's first success, there were still only eleven firms in the town that used his method.[180] And these firms operated on a small scale. The ingot produced from a single crucible rarely exceeded 20lb in weight during the eighteenth century (compared to the 50lb to 60lb that was standard in the mid-nineteenth century), and a dozen holes in a melting shop was considered a sizeable number.[181] Indeed, it was not until the 1820s and 1830s that crucible steel was made on a large scale, and not until the 1840s and 1850s that really large crucible plants, those boasting a hundred or more holes, came on stream.[182]

Cast or crucible steel was confined to those sectors where its superlative qualities were indispensable and to users who felt its premium price was therefore justified. Gabriel Jars, the French engineer, spoke of quite specific uses for crucible steel: the 'best razors are made from it... the best steel chains, the springs of watches and small watchmakers' files'. Indeed, Jars went so far as to assert that it was 'only used for those items requiring a fine polish'.[183] There was some exaggeration here—cast steel was much sought after by saw manufacturers, for example, not just by makers of fancy wares—but Jars's statement suggests something of the attraction that Huntsman's steel had for contemporaries. Chemically homogeneous and clear of foreign matter, it was a boon not just for those who sought technical precision in their steel, but for those who used steel for aesthetic purposes. Because the slag residues had been eliminated, cast steel could take a high, unblemished polish. Demand

---

[180] *A directory of Sheffield* (1787), p. 38.

[181] Barraclough, *Crucible steel*, pp. 41–42.

[182] J.G. Timmins, 'Concentration and integration in the Sheffield crucible steel industry', *Business History*, XXIV (1982), 61–78.

[183] Quoted in Barraclough, *Crucible steel* (1984), p. 214.

for it came from manufacturers who specialised in decorative items such as watch chains, jewellery, buckles, or the burnished steel buttons that enjoyed such a vogue in the 1770s. 'We have', Matthew Boulton reported from Birmingham, when the craze for highly polished buttons was at its peak, 'some button makers that order 2 or 3 Tons [of cast steel] at a time'.[184]

Indeed, it is likely that the demand for crucible steel was earliest and strongest amongst those who made items of personal adornment such as watch chains and buckles. Steel in the eighteenth century was more than a prosaic industrial input; it had many decorative purposes. The steel watch chains that became items of fashion in the early eighteenth century were a case in point. They were sufficiently robust to secure a pocket-watch, but they also had a polished glitter that announced to onlookers the taste and wealth of their owners. A Philadelphia merchant who visited Wolverhampton, one of the principle centres of their manufacture, much admired the chains that he saw: 'very neat', he scribbled in his journal. They were also expensive: 'some of them sold for 10 G[uinea]s Ea[ch]'.[185] Before the bulk production methods of the nineteenth century rendered steel a cheap, commonplace material, it enjoyed a prestige that has now been lost. An advertisement placed in the *Virginia Gazette* in the autumn of 1772 testified to that. A Williamsburg merchant announced the arrival of a 'neat Assortment of JEWELLERY and SILVER WORK', which featured 'PINCHBECK BUCKLES and FINE CUTLERY, such as Ladies Steel Watch Chains, Pocket and Penknives, and a Variety of Scissors and Spectacles, just imported from *London*'.[186] To the modern eye, there is a conceptual jumble here: items that are decorative and those that are functional are promiscuously thrown together. The hierarchy of value that seems appropriate to the post-Bessemer era, one that leads smoothly upwards from base metal to bullion, loses shape and definition when steel goods are spoken of in the same breath as precious metals or ersatz precious metals such as pinchbeck. To the eighteenth-century eye, however, there was no contradiction or confusion at work. Steel was routinely used in the production of high-status articles. A poet from Birmingham,

---

[184] Matthew Boulton, quoted in Barraclough, *Crucible steel*, p. 4.

[185] Winterthur Library, microfilm 296. Samuel Rowland Fisher journal 1767–1768.

[186] *Virginia Gazette*, 29 October 1772.

Illustration 2.19. 'Steel Buttons/Coup de Bouton'.

Courtesy of The Lewis Walpole Library, Yale University.
Caption: This print of the 1770s shows a lady of fashion dazzled by the ostentatiously large and reflective steel buttons sported by her companion. That the print appeared with a bilingual title indicates the European-wide market for these distinctively English articles.

the principal seat of steel jewellery manufacture, rhapsodised upon it, drawing extravagant comparisons with precious stones.

> What beauteous works from ORES refin'd arise,
> To grace the HEAD and NECK, and charm the eyes;
> To grace the HANDS, and FEET, the COAT, and VEST,
> And ornament our Belles and Beaux, full drest;
> The orient PEARLS, and blazing DIAMONDS, feel
> Their lustre, oft, outvied by polish'd STEEL.[187]

As Gabriel Jars recognised, however, Huntsman's steel was not restricted to the making of bijou goods. It was also sought after by those engaged in the horological trades—by those concerned with 'the springs of watches and small watchmakers' files'.[188] The catalogue issued by John Wyke of Liverpool, a prominent watch tool maker, illustrates the enormous range of files made. The first plate alone pictured 43 types of file, each of which came in six levels of abrasiveness (rough, rough bastard, bastard, smooth bastard, smooth and smooth-smooth). Some were no bigger than toothpicks.

John Wyke first set up as a tool maker in Prescot, in the heart of the busy watch and clock making district of south Lancashire, in the 1740s, at a time when the trade was dependent upon shear steel. Reinhold Angerstein, in visiting the workshop of Daniel Mather, a Liverpool contemporary of Wyke's, noted this dependence. Mather made 'all kinds of steel hardware required for a watchmaker's shop', specialising in 'a kind of grooved steel wire for pinions in small pocket watches'. The 'raw material for the pinion wire is Mr Bertram's Double Shear Steel costing 6 pence a lb [£56 per ton]'.[189]

In the later decades of the eighteenth century, however, cast steel began to infiltrate south Lancashire. Peter Stubs established a file cutting business in Warrington in the 1770s, advancing steel blanks to outworkers in the town and its hinterland. The artisans in his employ heated the steel, cut a sequence of minutely spaced teeth into its surface, then quenched and hardened the steel.[190] At first Stubs issued only blister

[187] James Bisset, *A poetic survey round Birmingham; with a brief description of the different curiosities and manufactories of the place...accompanied by a magnificent directory* (1800), p. 37.

[188] For background see M.L. Wayman (ed.), *The ferrous metallurgy of early clocks and watches: studies in post medieval steel* (British Museum Occasional Paper 136, 2000).

[189] *Angerstein*, pp. 313–14. William Bertram operated at Blackhall Mill in the Derwent valley, the historic centre for the manufacture of shear steel.

[190] E. Surrey Dane, *Peter Stubs and the Lancashire hand tool industry* (Altrincham, 1973).

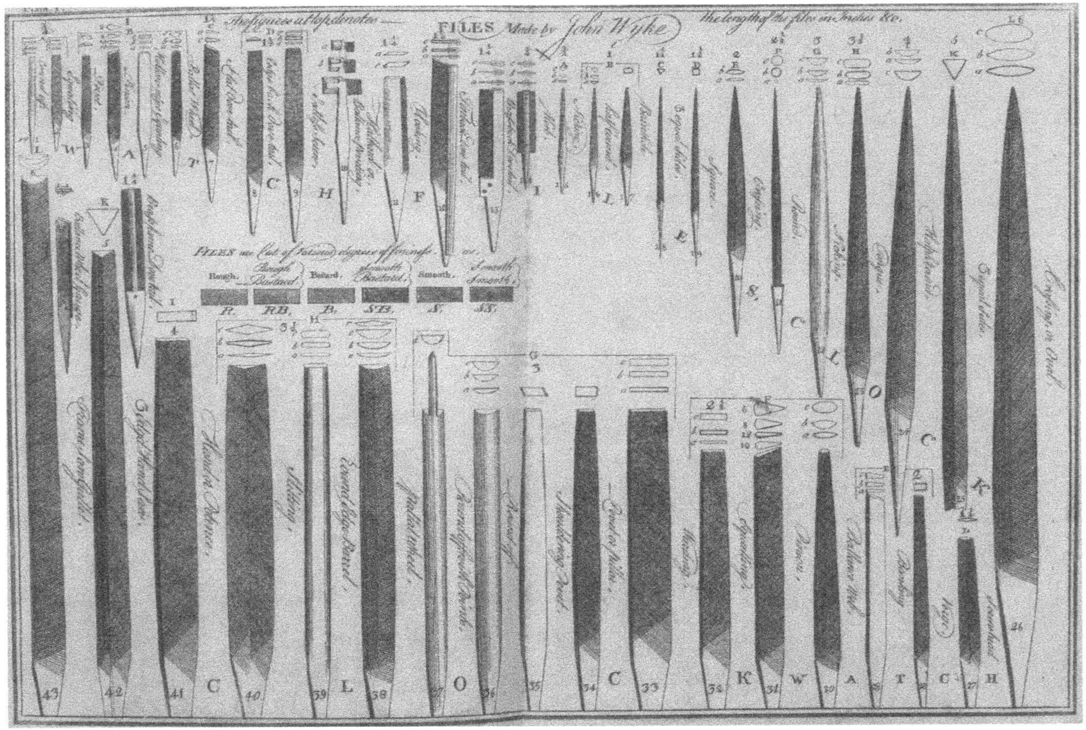

Illustration 2.20. Watch and clock files from the catalogue of John Wyke.

Courtesy of The Winterthur Library: Printed Book and Periodical Collection.
Caption: This first plate from the catalogue has been dated on stylistic grounds to the 1750s. See *A catalogue of tools for watch and clock makers by John Wyke of Liverpool, with an introduction and technical commentary by Alan Smith* (Charlottesville VA, 1978), p. 15.

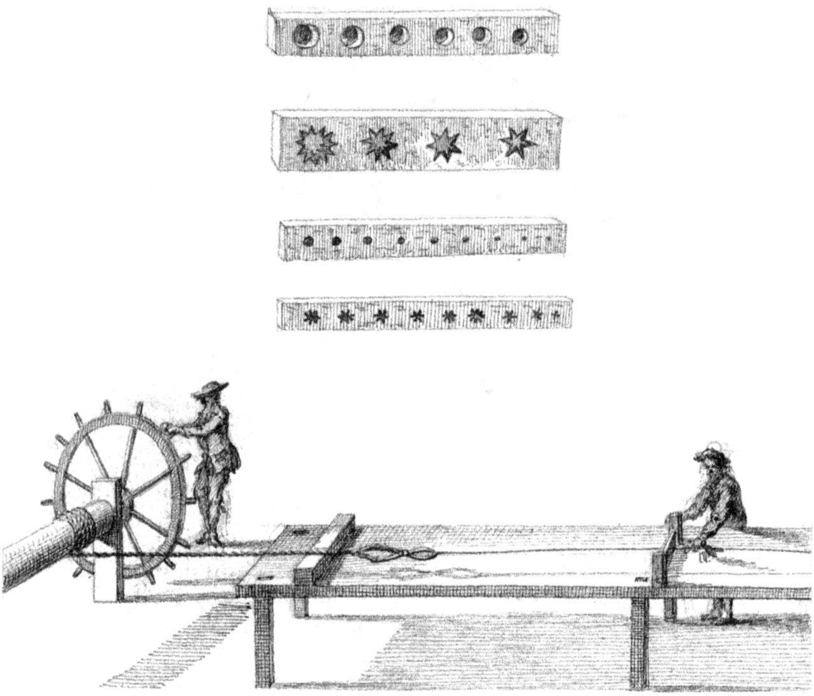

Illustration 2.21. Pinion wire drawing and dies as seen by the Swedish traveller Reinhold Angerstein.

Courtesy of Jernkontoret.
Caption: 'The steel is first drawn down under the hammer to thin bars that are subsequently cold-drawn to round wires of the diameter required... Drawing is then continued through dies with shallow grooves and through further dies with grooves of increasing depth until the wire is finished.'

steel and 'spur steel' (shear steel) to his outworkers, but in the 1780s 'cast steel' started to appear in his accounts.[191] Even so, cast steel failed to vanquish shear steel. Stubs was still buying shear steel in the 1790s and beyond. Indeed, the makers of shear steel poured scorn on the rival product. 'If you buy Cast Steel', Stubs was warned by Isaac Cookson of Newcastle, 'no wonder that you meet with great Impositions as the generality of Manufacturers of that article are subject to be imposed on with scrap steel of very bad quality, some of it from Russia iron converted into Steel and sometimes steel very badly converted'. It was little wonder, Cookson concluded, 'that your files often prove soft, others breaking in the teeth'.[192]

Like John Wyke, Peter Stubs made a quite extraordinary number of different files. By the end of the eighteenth century he could offer his customers 77 different categories of file, almost all of which were available in a variety of sizes. All in all, Peter Stubs was able to supply nearly 600 individual types of file.[193] Stubs's business extended into the making of all manner of tools and instruments: hammers, callipers, pliers, vices, screwdrivers, dividers, nippers and tweezers. These were so-called 'Lancashire Tools', specifically designed for the makers of watch and clock components.[194]

The south Lancashire watch district catered for two distinct markets.[195] One was domestic. Parts made in the low-wage Prescot district were sent south to high-wage London, where the parishes of Clerkenwell and St Luke's, on the northern edge of the city, teemed with specialised horological workers. A guide of 1747 defined the London watch trade as being composed of movement makers, wheel cutters, spring makers, chain makers, cap and stud makers, case makers, dial cutters, dial enamellers, gilders, and finishers.[196] The London 'watchmaker' whose

[191] Manchester Archives, L24/1 (Box 7), day book 1776–78 and workmen's day book 1788–91.

[192] Isaac Cookson & Co of Newcastle to Peter Stubs, 3 June 1799, quoted in T.S. Ashton, *An eighteenth-century industrialist: Peter Stubs of Warrington, 1756–1806* (Manchester, 1939), p. 48.

[193] Manchester Archives, L24/1 (Box 24), 'A List of the Prices of Lancashire Files, Manufactured by Peter Stubs, Warrington'.

[194] Manchester Archives, L24/1 (Box 24), 'A List of the Prices of Lancashire Tools. &c, &c, Manufactured by Peter Stubs, Warrington'.

[195] For background see F.A. Bailey and T.C. Barker, 'The seventeenth-century origins of watchmaking in south-west Lancashire', in J.R. Harris (ed.), *Liverpool and Merseyside: essays in the economic and social history of the port and its hinterland* (1969), pp. 1–15.

[196] Robert Campbell, *The London tradesman* (1747), pp. 250–51.

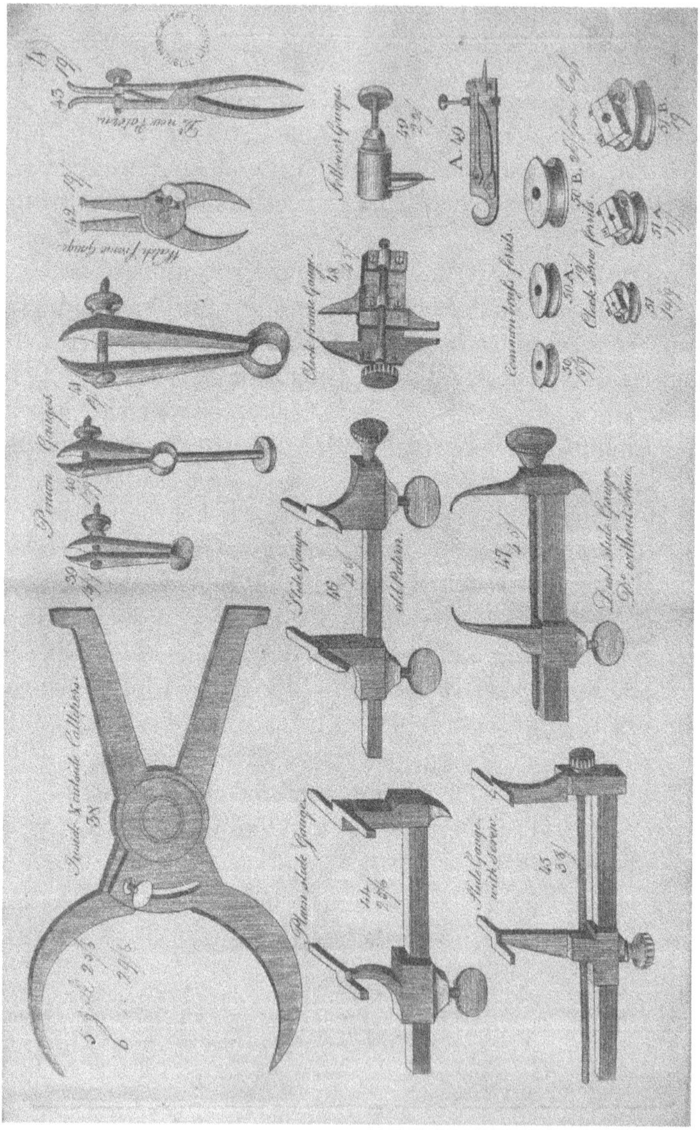

Illustration 2.22. Tools from the catalogue of Peter Stubs.

Courtesy of Manchester Archives and Local Studies (Central Library, L24/1, box 24). 'A List of the the Prices of Lancashire Tools, &c. &c. Manufactured by Peter Stubs, Warrington'.

name was engraved on the backplate was merely the entrepreneur who had set this small army of outworkers in motion.

The other market for Lancashire clockwork was transatlantic. Pinion sets and springs were shipped through Liverpool by the thousand, bound for New England. The components were sold to village craftsmen in Massachusetts and Connecticut.[197] These small town clockmakers added cases and dials of their own devising, but their clock movements were a matter of steel and brass prepared in Lancashire. In this way 'Orground' iron, transmuted into steel, flowed onwards into the wider Atlantic economy to govern time on a new continent.

'Orground' iron, transmuted into steel, also brought death to the Atlantic world, for steel or case-hardened Swedish iron was an essential element in gun making. In the mid-seventeenth century the manufacture of small arms had been centred upon London, home to the Gunmakers Company. By 1700, however, the trade had gravitated to Birmingham where steel, iron and brass were in plentiful supply and a skilled workforce was on hand. The ascendancy of Birmingham was signalled by the willingness of the Board of Ordnance to contract with West Midland gunmakers for muskets, the fulminations of the Gunmakers Company notwithstanding. In 1693, at the start of the state's quest for provincial suppliers, five Birmingham contractors combined to supply the Board of Ordnance with 200 pieces a month. Half a century later a single contractor, Samuel Galton, was capable of turning out 500 guns a week.[198]

Birmingham's earliest gunmakers were to be found in Digbeth, amid the densest concentration of metalworking in the town, but during the first half of the eighteenth century the trade migrated northward towards Steelhouse Lane where John Kettle had his two cementation furnaces. Samuel Galton and his brother-in-law James Farmer operated from 14 Steelhouse Lane in the 1750s, and as their business developed they took on additional premises in adjacent Weaman Street and Slaney Street.[199] The proximity to Kettle's furnaces could hardly

[197] Robert C. Cheney, 'Roxbury eight-day movements and the English connection, 1785–1825', *The Magazine Antiques*, (April 2000), 606–15.

[198] Wise, 'Birmingham and its trade relations', pp. 58f.

[199] BCA, 405/1–2 (Galton MSS); Wise and Johnson, 'The changing regional pattern'; B.M.D. Smith, 'The Galtons of Birmingham: Quaker gun-merchants and bankers, 1702–1831', *Business History*, IX (1967), 132–50; De Witt Bailey and Douglas A. Nie,

have been accidental, for Galton was a regular purchaser of steel converted from Leufsta and Åkerby iron. Nevertheless, if the organisation of the trade became centred on the northern edge of Birmingham, the manufacturing processes themselves were often dispersed, as both Reinhold Angerstein and Samuel Schröder made clear. The reports submitted by the two Swedes paid close attention to the business of gun manufacturing in the Birmingham district *c.* 1750. Together they provide a comprehensive overview of the four elements of the trade: barrel manufacture, lock making, stock making, and assembly.

Barrel manufacture began at a rolling mill. Angerstein, who watched iron being rolled at a mill between Bilston and Wednesbury, found that rolls produced a more even plate than the traditional battery hammers and did so expeditiously. The finished plates were then formed around a cylindrical die and welded longitudinally to make the barrel. Various brands of iron were used for this purpose. The best barrels, so Angerstein was assured, were made from Swedish iron or scrap. English iron was used only for 'trade' guns that were to be exported to Africa.[200] Schröder concurred. His informant Thomas Hadley used iron from Gammelbo *bruk* in central *Bergslagen*—a brand that Graffin Prankard dealt in extensively in the 1730s, as we shall see.[201]

The rough barrels had to be smoothed and shaped, inside and out. Angerstein witnessed barrels being brought to a red-heat ('cherry-red') and planed at the rolling mill itself. The boring of the barrel, on the other hand, was done at a specialised workshop. Barrels were fixed to a workbench and finished to the correct internal gauge by the application of a steel-tipped boring rod. As a rule, the process was water-powered, although the finest guns were bored with the help of a hand-turned crank. The larger Birmingham gun makers—Samuel Galton or the Grice family, for example—usually had their own boring mills. Such a facility hastened throughput, although as with all hydraulically-powered systems a period of drought could bring production to a stand. There were, as Angerstein noted, attempts to overcome this problem. John Willet (the gun maker who set up a cementation furnace at Tetbury in partnership with Sampson Lloyd) was the owner of a boring mill near Wednesbury that was designed to work off both water and wind power.

English gunmakers: the Birmingham and provincial gun trade in the eighteenth and nineteenth century (London and Melbourne, 1978).

[200] *Angerstein*, p. 320.

[201] Schröder I, fo. 185ff.

Yet the design was not a success: the Dutch-style windmill was 'called "Mr Willets Folly" by the people living in the district'.[202]

Once bored, the barrels—or at least those that were intended for the better quality guns—were ground to a smooth finish at a grinding mill, with the very best barrels being hand-polished with emery and oil. Grinding mills, unlike boring workshops, were seldom dedicated exclusively to the gun trade. Thomas Hadley, who owned a two-stone mill outside Birmingham, ground a variety of iron and steel items, often subcontracting work for other local manufacturers. Schröder watched 'saw blades, rapier blades, and assorted edge tools' being ground as well as gun barrels.

The making of gun locks was an entirely different affair. It was often in the hands of a single artisan, working in a single workshop with just one assistant. For all that, lock making was a highly complex procedure, involving the shaping and fitting together of a dozen or more metal components. This was more than brute hammering on an anvil. Schröder was impressed by the widespread use of dies to shape the different parts; this, he declared, allowed a lock to be fitted together with the absolute minimum of filing.[203] Once the lock maker was satisfied with the fit of the lock the components had to be hardened. This was especially so with the hammer plate, the surface against which the flint would snap when the trigger was drawn: a hammer plate that was too soft would produce no spark, rendering the weapon useless. Case hardening was a trade in itself, quite separate from that of the lock maker. First, old leather was charred on a coal fire, then the carbonised residue was used to line the fire-proof boxes into which the lock parts were packed. Once the vessels had been sealed with horse dung, they were fired in a furnace for two hours. It was the cementation process in miniature: carbon from the burnt leather infused into the surface of the hammer lock, giving it a steely, resistant quality. The wooden stock upon which the metal elements were to be mounted was prepared separately, passing from one set of specialised workers to another. 'The stocks', Schröder said, 'go through many hands before

[202] *Angerstein*, p. 49f.

[203] In other words, Birmingham lock makers were striving for interchangeability of parts, an aim more usually associated with French military engineers and passed on by them to American manufacturers: Ken Alder, *Engineering the Revolution: arms and Enlightenment in France, 1763–1815* (Princeton, 1997); David A. Hounshell, *From the American System to mass production, 1800–1932: the development of manufacturing technology in the United States* (Baltimore, 1984), pp. 25–28.

being ready...One planes and another files and cuts out the space for the lock. Another adds the brass fitting.'

The three component parts—the barrel, the lock and the stock—were assembled in Birmingham. It was here that major gun makers stockpiled iron and steel; here where their fitters put together the final product; and here where a firm such as Farmer & Galton maintained a 'packing chamber' in which the guns were wrapped in brown paper and boxed up for shipment. The output of the Birmingham gun trade issued from a complex production matrix that combined in-house workers with a dispersed body of artisans. Thomas Hadley, Schröder reported, employed a number of master workmen, 'some within and some outside his house, who, with all their apprentices and boys, work for him. He pays everyone of these masters per dozen, and they in turn pay the apprentices and boys'. The ways in which a gun maker might organise production were various. Thomas Hadley had his own grinding mill, but no boring facilities; Samuel Galton had a boring mill but no grinding troughs. Schröder implied that Hadley bought completed gun locks from independent masters in Wednesbury, whereas the records of Farmer & Galton suggest that the partners put out iron and steel to lock makers who depended upon them for materials.[204] The muskets for which Birmingham became famous emerged from streams of materials and credit that surged and eddied through the courts and cellars of the town, flowing outward to Wednesbury and Darlaston and returning to the warehouses of the great gun making firms.[205]

The biggest firms could turn out firearms in considerable quantity. The mobilisation at the start of the Seven Years' War revealed the formidable productive power of the Birmingham district. The Board of Ordnance issued warrants for 25,000 new land muskets in 1756, all of which were awarded to Birmingham contractors.

---

[204] BCA, 405/2, Samuel Galton to John Galton, 27 May 1755, in which an inventory of materials 'then in being & in Workmens hands' is mentioned.

[205] Our interpretation differs from that of Clive Behagg in his 'Mass production without the factory: craft producers, guns and small firm innovation, 1790–1815', *Business History*, XL, 3 (1998), 1–15. Writing of Birmingham in the Napoleonic era, Behagg identifies gun makers of Galton's or Hadley's type as merchants who interposed themselves between the master workmen who organised the actual making of guns and major customers such as the Board of Ordnance. Yet evidence from the mid eighteenth century shows gun makers having a substantial involvement in the productive process by operating substantial fixed plant (such as boring mills), employing workmen directly in their own premises, and organising outworkers.

Table 2.4. Contractors for land muskets to the Board of Ordnance, 1756.

Joseph Oughton	6000 barrels	–
Grice & Edge	–	6000 locks
Edward Jordan	4800 barrels	4800 locks
James Farmer	4400 barrels	4400 locks
Samuel Galton	3800 barrels	3800 locks
[Thomas?] Hadley	3600 barrels	3600 locks
[John?] Willet	2400 barrels	2400 locks

Source: adapted from De Witt Bailey, 'The Board of Ordnance and the small arms supply: the Ordnance System, 1714–1783', (unpublished Ph.D. thesis, University of London, 1988), p. 147. (The Board did not issue warrants for guns as such, only for components. These components were delivered to the Tower of London, inspected, and then handed over to London-based setters-up for final assembly.)

These were sizeable contracts, but they could be fulfilled with surprising speed. In December 1754 Samuel Galton grumbled to his partner that '500 or 600 Guns a week is no small Quantity'; but nor, by implication, was it unrealisable.[206] If Galton was to have devoted his entire workforce to the Ordnance contract of 1756 the 3,800 barrels and an equal number of locks could have been boxed ready for shipment within seven weeks.[207] This left plenty of time to attend to civilian markets. Indeed, non-military customers were crucial, despite the strong pulse of additional demand that accompanied the outbreak of every one of the eighteenth century's many wars. It was the non-state market that absorbed tens of thousands of firearms every year.

Sustaining an output of this order called for a ready supply of raw materials. This was not always possible. 'I am in great want of iron', Samuel Galton wailed to his Bristol-based brother John in September 1755. Only five days' supply remained in stock, and part of that was already in the hands of workmen. John Galton was to send new supplies up the Severn with the greatest despatch. Waggons would be waiting at Bewdley.[208] Despite recurrent shortages, Galton was always alert to the

[206] BCA, 405/1, Samuel Galton to James Farmer, 9 December 1754.

[207] Schröder reckoned that a barrel roller and two apprentices could turn out 30 plates a day; that a barrel maker's workshop could produce a dozen rough barrels daily; and that a locksmith and his boy could make a dozen locks. (Schröder I, fo. 174 and 185ff.) These were just the core workers. One contemporary authority ennumerated 21 separate branches to the gun trade, making it impossible to give a firm figure for those employed by a major gun maker. Even so, Samuel Galton could hardly have got by without a couple of hundred operatives or subcontractors working at his behest.

[208] BCA, 405/2, Samuel Galton to John Galton, 16 September 1755.

question of quality, specifying the marks he required. If, in the worst of circumstances, an inferior brand was all that was available, the fact should be disguised: 'I suppose its Gothenburg Iron', Galton muttered, '& if we take the whole that's made at that forge will it not be better to have no Mark fix'd on it'?[209]

The supply of steel was equally uncertain and the question of quality control just as prominent. There were occasions when John Kettle could not furnish Farmer & Galton with all that they needed. 'We are quite unsorted', Galton told Farmer in May 1752, 'and none but the [Leufsta and Åkerby marks] will do for our customers'.[210] On such occasions application might be made to other local manufacturers like Tibbits, the Wednesbury saw makers, who 'us'd last year 50 Tons of Steel'. (Alas, they had 'worked up almost all their [Leufsta and Åkerby]' and could spare none.)[211] Such makeshifts would not do for long. The growth of the gun trade and changes in gun design that called for a greater steel content in the finished weapon—such as the replacement of wooden rammers by more flexible steel equivalents—made an enhanced supply of steel imperative. This led several gun makers to consider moving back up the supply chain and becoming steel manufacturers in their own right. John Willet did so in 1739. Samuel Galton and James Farmer followed suit in the early 1750s.

Farmer & Galton acquired land at Belbroughton in Worcestershire and set about building a cementation furnace where 'we shall try to have the bar of iron converted'.[212] A forge was already *in situ*, so the firm had also acquired the capacity to manufacture its own bar iron; hence Farmer & Galton's importation of Bush River pig iron from the Chesapeake.[213] The acquisition of plant to make bar iron and steel did not, of course, necessarily resolve the problems of supply, it merely shifted the difficulty a link or two back in the commodity chain. The gun makers had now to acquire pig iron (and not just bar iron) and 'Orground' iron (rather than steel). This was no easy matter as the demand for guns marched upwards in the mid-eighteenth century.

---

[209] BCA, 405/1, Samuel Galton to James Farmer, 31 October 1751.

[210] BCA, 405/1, Samuel Galton to James Farmer, undated but in a context of May 1752.

[211] BCA, 405/1, Samuel Galton to James Farmer, 19 October 1751.

[212] BCA, 405/1, Samuel Galton to James Farmer, 17 July 1751.

[213] BCA, 405/2, Samuel Galton to Mr Parr, 20 November 1755. Hence too the dispute that the firm entered into with Edward Knight, the most powerful ironmaster in the Midlands, about charcoal supplies: Samuel Galton to James Farmer, 20 October 1751.

When Samuel Galton compared the gun trade of Birmingham with the textile industry in Manchester he found the comparison very much to the advantage of the cotton masters. 'I think there is very little affinity in the Gun Trade and Manchester, as the manufacturers in those goods [i.e. cotton] keep severally a stock on hand and can readily supply another whereas... each manufacturer in Guns hath orders for more than [he] can supply.'[214] Gunmakers competed for a very finite supply of raw materials and the loyalties of a never quite adequate workforce. The gun trade, in other words, was continually straining at the limits of the human and material resources at its disposal. Not the least of the reasons for this was the incessant demand for muskets in Africa.

Samuel Galton's despairing contrast between the elasticity in the supply of Manchester-made cotton goods and the tardiness of his own supply network was prompted by his inability to complete an order for some Liverpool Guinea merchants. The proportion of Farmer & Galton's output that was directed to slavers cannot be determined, but it must have been considerable. It has been estimated conservatively that the Slave and Gold Coasts alone were absorbing 180,000 firearms annually by 1730.[215] It would have been extraordinary if any of the leading Birmingham partnerships had abstained from such a trade. They did not. Schröder noted that a large part of the output of Thomas Hadley was 'shipped to the Coast of Guinea in Africa, where they are bought by the Barbarians'. The making of 'Angola' muskets also featured heavily in the correspondence of Samuel Galton, and the west coast slaving ports received large and regular consignments of guns for the African market. Galton had a warehouse at Bristol under the charge of his brother and an influential agent in Liverpool in the person of John Parr. In early 1772, at one of the peak moments in the African trade, a single order from Liverpool had Farmer & Galton boxing up 6,410 pieces.[216] Guns for the Guinea trade were also sent to London where Farmer & Galton counted Grant, Oswald & Co, proprietors of the great slaving depot on Bance Island at the mouth of the Sierra Leone River, among their customers.[217]

---

214 BCA, 405/1, Samuel Galton to James Farmer, 9 December 1754.

215 W.A. Richards, 'The import of firearms into West Africa in the eighteenth century', *Journal of African History*, XXI (1980), 46.

216 Joseph E. Inikori, *Africans and the Industrial Revolution in England: a study in international trade and economic development* (Cambridge, 2002), p. 464.

217 See David Hancock, *Citizens of the world: London merchants and the integration of the British Atlantic community, 1735–1785* (Cambridge, 1995), chap. 6.

Firearms had circulated in West Africa since the fifteenth century, but it was only in the late seventeenth century, with the introduction of the flintlock musket, which functioned better than matchlock weapons in humid equatorial conditions, that they became a major item of commerce.[218] The impact of guns on African society was often profound. The Dutch commander at Elmina on the Gold Coast identified a bitter, destructively self-sustaining relationship between the import of firearms and the extension of the local slave trade. 'The great quantity of guns and powder which the Europeans have brought', he reported in 1730, 'have caused terrible wars between the Kings and Princes and Caboceers of these lands, who made their prisoners of war slaves; these slaves were immediately bought up by Europeans at steadily increasing prices, which in its turn, animates again and again these people to renew their hostilities.'[219] The relationship was not automatic, however. The waging of war on the Gold Coast depended upon more than the inflow of weaponry; it rested as well on the existence of strong states with centralised armies that were capable of exploiting European firepower.[220] Elsewhere, guns were put to non-battlefield uses. They might be used for festive or religious purposes, or amassed by chiefs as an expression of royal prestige. It was this kind of conspicuous consumption that Schröder alluded to when he suggested that 'the Barbarians bury them in the ground...as their wealth consists in having a large number of guns'.

Because the employment of firearms in Africa varied so widely, so did the models manufactured in Birmingham. Some were poor stuff from a technical point of view, quite incapable of passing proof for the Board of Ordnance. Schröder claimed that Thomas Hadley had trade guns bored 'only about 2 to 3 inches at the muzzle' and did not trouble much over grinding the barrels. As a result, the guns were often as much a threat to their users as anyone. Some of Farmer & Galton's pieces were little better. 'What is shocking to humanity', Lord Shelburne wrote after visiting the Steelhouse Lane premises, 'is that above half of them, from the manner they are finished in, are sure to burst in

[218] H.A. Gemery and J.S. Hogendorn, 'Technological change, slavery and the slave trade', in C. Dewey and A.G. Hopkins (eds), *The imperial impact: studies in the economic history of Africa and India* (1973), pp. 243–59, especially pp. 248–50.

[219] Quoted in Richards, 'The import of firearms', 46.

[220] See John Thornton, *Africa and Africans in the making of the Atlantic world, 1400–1800* (2nd edn, 1998), pp. 120–24.

the first hand that fires them.'[221] But this was to assume that the guns were intended as battlefield weapons. Many were not, or they would scarcely have been saleable on the Guinea coast. Farmer & Galton sold fourteen types of trade gun in the 1750s, each meant for a different market. The cheapest, the 'Catch Trading' musket, which sold for 6s 8d, cannot have amounted to much, and 'Bonny' and 'Angola' muskets were little better. 'Danish' muskets, on the other hand, cost 12s 6d apiece.[222]

It was this attention to African conditions and African tastes that allowed Birmingham makers to meet the growing demand for firearms that swept eastward along the Guinea coast. Guns were being traded in large numbers along the Slave Coast by 1680. A generation later, they had appeared in Benin. A generation later still, in the 1730s, box after box of muskets was being brought ashore in the Bight of Biafra.

Calabar

At the start of the eighteenth century the Bight of Biafra was of small consequence for English slavers. The Royal African Company had its headquarters at Cape Coast castle on the Gold Coast, hundreds of miles to the west. But in the 1730s the trading towns of the Niger and Cross River deltas assumed a major importance as Bristol merchants strengthened their links with Bonny and Calabar. Slave shipments from the Bight of Biafra rose fourfold between the 1730s and 1760s, from 34,100 in 1731–40 to nearly 152,100 in 1761–1770.[223] Calabar, an important node in the trading networks that snaked up and down the rivers and estuarial creeks of the region, now became more intimately involved in the wider Atlantic economy.[224]

---

[221] Quoted in Hugh Thomas, *The slave trade: the history of the Atlantic slave trade, 1440–1870* (1997), p. 325.

[222] Richards, 'The import of firearms', 53. In the early 1680s the Royal African Company distinguished between eight varieties of gun for the West African markets, varying in price from 8*s* 8*d* to 20*s* 8*d*: R.A. Kea, 'Firearms and warfare on the Gold and Slave Coasts from the sixteenth to the nineteenth centuries', *Journal of African History*, XII, 2 (1971), 196–97.

[223] David Eltis, Stephen D. Behrendt, David Richardson, and Herbert S. Klein (eds), *The transatlantic slave trade: a database on CD-ROM* (Cambridge, 1999).

[224] See Efiong U. Aye, *Old Calabar through the centuries* (Calabar, 1967); Stephen D. Behrendt and Eric J. Graham, 'African merchants, notables and the slave trade at Old Calabar, 1720: evidence from the National Archives of Scotland', *History in Africa*, XXX

The coastal areas, with their sandy spits and saltwater swamps, did not support an intensive agriculture. The Efik people of the coast obtained yams and other staple foods by trading salt and dried fish with the Ibo people of the interior. By supplying European goods they were also able to obtain slaves. Calabar's trade with the interior was controlled by a small group of African merchant dynasties, known to their English counterparts by anglicised versions of their local titles. It was they who made up the ruling elites of the different 'wards' into which Calabar was divided: the Robin family, for example, was active in Old Town, whilst the Duke clan was the dominant force in New (or Duke) Town. These powerful lineages developed a polyglot cosmopolitanism to ease their integration into the Atlantic economy. 'The Black Traders of Bonny and Calabar' were said to be 'very expert at reckoning and talking the different Languages of their own Country and those of the Europeans'.[225] English, or a pidgin thereof, became the language of commerce. Some Efik traders affected a European mode of dress: they 'Drisht whit men', as one of them put it.[226] Others built two-story wooden houses in the European style, employing visiting ships carpenters for the purpose. Egbo Young of Duke Town called his 'Liverpool Hall' in honour of his trading partners from the Mersey. So strongly was Calabar's elite imbued with the spirit of circum-Atlantic enterprise that by the second half of the eighteenth century it was not uncommon for the sons of the most eminent families to be sent to England for their education. Robin John Otto Ephraim, the son of 'King George' of Old Town, was one such, sent to Liverpool in 1767.[227] He retained a vivid impression of his time there. Years afterward he added a postscript to a letter to

---

(2003), 37–61; A.J.H. Latham, *Old Calabar, 1600–1891: the impact of the international economy upon a traditional society* (Oxford, 1973); David Northrup, *Trade without rulers: pre-colonial economic development in south-eastern Nigeria* (Oxford, 1978); Randy J. Sparks, *The two princes of Calabar: an eighteenth-century Atlantic odyssey* (Cambridge MA, 2004), chapter 2; and Lorena S. Walsh, *From Calabar to Carter's Grove: the history of a Virginia slave community* (Charlottesville, 2001), pp. 67–76.

[225] *Report of the Lords of the Committee of the Privy Council* (1789), quoted in Elizabeth Donnan (ed.), *Documents illustrative of the slave trade to America* (4 vols, New York, 1965), II, p. 598.

[226] 'The diary of Antera Duke', in Daryll Forde (ed.), *Efik traders of Old Calabar* (1956), p. 84.

[227] Paul E. Lovejoy and David Richardson, 'Trust, pawnship, and Atlantic history: the institutional foundations of the Old Calabar slave trade', *American Historical Review*, CIV, 2 (1999), 342. See also Paul E. Lovejoy and David Richardson, '"This horrid hole": royal authority, commerce and credit at Bonny, 1690–1840', *Journal of African History*, XLV (2004), 363–92.

Ambrose Lace, the Liverpool slave merchant: 'Remember me to your Wife and your son Joshua [and to] Ambrose[,] William and Polly'.[228]

The arrival of European ships was a matter for celebration among the Efik trader chiefs. Guns would be fired in salute as slaving vessels nosed around Seven Fathom Point to drop anchor in the turbid, mangrove-fringed waters of the Cross River. Slave trading usually began in the late summer or early autumn. Spring was the yam planting season, when the movement of slaves was suspended, but once the harvest had been brought in shipments could begin in earnest, not least because yams were now available as provender for the human cargo during the Middle Passage.[229] To set the trading cycle in motion, European articles were advanced to the merchant dynasts of Calabar. As a guarantee that the credit placed at their disposal would be repaid the merchants would hand over 'pawns' to the slave captains, usually personal slaves but sometimes family members. These human pledges would be kept on board ship until slaves equivalent to the value of the goods advanced had been supplied. If the Calabar merchant failed to fulfil his obligations, as sometimes happened, his unfortunate pawns would themselves be shipped to the Caribbean.

The European manufactures would be entrusted to lesser merchants in marketing centres in the interior. They would buy up captives at the monthly fairs at Bende or Uburu and send them down-river.

> Twenty or Thirty Canoes, sometimes more and sometimes less, come down at a Time. In each Canoe may be Twenty or Thirty Slaves. The Arms of some of them are tied behind their Backs with Twigs, Canes, Grass Rope, or other Ligaments of the Country; and if they happen to be stronger than common, they are pinioned above the Knee also. In this situation they are thrown into the Bottom of the Canoe, where they lie in great Pain, and often almost covered with Water. On their landing they are oiled, fed, and made up for Sale.[230]

---

228 Quoted in Gomer Williams, *History of the Liverpool privateers* (1897), p. 549. See also Paul E. Lovejoy and David Richardson, 'Letters of the Old Calabar slave trade, 1760–1789', in Vincent Caretta (ed.), *Genius in bondage: literature of the early Black Atlantic* (Louisville KY, 2001), pp. 89–115.

229 For the seasonality of slaving in the Bight of Biafra see Stephen D. Behrendt, 'Markets, transaction cycles, and profits: merchant decision making in the British slave trade', *William and Mary Quarterly*, 3rd ser. LVIII, 1 (2001), 184–85.

230 *Report of the Lords of the Committee of the Privy Council* (1789), quoted in Donnan, *Documents illustrative*, II, p. 598.

Slaves were sold in small parcels, sometimes individually. The 566 captives that were taken on board the *Dobson* of Liverpool between July 1769 and January 1770 arose from no fewer than 326 transactions. One supplier, Antera Duke, furnished the *Dobson* with 37 slaves over a six-month period. Duke's first sale, on 31 July 1769, was of two males for whom he received eight iron bars, fifteen copper rods, four kegs of gunpowder, two basins, two trade guns, four pounds of beads, and an assortment of cloths.[231]

This basket of goods is worthy of note, for the goods traded for slaves on the Cross River differed from those used on the Gold Coast or in Senegambia. Each sector of the African coast had its distinctive pattern of demand, as one English commentator explained. 'Brass-mounted Cutlasses are peculiar to the Windward Coast', he wrote, 'as are brass Pans from Rio Sesthos to Apollonia.' At Ouidah it was cowry shells that were most sought after, but at Calabar it was 'Copper and Iron Bars'.[232] These broad claims are borne out by the experience of Bristol and Liverpool slave ships that sailed south during or immediately after the Seven Years' War. Bar iron accounted for just 1.8 per cent of the cargoes shipped to the Windward Coast, but 11.7 per cent of cargoes for Calabar, and 18.8 per cent of cargoes sent a little further east along the Bight of Biafra, to the Cameroons.[233] This thirst for metals did not arise from an absence of iron along the Bight of Biafra. Quite the contrary, there was a flourishing tradition of iron making in Africa. 'The basic smelting process diffused from the Middle East to West Africa (as it had to northwest Europe) during the last half-millenium before Christian era.'[234] The savanna zone that extended between latitudes 10° and 15° north was rich in ore and dry woodland. From here iron was brought south to the forest belt. Iron was therefore a very familiar commodity in Calabar's hinterland, where it was worked up by the Awka, itinerant smiths who were a conspicuous feature of Ibo society.[235] In fact, iron

---

[231] P.E.H. Hair, 'Antera Duke of Old Calabar—a little more about an African entrepreneur', *History in Africa*, XVII (1990), p. 361.

[232] John Atkins, *A voyage to Guinea, Brasil, and the West-Indies* (1735), quoted in Donnan, *Documents illustrative*, II, p. 274.

[233] Data for the Windward Coast (five observations 1760–1771), Calabar (six observations 1757–1770) and the Cameroons (eight observations 1758–1769) taken from Richardson, 'West African consumption patterns', table 12.2, pp. 312–14.

[234] Philip D. Curtin, *Economic change in precolonial Africa: Senegambia in the era of the slave trade* (Madison WI, 1975), p. 208.

[235] Lars Sundström, *The exchange economy of pre-colonial tropical Africa* (1974), p. 188.

tokens were used as a currency. It was a demand for additional iron, not a lack of metallurgical knowledge in African society that drew down European imports.[236] It was this that brought the Ibo people into a relationship with forest communities in midland Sweden.

The *Amoretta* cleared Bristol under the command of David Jones on 24 November 1735, riding the ebb tide down the Avon to the open sea. A slaver of 85 tons, carrying eight guns, she had been fitted out by Joseph Iles & Co, a partnership of some of Bristol's leading slave merchants. The *Amoretta* was bound for the Bight of Biafra, where she would take on a full cargo of slaves, 224 of whom would survive the Middle Passage to be disembarked in South Carolina in July 1736. She carried a range of trade goods to exchange for slaves. Typically, these would have included textiles, hats, guns, cutlasses, rum, and prestige articles such as glassware and beads.[237] The *Amoretta* was also laden with metal. Copper rods were much in demand along the Bight, and Thomas Coster, the Bristol copper and brass manufacturer, who was an investor in the voyage, was perfectly placed to supply them.[238] The ship carried iron as well: 1,186 bars of Swedish voyage iron furnished by Graffin Prankard.[239]

Voyage iron—'the only sort and size used throughout all *Nigritia*, *Guinea*, and *West-Æthiopia*, in the way of trade'—had long been a component part of slaving.[240] A Gothenberg merchant who shipped voyage iron to the English market in 1670 claimed to have been supplying Dutch

[236] But note that it has been suggested that climatic change and desertification, by raising fuel costs, raised the price of indigenously made iron, opening the way for European imports. See Candice Goucher, 'Iron is iron 'til it rust: trade and ecology in the decline of West African iron-smelting', *Journal of African History*, XXII (1981), 179–89.

[237] We have no manifest for the *Amoretta*, but the goods supplied on the *Fly*, which sailed from Bristol in 1787, included 140 bars of Swedish iron, 417 pewter bowls, a quantity of earthenware mugs, rolls of linen, 200 brass kettles, satin and chintz to the value of £210, 100 'Bonny' musquets, 400 lead bars, 20 barrels of gunpowder, over 200 'Negro' hats, and a large selection of Birmingham goods from William Gibbons & Co—razors, padlocks, mirrors, japanned snuff boxes, pen knives, 'women's scissors', silver manillas, gilt earrings, and cutlasses. TNA: PRO, C 107/1, bundle A, 'Flys's Insett from Africa & Antigua', 28 June 1787. See also Stanley A. Alpern, 'What Africans got for their slaves: a master list of trade goods', *History in Africa*, XXII (1995), 5–43.

[238] Richardson, *Bristol, Africa and the eighteenth-century slave trade... Volume 2*, p. 60. For Coster, see Madge Dresser, *Slavery obscured: the social history of the slave trade in an English provincial port* (2001), p. 104, and Day, *Bristol brass*.

[239] Eltis *et al.*, *The transatlantic slave trade*; SA, DD/DN 439, 20 November 1735.

[240] John Barbot, *A description of the coasts of North and South Guinea* (1746), p. 44.

slavers for 'over 50 years'.[241] English involvement in the slave trade, hitherto secondary to that of the Portuguese and the Dutch, mounted with the expansion of sugar production on Barbados and Jamaica in the third quarter of the seventeenth century, and received official endorsement with the incorporation of the Royal African Company (RAC) as a monopoly trading concern in 1672. As the English slave trade grew, so did the demand for iron amongst the London merchants who furnished the RAC with trade goods. By the early 1680s the RAC was exporting about 10,000 bars annually, sufficient for 830 slaves or 10 per cent of the number embarked yearly by the Company.[242]

In 1698 the monopoly of the RAC was rescinded and the slave trade was opened up to private traders. With this, the English slave trade grew in scale and shifted in focus. London, headquarters of the RAC, diminished in importance, whilst Bristol, with its command of the western approaches, emerged as England's premier slaving port. Bristol merchants entered the slave trade as soon as the RAC's monopoly powers were surrendered in 1698, although their involvement was at first tentative. Only nine slave ships left Bristol for the Guinea coast in 1701–1705. Progress thereafter was more rapid. Forty-two slave voyages cleared Bristol in the quinquennium 1706–1710, then 75 voyages in 1711–1715, and then 117 in 1716–1720. By the late 1720s the Bristol slave trade was at its zenith. Bristol merchants fitted out 203 ships for the African trade in the years 1726–1730, landing 57,862 captives in the New World. (A further 10,585 unfortunates were stowed on board Bristol vessels but did not survive the Atlantic crossing.)[243]

As Britain's slave trade grew, spearheaded by Bristol men, so did the demand for voyage iron. Bar iron exports from Britain to the African coast crept steadily upwards in the first half of the eighteenth century, from an annual average of 360 tons in the years 1701–03, to 536 tons

[241] J. Van Savelant to Leonora Marescoe, 18 October 1670, in Henry Roseveare (ed.), *Markets and merchants of the late seventeenth century: the Marescoe-David letters 1668–1680* (Oxford, 1987), p. 341. It has been said that the Dutch 'introduced the standard iron bar, probably in the 1630s'—a date that would coincide with the rise of Dutch influence within the Swedish iron industry: A.F.C. Ryder, *Benin and the Europeans, 1485–1897* (1969), p. 98.

[242] Thomas, *The slave trade*, p. 321. The exchange rate used here, that of 12 iron bars per slave, is taken from an inventory of goods at the RAC's factory at Ouidah in April 1681: Robin Law (ed.), *The English in West Africa, 1681–1683. The local correspondence of the Royal African Company of England 1681–1699. Part 1* (British Academy and Oxford University Press, 1997), p. 222.

[243] Data extracted from Eltis *et al.*, *The transatlantic slave trade.*

in 1727–29, then to 990 tons in 1737–39.[244] This was a market that Graffin Prankard, newly emerged as Bristol's leading Baltic merchant, was in a position to dominate. The market for voyage iron required careful management, however. For one thing, careful attention had to be paid to the requirements of African consumers. 'The Blacks of the Gold Coast', it was said, 'examine and search very narrowly all our merchandize, piece by piece, to see each to be of the quality and measure contracted for by samples'.[245] As factors at the RAC's West African forts discovered in the early days of English slaving, local merchants were particular about the provenance of iron and looked for brand marks on the bars before buying. 'These people begin to aske for iron bars', Robert Thelwall reported from Anamaboe in July 1683, 'and I have a great many but they doe not like them, for they must be all marked and noe flau's in them.'[246] The weight and dimensions of the bars were also of critical importance. Yet these specifications tended to change over time. Bars supplied to the Dutch West India Company in the mid seventeenth century were 32lb apiece, making 70 bars to the ton.[247] Voyage iron bought by the RAC later in the century came rather lighter, at 28 to 30lb per bar, or from 75 to 80 bars to the ton.[248] In the 1720s the bars required by Graffin Prankard's customers were lighter still: they wanted bars that 'run neare about 92 to ye ton', that is, about 25lb apiece, and be '10 foott 6 Inch or 10 foott 8 long'.[249]

---

[244] TNA: PRO, BT 6/241. Statistics have only survived for selected three-year periods. The overall trend is upwards, but exports fluctuated considerably, often in response to political conditions. Voyage iron exports slumped to an average of 370 tons in the war years 1745–47, then recovered to 900 tons annually in 1752–54.

[245] Barbot, *A description*, p. 273.

[246] Law, *The English in West Africa, 1681–1683*, p. 137.

[247] Roseveare, *Markets and merchants*, p. 341: 'before I came to Sweden I supplied large amounts in Holland, at 34 bars per 1,000lb—Holland weight'. The Dutch pound was equivalent to 1.09 English pounds.

[248] The contract made by the Company with Peter Joye of London in 1685 stipulated that the bars 'be of the usual lenght [sic] with Mark or Marks on each Barr and the Number to be from 75 to 80 Barrs at the least in each tun…' K.G. Davies, *The Royal African Company* (1957), p. 171. Bars exported by the French Compagnie du Sénégal in the 1690s were of similar dimensions: 'about eighty of these bars weigh a ton, or twenty hundred weight English'. Barbot, *A description*, p. 44. Voyage iron made at Bassaleg in south Wales ranged from 29 to 30lb per bar in 1711–1713: NLW, Tredegar 76/27–28, Bassaleg forge accounts.

[249] GP to FJ, 1 December 1731. Later in the century Sven Rinman, the great Swedish metallurgist, gave another definition. 'Voyage iron is bar iron that is commonly forged 5 to 5½ aln in length, 1½ tum flat, and 3/8 tum thick, and that is bent double two or four times, so that in foreign places it can be carried on a donkey'. Sven Rinman,

Bristol slavers had to be attentive to the demands of their African counterparts, which caused them to be demanding in what they asked of Baltic merchants. As an exasperated Graffin Prankard told his Rotterdam correspondents in 1729, 'our people here are become very nice in their voyage Iron'.[250] Just as the making of 'Orground' iron required Walloon forgemen to be responsive to the concerns of steel makers in England, the fabrication of voyage iron called for Swedish forgemen to track the changing preferences of African consumers. In effect, the bars were being used as a currency on the African coast. If they were too heavy they were devalued as a unit of exchange for Bristol merchants and Prankard's customers would insist upon a rebate.[251] If they were too light, African merchants in Bonny and Calabar would reject them. Because of this, Prankard was anxious to have a regular and assured supply of voyage iron, made by a workforce that understood the specifications of the product. The *bruk* at Gammelbo fulfilled this role for him throughout the 1730s. As was the case with 'Orground' iron, credit was advanced to the proprietor of the works, allowing Prankard to dictate the form that the final commodity would take: 'press hard on Feoffe [Jacob Feiff, the merchant who handled the sale of Gammelbo iron in Stockholm]', he told Jennings in February 1733, 'for Striking the Voyage of [Gammelbo] much wider & to run about 90 to ye Ton'.[252]

Gammelbo *bruk* lay in the parish of Ramsberg in central *Bergslagen*. The context of iron making here was very different to that prevailing in Uppland, where *Leufstawerken* encompassed all aspects of the production chain from mining to bar iron making. As has been noted, the *Vallonbruk* of Uppland inhabited a specialised enclave. Elsewhere in *Bergslagen* the Swedish state had imposed a spatial, technical and social division of labour. Mining and smelting were assigned to *bergsmän* in the central parts of the mining district, while bar iron manufacture was left to *brukspatroner* on the fringes of *Bergslagen*. This model was visible in Ramsberg, but not as clearly as state officials might have wished. Local *bergsmän* did smelt ore and sell pig iron to outlying *bruk*, but some

---

Bergwerks Lexicon (2 vols, 1788–89), II, p. 1180. An aln equals 59 centimetres, so the bar would have been from 9 feet, 8 inches to 10 feet, 7 inches in length.

250 GP to Coysgarne & Lloyd, 1 March 1732.

251 Isaac Hobhouse & Co were rebated £1 per ton on a consignment of voyage iron in June 1738 because of 'ye heaviness of ye barrs': SA, DD/DN 439, 22 June 1738.

252 GP to FJ, 28 February 1733.

operated their own forges as well. Moreover, there were licensed *bruk* within the parish. As a result, the pattern of iron making in the Linde and Ramsberg district was somewhat motley, with 27 furnaces in blast in the first half of the eighteenth century and 20 forges at work.[253]

Gammelbo *bruk* was the largest of the 'non-*bergsmän*' production facilities, with a blast furnace and four forges. The furnace was adjacent to the manor house at Gammelbo itself. The forges lay beside the river that flowed southward through the forest: Hägernäs, Nyhammar, Bergshyttan and Sundbo. This made Gammelbo a large *bruk*, perhaps the largest in the region, with an authorised output of 2,875 *skeppund* (430 tons) in 1748.[254] As a *bruk* with its own pig iron making capacity it was something of an anomaly, yet the output of the furnace at Gammelbo was not sufficient to supply four forges. Additional pig iron had to be brought in from outside. Regular consignments came from two nearby furnaces in which the Gammelbo estate may well have held shares, but large quantities of pig iron were also purchased on the open market, either directly from *bergsmän* or from merchants in the local market towns of Lindesberg and Arboga. Unlike Leufsta, which drew upon a single, self-managed source of pig iron, Gammelbo depended upon a wide variety of suppliers.

Gammelbo's forges were also quite distinct from their counterparts at Leufsta. The four forges at Leufsta were gathered at the centre of a planned production landscape. In Ramsberg there was no unity of design; the Gammelbo forges were scattered at some distance from one another. Most important of all, the Gammelbo forges employed the German forging method, not the Walloon technique that was the specialism of the Uppland forges. In German forging no distinction was made between a finery hearth and a chafery. The same hearth served both for fining the metal and for reheating it during the drawing out of the bars, even though most German forges contained two hearths. Furthermore, the same workmen performed both functions. There was no division of labour between finers and hammermen; a single hammer crew refined the iron *and* shaped the bars.

---

[253] Ture Omberg, *Bergsmän i hyttelag. Bergsmansnäringens utveckling i Linde och Ramsberg under en 100–årsperiod från mitten av 1700-talet* (Uppsala, 1992), p. 45, and Anders Florén, 'The making of the forgeman. Social relations and bar iron production in Sweden *c.* 1650–*c.* 1750', in Hans-Jürgen Gerhard, Karl Heinrich Kaufhold and Ekkehard Westerman (eds.), *Europäische Montanregion Harz* (Bochum, 2001), p. 199. Nine of the forges were operated by *bergsmän* and eleven were attached to *bruk*.

[254] Hammarskattelängden 1748, Bergskollegiets arkiv, RA.

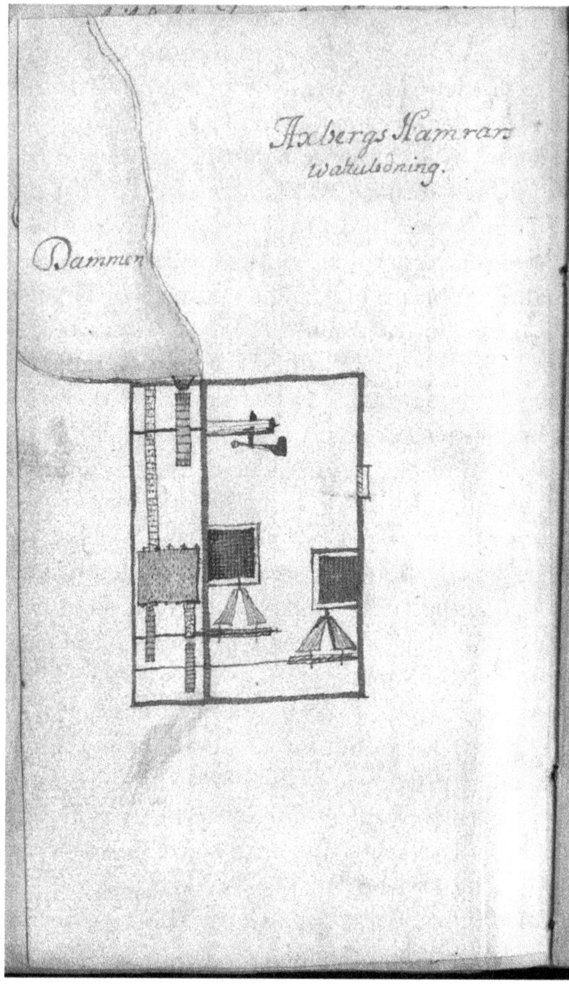

Illustration 2.23. The German forge at Axberg.

Source: Daniel Tilas, Diarum pro Anno 1733, UUB, X 300.
Caption: The *Bergscollegium* official Daniel Tilas kept a detailed diary of his movements in 1733, a year in which he visited Gammelbo, the *bruk* owned by his aunt Greta Tilas and his birthplace. Whilst at Gammelbo he inspected the nearby forge at Axberg. His sketch plan shows a layout typical of a German forge, the type most commonly found in Sweden. The forge pond (top-left) fed the large water wheel that drove the hammer, and (via a launder and reservoir tank) the two smaller wheels that powered the bellows. The two hearths were identical; they served for both the fining of the metal and the reheating of the blooms as the latter were drawn out into bars.

The proper composition of a German forge crew had been defined in a decree of 1637. There should be three forgemen: a master (*mästersmed*), a forgehand (*mästersven*), and an apprentice (*smedsdräng* or *koldräng*). In fact, this law was honoured more in the breach than the observance. A study of forge crews in the Nora and Linde district, of which Ramsberg was a part, has revealed that few crews adhered to the legal model. It was quite common for just two men to share a hearth.[255] On other occasions one master forgeman took responsibility for an entire forge, overseeing work at two hearths. It was this pattern that was to be found at Gammelbo in the 1730s, despite a decree in 1703 that had reiterated the old regulations. At Bergshyttan in 1736, for example, the *mästersmed* (Hans Hansson Palt) worked with three *smedsdrängar* (Nils Krabbe, Anders Brace, and Nils Hansson), and a *koldräng* (Carl Carlsson). Palt worked with Nils Hansson at one hearth, while Krabbe and Brace laboured at the other. Young Carlsson supplied charcoal to both hearths. *Mästersmed* Palt was paid for the entire output of the forge at a set rate per *skeppund*. He then paid the other members of the forgecrew according to their level of experience.

The situation can be contrasted with that at Leufsta. A common piece rate was also paid to forge crews at Leufsta, whether they were finers or hammermen. But at Gammelbo the rate would vary according to what type of bar was being made: ordinary bars brought 3 *kopparmynt* per *skeppund*, voyage iron 3½ *kopparmynt*, and 'extra voyage' 4 *kopparmynt*. The piece rate at Leufsta never varied, no matter what sort of bar the forgemen produced. Work at the *Vallonbruk*, as we have seen, was at a high tempo. The purity of the product was paramount, not the dimensions in which the bars came, so hammermen were not required to linger over the bars. At German forges the reverse was true. German forgemen did not have access to the exceptional ores of Dannemora, but they could make bars to very precise specifications, provided that the wage structure was adjusted to reward the greater care taken in shaping the bars. Output was lower at German forges: production at Bergshyttan in 1736 was just 136 tons, for example,

[255] Anders Florén and Göran Rydén, 'The social organisation of work at mines, furnaces and forges', in Maria Ågren (ed.), *Iron-making societies: early industrial development in Sweden and Russia, 1600–1900* (Oxford, 1998), pp. 110–13; Florén, 'The making of the forgeman', pp. 201ff. The frequency with which two-man crews were to be found probably relates to the length of the working day. A three-man team was necessary to facilitate round-the-clock working, but was top-heavy if work went on only during daytime hours.

only a third of what was being made at the Leufsta forges. This was not because production had to be halted due to water shortages or other natural impediments—the forge was at work for 48 weeks in the year—but because the forgemen lavished more attention on the form of the bar. Bars at Leufsta weighed about 20 kg on average; those made at Bergshyttan were far finer, averaging 15 kg.

The output at Gammelbo *bruk* was far more varied than that at Leufsta. Hans Hansson Palt and the forgemen at Bergshyttan made voyage iron, 'extraordinary' bars, squares, and *schampluner* in the course of 1736. (*Schampluner* denoted any sort of bar made to irregular specifications at the bidding of particular customers; five different batches were drawn out in 1736.) The dimensions of bars could vary radically. The 4¾ inch bars made at Bergshyttan weighed 35 kg on average, whereas voyage iron was a mere 10 kg. Of all these sorts, voyage iron was the most important. It often accounted for one-third of production at Gammelbo in the 1730s, and in 1738 made up over half of production.

Voyage iron manufacture had its own season. The Gammelbo forgemen began making it in the spring and usually made nothing else until the early autumn. This was in response to instructions from Bristol. Graffin Prankard would transmit his yearly order for voyage iron to Francis Jennings in January or February.[256] It would be passed on via Jacob Feiff in Stockholm to Greta Tilas, the widow of the *brukspatron*. Her forgemen had spent the winter, when communications with the outside world were slow and the difficulties in transporting iron at their most extreme, in making standard bars. For these, a market of some sort would always exist and they could be safely stockpiled. Demand for voyage iron, on the other hand, was conditional upon developments in a volatile branch of the Atlantic economy. It was best, therefore, to wait for signals from the international market before commencing work.

Yet the signals were not always interpreted correctly, and Graffin Prankard was quick to complain if the forgeman erred. 'I know not what to do with [Gammelbo iron]', he declared in 1736, 'it being to [sic] light...wch [has] rendered it unsalable'.[257] His plaints were heeded, however. The width of the undersize Gammelbo bars was increased by ¼ inch in 1737. Prankard was in a position to insist. After all, he made the financial advances, routed through Amsterdam and Stockholm,

---

[256] See, for example, GP to FJ, 18 January 1731, or 18 January 1732.
[257] GP to FJ, 27 September 1736.

which made production possible. Jacob Feiff seems to have relayed instructions to Gammelbo throughout the summer, ensuring that the bars met with Prankard's approval.[258]

Yet there were difficulties facing a dealer in voyage iron quite separate from those of quality control. Although voyage iron tended to sell in bulk—Prankard sold an average of 9.58 tons per parcel as against 2.68 tons for Swedish 'common sorts'—it was a branch of trade that could tie up capital for long periods. Credit had to be extended to *brukspatroner* in order to guarantee the supply. Moreover, the arrival of voyage iron from the Baltic coincided a little too closely with the fitting out of slave vessels at Bristol. Most slaving expeditions left between June and October, the very period at which Prankard's ships started to arrive from the Baltic. This, Prankard complained, forced him to put voyage iron into store over the winter. He was, he told his correspondents in Gothenberg, 'almost Sick of ye Trade[,] being kept out of our money for ys Commodity on an averidge 9 to 10 months'.[259] Yet Prankard's lamentations over 'ye long loose uncertain pay' associated with voyage iron were surely overdone. Voyage iron supplied to slavers that cleared Bristol in the spring must indeed have been landed the previous year, but by August, the busiest month for clearances to Guinea, fresh supplies, newly arrived from Sweden, would have been available.[260]

The market for voyage iron was very volatile. Trading conditions on the Guinea Coast were not predictable, demand in the Americas was given to fluctuations, and like any branch of maritime trade slaving was affected by hostilities between European powers. The early 1730s saw a slump in slaving voyages out of Bristol, reflecting the depression in sugar prices of that time. From an all-time peak in 1729, when 17,750 slaves were crammed into Bristol ships, the trade reached a low of just 7,039 slave embarkations in 1734. Prankard's sales fell in tandem. Whereas 346 tons were sold in 1732, just 148 tons of voyage iron left Prankard's warehouse in 1733: 'our Guinea Trade is wholly at a Stand'.[261] Two lean years followed. It was only in 1736 that sales regained their earlier buoyancy: a 'famine or near to it In Affrica very

---

[258] Gammelbo bruksarkiv, letterbook 1733, Greta Tilas to Jacob Feiff, 1 June 1733.

[259] GP to Maisters & Grundy, 14 December 1731.

[260] See Behrendt, 'Markets, transaction cycles, and profits', for the difficulties merchants faced in coordinating the supply of shipping, seamen, trade goods, and slaves with the demand for slaves in the plantations.

[261] GP to FJ, 28 April 1733.

lately have Caused a Great Plenty of Slaves...& gave a Life to the Trade', Prankard announced, ordering additional supplies of iron.[262] Indeed, the next two years saw a boom in his sales of voyage iron, which topped 616 tons in 1738. There was then a sharp contraction as the outbreak of the War of Jenkins' Ear with Spain curtailed the slave trade. Only 6,249 Africans were embarked on Bristol ships in 1740, compared with 14,714 in 1738.[263]

Despite all these uncertainties, the rationale for Prankard's involvement in the trade in voyage iron was clear enough. Bristol was Britain's premier slaving port, and Prankard's customers included every major slave merchant in the city: James Iles, Isaac Hobhouse, James Laroche, Henry Dampier, and others. Prankard guarded this clientele jealously. 'I dont pretend to be so vain as to think yt no one should Sell Voyage but my Self only', he told Jennings in 1730, but this was disingenuous.[264] He had every intention of excluding rival traders. The success with which Prankard engrossed the Bristol market is not easily determined. His account books do not always specify the vessel for which a consignment of voyage iron was intended, so his sales to slave merchants cannot always be aligned with known sailings to West Africa. Nevertheless, in 1738, his peak year in this branch of the iron trade, Graffin Prankard sold bar iron to 19 of the 20 Bristol partnerships that are known to have fitted out ships in the Guinea Trade.[265] Clearly, Prankard was a major player in the market, not an occasional dabbler. The full scale of Prankard's achievement can be grasped when it is realised that in 1738 he alone handled 54 per cent of the voyage iron that left British quays.

The surge in demand in 1736 exhausted the immediately available iron from Gammelbo. Francis Jennings was instructed to scour the Stockholm *järnvåg* for additional supplies and Prankard's Gothenberg correspondents were put on alert. As the recovery continued through 1737 Prankard's demands became more and more insistent. His Dutch correspondents were ordered to pick up whatever Swedish voyage iron appeared on the Rotterdam market and to bring German voyage iron down the Rhine. Prankard also recognised that the suddenly bouyant

---

262 GP to FJ, 30 September 1735.

263 Eltis *et al., The transatlantic slave trade.*

264 GP to FJ, 22 June 1730.

265 A conclusion based upon cross-referencing entries in Prankard's waste books with data in Richardson, *Bristol, Africa and the eighteenth-century slave trade... Volume 2.*

African market might be a way of disposing of the surplus 'Orground' iron he had on his hands after Samuel Shore and he had contracted to take '2nd orgrounds' in addition to Leufsta and Åkerby. There was always the possibility of having 'Orground' iron that had been struck into 'narrow flats'—bars that went '60 or 61 to ye Ton full out 15 foott & ½ long'—re-cut as a serviceable voyage iron.[266] When Graffin Prankard found himself hopelessly overloaded with 'Orground' iron in 1736 and 1737, this is what he did.[267] Bristol smiths reprocessed the narrow flats at Prankard's behest. The re-cut bars were loaded into the holds of slaving vessels. The 'bits of bars' that were left were sold off to slitting mill proprietors to be turned into nail rods: 'ye 792 Pieces is what I Cut of ye Guinea Iron', Thomas Lewis, the Swansea ironmonger was told, 'and I know will answare your purpose very well for Slitting into rods for nailes.'[268]

The 616 tons of voyage iron that Graffin Prankard sold in 1738 was never to be surpassed. The outbreak of war in 1739 war disrupted American markets and brought Spanish privateers swarming into the sea lanes that connected Britain to west Africa. Slaving was never a risk-free activity, but the coming of war persuaded many Bristol merchants that the Guinea trade was just too insecure an investment for the time being. The number of slavers clearing the port dropped sharply, from 53 in 1738 to 28 in 1741. The reverberations were felt far off in Ramsberg parish. The Gammelbo forgemen had drawn out 1408 *skeppund* of voyage iron in 1738; in 1741 they made just 122 *skeppund*. As the number of slaving ships arriving in the Cross River dwindled, Hans Hansson Palt and his forgehands at Bergshyttan turned their hand to *schampluner* of a different sort, for a different market.

---

266 GP to FJ, 9 June 1729.

267 GP to FJ, 9 February 1737: 'as to ye other orgrounds of wch I will Strive to take as much as possible I wish thee coud prevail to have a quantity of ¾ Squares Struck of it as also narrow flats fit to Cut for Voyage for of Voyage I shall want 300 tons at least ys year & if coud have but 80 or 100 tons Orgrounds drawn fine Enough to Cutt woud help mee out with ye Common Orgrounds'.

268 GP to Thomas Lewis, 11 January 1737. Over 3 tons of such 'pieces' were sold to Lewis at the bargain rate of £12. 15s. per ton: SA, DD/DN 439, 5 January 1737. Lewis was, presumably, the proprietor of the slitting mill at Ynyspenllwch mentioned in E.H. Brooke, *Chronology of the tin plate works in Great Britain* (Cardiff, 1944), p. 163. A commission of bankruptcy was issued against the partnership of John Morse and Thomas Lewis in September 1737. Prankard acted as an assignee on behalf of other creditors, and papers concerning the disposal of the firm's assets can be found in SA, DD/DN 454.

St. Petersburg

In the autumn of 1746 Joseph Baker, a London merchant, waited anxiously for news of the wartime convoy that was homeward bound from St Petersburg. Baker, like Graffin Prankard and Josias Wordsworth, dealt in the commodities of both the Atlantic and the Baltic. He traded in sugar and rum from the West Indies, and in hemp, tar, flax and iron from northern Europe. Baker's ship, whose arrival from the Russian capital was of such concern, was laden with 79 tons of hemp and '376 Bars of Siberia Iron in Flat Bars'. Encouraging intelligence arrived in early October. The Petersburg convoy had passed the Sound. 'By the last Holland mail we learne the Baltick fleet was sailed ye 6th N[ew] S[tyle] from Elsenore consisting of 3 men of war & 108 Merch[ant] Ships Viz 65 Lond[on], 7 Plym[outh], 1 Yarm[outh] 18 Hull 1 Woolwich 1 Cow[e]s 6 Port[s]mouth 3 Lyn[n] 1 Poole 2 Bristol Viz Rich Thompson & Edw Hill 1 Dublin 1 Leith & 1 Aberdeen.' Britain had been engaged in the War of Austrian Succession for six years, and merchants had learnt the advisability of sailing under the protection of the Royal Navy. 'I hope in God', Baker confided, 'this will prove a fortun[at]e Bargain'.[269]

That a convoy of over one hundred ships should leave St Petersburg for Britain in 1746 was an index of the astonishingly rapid political and commercial ascendancy of that city. It had not existed in 1700. The marshy delta on which it was to be built was part of the Swedish territory of Ingermanlandia. Nyen, a small Swedish fortress and trading place, lay upstream on the Neva; the coastal area through which branches of the Neva and its tributaries flowed was as yet home to isolated farmsteads and small fishing settlements, nothing more.

The Great Northern War transformed the situation. Nyen was besieged, then razed by Russian forces. Peter the Great, bent on regaining that access to the Baltic that Russia had lost in the wars of the early seventeenth century, gave orders for a new military base to be built further downstream, where the Neva met the Gulf of Finland. Work on the Peter and Paul Fortress, the kernel of a new city, began in May 1703.

---

[269] 'Letterbook D, by Joseph Baker', 1746–9, Corporation of London Record Office.

The Neva delta soon became a gigantic building site. Thousands of labourers and artisans were drafted in from across the Russian empire. Large numbers of prisoners of war were added, creating a workforce that numbered 40,000 within a year of the city's foundation. During these first years construction workers toiled through thc malarial summers and the brutal winters to create a giant military citadel. Naval dockyards and the Admiralty headquarters were built on the left bank of the Neva, opposite the Peter and Paul Fortress, while the offshore island of Kronstadt was fortified to guard against any descent on the city from the open sea.

After the calamitous Swedish defeat at Poltava in 1709 the outcome of the Great Northern War became more certain and the future of the Tsar's new city more secure. 'With God's help the last foundation-stone of St Petersburg has now been laid', a triumphant Peter wrote from the battlefield.[270] Post-Poltava there was some relaxation in the martial atmosphere that had marked St Petersburg's formative period. Civilian buildings now began to rise alongside the redoubts and barracks. Peter's announcement in 1712 that St Petersburg was to be his imperial capital was the signal for an epic new building programme to get underway, following plans largely drawn up by the Swiss architect Domenico Trezzini (1670–1734). Trezzini was to design the Tsar's winter and summer palaces (1710–12), the State Offices (1714–19), the twelve administrative Collegia (from 1722), and the Peter-Paul cathedral (1712–33). Stone now began to replace the earlier wooden-built structures. Indeed, the Tsar was so determined that his capital should present a smooth, stone-clad face to the world that in 1714 he outlawed the erection of stone buildings elsewhere in the empire in an effort to draw stonemasons to the banks of the Neva.

The emerging urban landscape was mapped in 1722 by Carl Fredrik Coyet, a Swedish prisoner of war, a year after the Treaty of Nystad had confirmed Russian sovereignty over the region (Illustration 2.24). The Peter and Paul fortress, marked 'A' on the map, was separated from the main body of the 'Town Island' (B) by a small canal. On the 'Admiralty side' (D) was the Tsar's summer palace. The palaces of other members of the imperial family and the nobility extended along the embankment to the east of the Admiralty. By the 1730s this

[270] Quoted in Peter Englund, *The battle that shook Europe: Poltava and the birth of the Russian empire* (2003), p. 209.

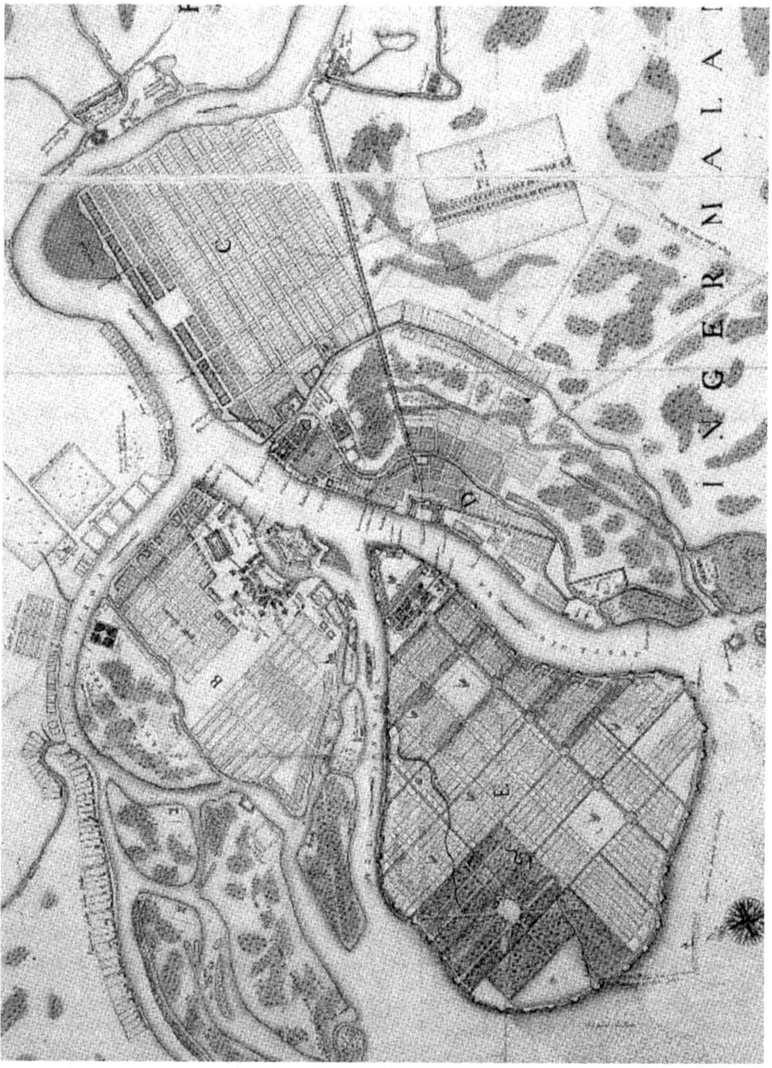

Illustration 2.24. St Petersburg in 1722.

Courtesy of Krigsarkivet, Stockholm.

was the social and political hub of the city. The commercial centre lay elsewhere. The first specialised marts, the *Gostiny Dvor*, through which all export trade had to pass, had been established on the Town Island, but by the time Carl Reinhold Berch (the uncle of the Christer Berch who we have met at Leufsta) visited St Petersburg in 1735 they were in decay. 'The Shops, (Gostiny Dvor) which are built in a large square on the Peterburg Island, have deteriorated considerably through age, having been built merely in half-timbered style.' A new Exchange, completed in 1733, stood on Vasilyevsky Island (E), which Peter had determined should be 'the most distinguished and best part of the city'. Coyet shows the island with a grid of avenues and canals, but this layout was as yet unrealised, as was the fortified perimeter dyke. Large parts of Vasilyevsky Island remained forested in the mid-eighteenth century, only the southern shore and the eastern spit were fully developed.

By the middle of the eighteenth century St Petersburg had grown into a European city of some stature. By 1740, when the city's population reached 75,000, it had surpassed Stockholm in size. Although visitors were apt to grumble about the crudity of the wooden dwellings, few could fail to be impressed by the italianate splendour of the public buildings and palaces, or the sweep of the Neva. 'It cannot be denied', a German visitor reported in 1737, writing of the new Exchange and adjacent administrative Collegia on Vasilyevsky island, 'that the island along the bank of the river... has been superbly built.'[271] Jean de Bedoire, a Swedish visitor of the early 1750s, admired the spaciousness of the city, with its 'regulated and broad streets'. The 'beautiful canals' were another characteristic of the city of which Bedoire approved.

St Petersburg's canals were of great utility as well as beauty. They drained the marshy land and allowed the smooth movement of goods through the city. This was of great significance, for it was Peter's intention that his city should be the commercial outlet of his empire, not just his capital.

Peter the Great's drive to the Baltic must be understood within the much wider objective of modernising Russia along western lines.

[271] Peter von Haven, quoted in Konstantin V. Malinovskij, 'En stad växer fram. Sankt Petersburg 1703–1740', in *Vattenstäder Sankt Petersburg—Stockholm* (Stockholm, 1998), p. 105. For population figures see James Cracraft, *The Petrine revolution in Russian architecture* (Chicago, 1988), p. 228. For other figures see Anthony Cross, *By the banks of Neva: chapters from the lives and careers of the British in eighteenth-century Russia* (Cambridge, 1997), p. 16.

The rise of St Petersburg was one facet of a general programme of reform—military, administrative, and economic.[272] Modernising initiatives were especially conspicuous in the field of industrial development and trade. At the start of Peter's reign monetary exchange was not a common feature of the Russian economy. 'Russia abounds in Merchandize', a Hanoverian envoy remarked, 'but not in ready Money'.[273] The feudal estate provided the framework within which the production of both foodstuffs and industrial commodities such as hemp took place. Nevertheless, economic growth was discernible from the 1720s, spurred on by closer links to the international economy and the availability of international credit. This growth was not limited to the agrarian world. Industrial development was spectacular, especially in the metallurgical sector.[274]

Large-scale iron making in the Urals began in 1701 with the building of the Neviansk works, inaugurating a phase of massive expansion. Nearly 200 iron or copper works were established in the Ural region in the course of the eighteenth century. The initial impetus was military. Munitions were needed for the Great Northern War, and they were hardly to be obtained from Sweden. When Wilhelm de Hennin, one of key figures in the development of the Russian iron industry, first headed for the Urals in 1714, he was charged with improving the quality of the cannon made there. De Hennin's brief was to make cannon equal to the Swedish, which were considered the finest in Europe at the time.[275]

A new phase of development started after the Peace of Nystad. De Hennin undertook a second mission to the Urals in 1722. This time he was to concern himself with far more than armaments; he was to create

[272] See Lindsey Hughes, *Russia in the age of Peter the Great* (New Haven CT, 1998) and Simon Dixon, *The modernisation of Russia 1676–1825* (Cambridge, 1999) for an overview of Russian development.

[273] F.C. Weber, *The present state of Russia* (1723) quoted in Dixon, *Modernisation of Russia*, p. 231.

[274] Dixon, *Modernisation of Russia*, pp. 221–52, and Hughes, *Russia*, pp. 145–58. For economic growth see Ian Blanchard, 'Eighteenth-century Russian economic growth: state enterprise or peasant endeavour?', *Jahrbücher für Osteuropas*, XLV (1997), 541–51.

[275] 'General Liutenanten Hennings Relation om Ryske Bergwerkens upkomst och början, In Septembr 1728, utaf Tyskan öfwersatt', Bergskollegiets Arkiv, vol. D VI:8 Tilas Samlingar, RA. This text, in Swedish, has been translated from German, and is an attempt by De Hennin to tell his side of the story of how the Russian iron industry was founded. For a general discussion of Russian iron making during this period see Ågren, *Iron-making societies*.

an iron industry capable of producing bar iron for the west European market. It was a task for which De Hennin was well equipped. In 1719 he had been sent on a Europe-wide odyssey to acquaint himself with the latest developments in iron making. He returned with an enhanced knowledge of new 'equipments and techniques', together with a large retinue of foreign workers. De Hennin, who already held rank as an officer in the Tsar's artillery corps, was now appointed director of the Board of Mines. This was a new body, modelled on the Swedish *Bergscollegium*. As director, De Hennin was handed 'all authoritative power' to realise Peter the Great's vision of a Urals iron industry. He lost no time in setting about the task. By 1725 he had completed the giant fortress-cum-factory at Ekaterinburg. Shielded by fortifications from Tartar raiders was a combined iron and copper works, consisting of blast furnaces, a forge for bar iron making, a wire-mill, a sheet-making forge, and a copper furnace.[276]

The construction of Ekaterinburg exemplified certain architectural trends in Petrine Russia. The town was enclosed within a square bastion, each side of which was about 700 metres in length. One side of the bastion was breached so as to admit the pond whose waters, penned back by an immense dam at the very centre of the community, would power the industrial processes. In the shadow of the dam stood the workshops, and around them were the school, the hospital, the churches and the master's mansion. The workers' housing stretched out in long, straight rows beneath the perimeter walls of the fortress. There are clear parallels with contemporary developments at Leufstabruk, or indeed with the proposed layout of Vasilyevsky Island in St Petersburg.[277] 'Regularity' (*reguliarnost*) was the most sought-after quality in Russian town planning at this time, both for new military towns and for industrial settlements on the Petrine frontier. Ekaterinburg, which was both an industrial community and a military outpost, embodied this fully. As the headquarters of the Siberian mining administration it mirrored distantly the architectural ambitions that were to be seen on a far grander scale in St Petersburg.[278]

Reliable statistics for bar iron production in the Urals, as opposed to Russia as a whole, are hard to come by, but the trend is reasonably clear.

---

[276] 'General Liutenanten Hennings Relation'.

[277] R. Lotareva, 'Factory-towns of Russia and C.-N. Ledoux', in *Claude-Nicolas Ledoux and Russian architecture* (Ekaterinburg, 2001).

[278] Cracraft, *Petrine revolution*, pp. 257ff.

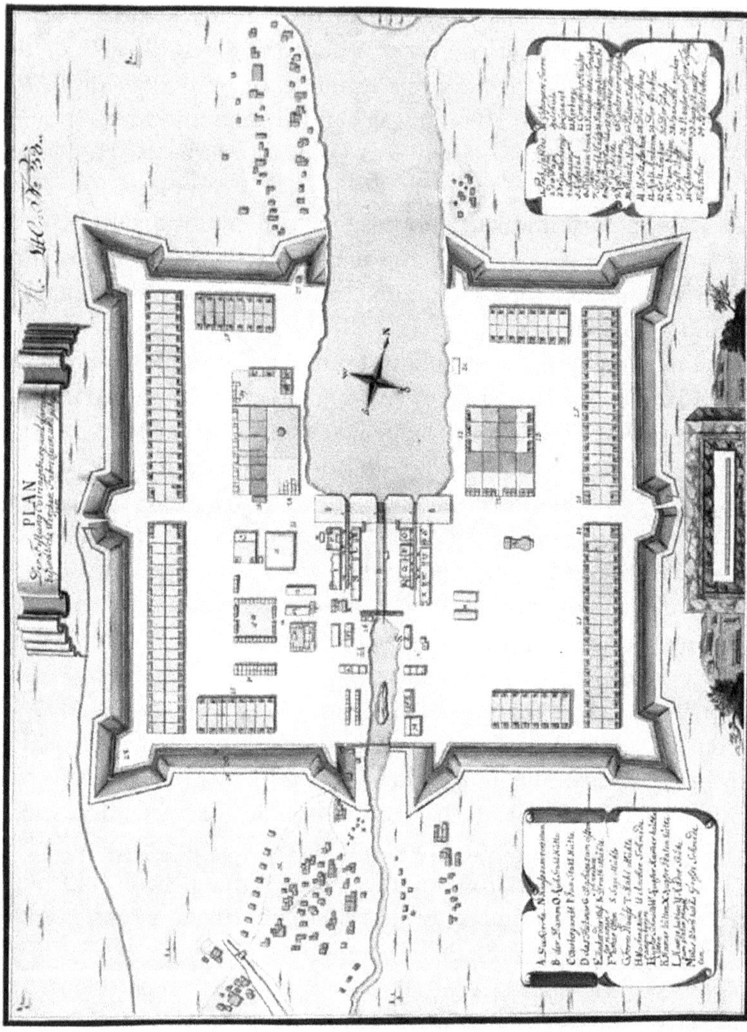

Illustration 2.25. Ekaterinburg.

Courtesy of Krigsarkivet, Stockholm.

There were two waves of investment in state-directed ironworks, the first in the mid-1720s, the second in the 1740s. The development of private works followed a slightly different trajectory. Entrepreneurial families such as the Demidovs had been active in the region since the first years of the century. Nikita Demidov, the founder of the family's fortunes, had moved from the old metalworking centre of Tula to take over the Neviansk works in 1702. Two new works were built in the 1710s, but the next major surge in investment by the Demidovs coincided with the arrival of De Hennin at Ekaterinburg. Their famous Nizhnii Tagil works, for example, dates from 1725. By 1736 the combined output of state-owned and private works amounted to 8,000 tons, according to Nils Psilanderhielm, a Swedish prisoner of war, with the private works contributing more than half of the total.[279] And in the late 1730s, while the output of state-owned works marked time, output at the Demidov works started to soar.[280]

Technologically, the iron industry in the Urals had much in common with its Swedish counterpart. Both made malleable iron by the indirect technique, smelting ore in a blast furnace and then refining pig iron at forges. Both used the German forging method (the Walloon forges of Uppland being a technological aberrant in Sweden). Russian forgemen would, therefore, have worked in much the same way as the forgemen at the Gammelbo forges. However, there was a striking disparity in scale, both with respect to individual works and individual workshops. Russian iron making was undertaken on a much grander scale. Urals ironworks were often equipped with two blast furnaces rather than one, and the Russian forges were larger than the Swedish. The average number of hearths per forge in the Urals was five in the mid-eighteenth century, compared to the two hearths that were common in Sweden. Nils Psilanderhielm's report also revealed that Russian ironworks often had more than one forge. The Russian workforce was correspondingly larger. Another Swedish prisoner of war, Petter Schönström, noted in the 1720s that a master forgeman worked with a journeyman and a forge hand, and that two such crews interchanged, allowing around-the-clock production. The number of workers per hearth was thus

[279] Nils Psilanderhielm, 'Berättelse om Ryska och Sibiriska Jernwerken Ingifwen til Kongl. BergsCollegium den 8 Decembr. 1743', Bergskollegiets Arkiv, vol. D VI:8 Tilas Samlingar, RA.

[280] Hugh D. Hudson, Jr., *The rise of the Demidov family and the Russian iron industry in the eighteenth century* (Newtonville, 1986), chapter 3.

double that found at Gammelbo. The additional presence of a couple of supervisors per forge and some ancillary workers responsible for maintenance made for a far more elaborate division of labour in Russia.[281] The output at Russian forges, on the other hand, was very uniform. Every bar, Psilanderhielm noted at Contzoner, was 2½ inch wide and ½ inch thick, although of varying length. This was a system geared towards the production of large volumes of iron, but iron that came in only a limited range of sizes.[282]

Yet the most fundamental difference between Russian and Swedish iron making was not scale, but the different social foundations upon which production was based. In Sweden all parts of the productive process were undertaken by peasants or workers who enjoyed personal freedom. In Russia this was not the case. The making of bar iron for the international market was embedded within a feudal economy. The dramatic take-off of the Urals iron industry depended upon the mobilisation of a large workforce in a region that was sparsely populated. This was achieved through coercion. Although the possibility of employing free labour was sometimes discussed, compulsion was the central feature of the labour regime in the Urals. Everything from the felling of timber to the making of the finished bars was carried out by peasants or industrial workers who were legally bound to an ironworks estate. The Demidovs, like other private proprietors, relied upon serf labour, operating within what has been aptly termed a 'fear-factory', to make their iron.[283]

Until the 1730s the new metallurgical complex in the Urals was only loosely articulated with the international market for bar iron. The geographical barriers were formidable. The journey from the Urals to St Petersburg commonly took a year, sometimes longer. Iron

---

[281] Ågren, *Iron-making societies*, pp. 96–113, and Psilanderhielm, 'Berättelse om Ryska och Sibiriska Jernwerken'. For detailed description of work in Russian forges G.W. de Hennin, *Description of Ural and Siberian factories in 1735* (Washington D.C., 1992), pp. 190–230, should also be consulted. This is a translation of a Russian text of the 1730s that differs from 'General Liutenanten Hennings Relation om Ryske Bergwerkens' cited above.

[282] Psilanderhielm, 'Berättelse om Ryska och Sibiriska Jernwerken'.

[283] Hudson, *Demidov Family*, p. 57f.; Hugh D. Hudson, Jr., Bruce J. DeHart, and David M. Griffiths, 'Proletarians by fiat: the compulsory Ural metallurgical work force, 1630–1861', *International Labour and Working-Class History*, XLVIII (1995), 94–111; Hugh D. Hudson, Jr., 'Religious persecution and industrial policy in the reign of Anna I: V.N. Tatishchev and the Old Believers reconsidered', *Jahrbücher für Osteuropas*, L (2002), 22–36.

was transported by barge, using the vast river systems of Russia, but because the main river systems in Russia flowed along a north-south axis, rather than east-west, this involved a protracted, crab-like movement in which iron had to be disembarked and taken by wagon across watersheds. Many complaints were made about the 'Tediousness of the Way', and the necessity of having to 'wait for Floods and Rains at several shallow places'.[284] Much depended upon iron being ready at the riverside magazines when the spring thaw allowed water-born transport to get underway. If the moment was lost, Joachim von Ditmer, the Swedish envoy in St Petersburg noted in 1729, iron might have to remain in store until the following season.[285] A memorandum by von Ditmer in 1730 reported that 8,872 tons of bar iron had been brought to St Petersburg in that year. This amounted to two year's production, as no iron at all had been shipped to the capital in the previous year because of transport difficulties. Matters improved markedly in the 1730s as a giant canal network, linking St Petersburg to Novgorod and thence, via the river Msta, to the Volga basin, was finally brought to completion.[286] With this, the export of iron to western markets could begin in earnest.

The first record of Russian iron passing the Sound comes from 1716 when a parcel of just 36 tons was shipped to Britain. Shipments from St Petersburg remained tiny until the mid-1720s, and it was not until the 1730s that iron exports from the capital achieved real significance. They averaged 2,612 tons annually between 1730 and 1739, as opposed to 711 tons annually in 1720–29. Then, after some slackening in the pace of growth in the 1740s, came a fresh surge in the 1750s, one that continued through the 1760s, so that by the 1770s the annual export of bar iron from St Petersburg averaged 27,840 tons.[287]

The export of iron was handled by foreign merchants. They had the knowledge of overseas markets that native Russians lacked. In the early phases of St Petersburg's development it was Dutch factors who were dominant, but by 1730 the British had assumed control; so much so,

---

[284] Quoted in S.J. Newman, 'Russian foreign trade, 1680–1780: the British contribution', (University of Edinburgh Ph.D., 1985), p. 48.

[285] 'Extraordinarie Envoyen Joachim von Ditmers Swar til Kongl. BergsCollegium d. 21 aug 1729', Bergskollegiets Arkiv, vol. D VI:8 Tilas Samlingar, RA.

[286] 'Extraordinarie Envoyen Joachim von Ditmers Bref til Riksrådet och Presidenten Grefwe Arved Horn 3 dec 1730', Bergskollegiets Arkiv, vol. D VI:8 Tilas Samlingar, RA.

[287] Small quantities of bar iron were also shipped via Narva and Riga.

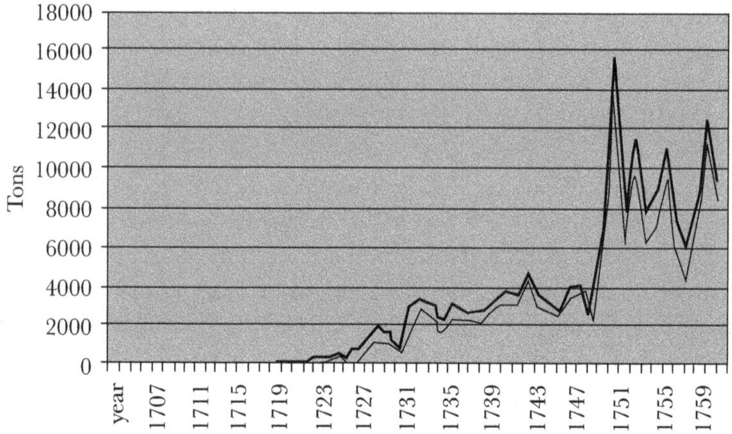

Source: Nina Ellinger Bang & Knud Korst, *Tabeller over Skibsfart og varetransport gennem Öresund 1661–1783* (2 volumes, Köbenhavn 1930–1945).

Figure 2.13. The export of iron from St Petersburg through the Sound, 1710–1760.

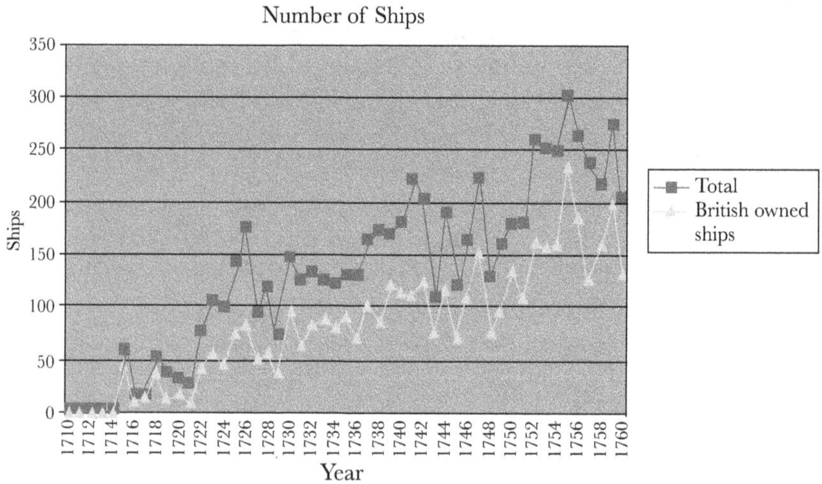

Source: Nina Ellinger Bang & Knud Korst, *Tabeller over Skibsfart og varetransport gennem Öresund 1661–1783*, Vol 1–2, Köbenhavn 1930–1945.
Note: In this graph figures for registered British ships have been used instead of ships destined for Britain. These figures match almost exactly; the discrepancy is only a couple of ships a year.

Figure 2.14. The number of ships leaving St Petersburg for the Sound, 1704–1750.

that the merchant quarter to the west of the Admiralty became known as the 'English Embankment'. As the English (and Scots) took control so the destination of Russian iron shifted. The Dutch market had once been of prime importance, and the Mediterranean not negligible, but it was the British market that reigned supreme after 1730. English ports had received just 28 per cent of St Petersburg's iron exports in 1720–24, and 44 per cent in 1725–29, but the English share reached 75 per cent in 1730–34 and then 80 per cent in 1735–39. That share was to remain above 80 per cent for the remainder of the eighteenth century.[288]

Some trade routes are very ancient. That taken by the 'Baltick Fleet' in 1746 was not. It was an eighteenth-century novelty, not the Silk Road. It was the wilful creation of Peter the Great. It was he who wrenched Russia's trade with the west away from Archangel, and he who in 1723 ordered the old-established British merchant community in Moscow to decamp to his new city on the Neva. It was some time, however, before Peter's city achieved a degree of commercial maturity. It was not until the 1730s that the export of iron from the city's quays became sizeable, or that shipments of flax and hemp achieved parity with those passing through Riga. Peter's iron industry in the Urals showed a similar pattern of development. It was founded at the dawn of the eighteenth century, but it was not until the second quarter of the century that Russian iron was drawn into the whirl of commerce that found its centre in the British Isles.

Graffin Prankard was always careful to specify the physical form that 'Orground' iron or voyage iron was to take. If voyage iron was not of the correct dimensions it was effectively unsaleable; and if 'Orground' was to be acceptable to steel makers it had to be struck into broad bars. With Russian iron he was less particular. He could not, after all, do much to affect production patterns in the Urals. The extension of credit to Swedish *brukspatroner* allowed Prankard to impose certain conditions upon them, but the feudal magnates who presided over the massive *usines* of the Urals were impervious to such pressure. Moreover, Russian iron took such a long time to arrive on the British market that any attempt to convey consumer preferences to Russian forgemen would be fruitless.

---

[288] Calculated from figures given in Newman, 'Russian foreign trade', pp. 283–85.

The Urals ironworks did not produce voyage iron, nor did they make the finer sorts of squares; they made broad bars. 'Russian iron is generally in bars, about 3 inches broad & ½ inch thick—not square at the end, but swelled and rounded'.[289] Iron from the older centres of iron making in central Russian such as Tula might not even come in the form of bars. A parcel of Russian iron forwarded to Prankard in 1730 was of 'but short lengths ye greatest part but abt 6 foott long & some of it in pieces about 1½ foott or 2 foott long'.[290] Such irregularity was not necessarily a drawback, however, for most of the Russian iron that was imported to Britain was destined for slitting mills, to be rolled and slit into nail rods. It was the character of Russian iron, not the crudity with which some of it was shaped, that interested iron merchants. Some brands, like the 'Old Sable' made by the Demidovs or the 'Government Siberia' that carried the imperial double eagle as its mark, were tough in the manner of Swedish iron. 'Hard stubborn iron', was how one assessment of these top brands ran, that 'works hard when hot, bad for joining or welding to steel—[but] durable, used chiefly for horse shoes and coach tire[s].'[291] Most brands of Russian iron, however, were made from phosphoric ironstones and therefore tended to brittleness. They were, in the parlance of the trade, 'coldshort'. Coldshort irons were not suitable for smithing purposes where durability was a desideratum, but they were eminently suitable for the manufacture of nails. Tensile strength was highly desirable in an anchor, but of far lesser value in a nail. Besides, the effort of fashioning a nail from so resistant a material as 'Orground' iron added materially to the cost of production. Coldshort iron, which was far more readily cut and trimmed, made cheap nails.

The nail trade was of colossal importance in eighteenth-century Britain. In a world where timber was a ubiquitous building material, nails were required by the million. Nails were also required in huge variety. 'The Wholesale Dealers in Nails have found it necessary to distinguish them into GENERAL and SPECIAL', said one authority. 'Under the *General* Sorts of Nails, they comprehend, 1. *Brads*; 2. *Hobbs*;

---

289 CCL, MS 3.250/4. Compare Prankard's assessment: 'Siberia Iron is Gennerally Stout drawn about 2½ to 2¾ & 3 Inch [wide]'. GP to David Skinner, 28 February 1730. Daniel Tilas reported that the state-owned ironworks made only four different sorts: 3 × 1½, 2½ × ½, 2 × ¼ and 1¾ × ⅜ inch. Bergskollegiets Arkiv, vol. D VI:8 Tilas Samlingar, RA.

290 GP to Francis Homfray, 3 November 1730.

291 CCL, MS 3.250/4.

and 3. *Nails*.' There were three basic types of brad, five sorts of hobb, and twelve varieties of nail.[292] Each of these came in several subvarieties. Standard nails were differentiated by their weight per thousand: a thousand '4 pundy' nails, for example, weighed 4lb.[293] In the Dudley district, so Reinhold Angerstein noted, the range of ordinary nails stretched from the delicate 2lb nail, which earned the nailer 1*s.* per thousand, to the weighty 20lb nail, which brought a nailer 5*s.* 3½d per thousand.[294] Then there were the special nails, a category that included such monsters as the ribbing-nail used in shipbuilding, 'from 5 to 10 Inches long', that were so laborious to make that they were priced by the hundred rather the thousand. The Crowley nailing works at Winlaton near Newcastle upon Tyne, the most extensive of its kind, no doubt made the widest selection. When inventoried in 1728 there were 154 varieties of nail in stock.[295]

Much of this profusion was absorbed by the domestic market, one bouyed by strong urban growth, but it was overseas markets that were gaining in prominence in the first half of the eighteenth century. The export of nails from England and Wales grew from 542 tons in 1700 to 1848 tons in 1750.[296] European sales were of negligible importance; it was transatlantic markets that were critical. The Thirteen Colonies and the Caribbean sugar islands took between 85 and 95 per cent of exported nails in the second quarter of the century, and it was these markets that received the attention of nail manufacturers in the specialised production zones of the English Midlands and North. In the south Yorkshire nailing district, where workers divided their time between nailing and agriculture, specific seasons were dedicated to different colonial markets: 'the men worked from March to August on making clasp nails for London. During the harvest, nailmaking stopped, but then during the autumn flat points were made for Virginia until

---

[292] Hoppus, *Practical measuring*, p. 187. See also Richard Neve, *The city and country purchaser's and builder's dictionary: or, the complete builder's guide* (3rd edition, 1736), *sub* 'Nails'.

[293] In fact, this would have been the weight of 1200 nails, as the nail trade used the 'long thousand'.

[294] *Angerstein*, p. 179.

[295] Suffolk Record Office (Ipswich), HAI/GD/5/14: 'Nailes in Robt Walls hands'. The Crowley works used 37 diffferent symbols when bagging nails just to indicate the type, many of which came in several sizes: BL, Add. MS 34555, pp. 72–75, 'Order No. 27'.

[296] Elizabeth B. Schumpeter, *English overseas trade statistics 1697–1808* (Oxford, 1960), table xxv.

Martinmas, then sharp points were made for the Leeward Isles and Jamaica until it was time to till the soil again'.[297]

Grafffin Prankard, from his vantage point at Bristol, was ideally placed to supply these transatlantic markets. He had close links with West Midland ironmongers like Francis Homfray, who put out rods of Russian iron to domestic nailers in his district on Prankard's behalf. Prankard, with his experience in the Atlantic trade, was able to keep Homfray abreast of colonial demand: 'as to the 4d nails send me as many as the[e] Canst this Spring now Coming on... also keep on the Hands on 14 [pundy] and 22 [pundy] of Clasp Nails to Send me as many as thee Canst of those Sorts with 10, 15 or 20 Baggs of any other Sizes, Clasp Nails that the[e] know to be Saleable at Carolina'.[298] When the *Parham* or the *Baltick Merchant* sailed for Charleston they invariably carried a heavy load of nails. Indeed, the *Baltick Merchant* was loaded with over 2 million nails when she sailed for South Carolina in 1736, together with gunpowder, English and German steel, hoes, ox chains, whip saws, files, and 10 tons of bar iron.[299] For those who were attuned to colonial demand this was a lucrative market. (For those who were not, it could be less rewarding: 'You have sent too Large a Quantity of sheathing nails & some sorts which are but Little Us'd here', one Charleston merchant told his London correspondent, '& [they] therefore lye on hand Unsold.')[300]

The technology of nail making was varied. At one extreme was the slitting mill, a water-powered facility of some complexity; at the other, the simple stone anvil of the common nailer. Of the two, it was the slitting mill that held the key to Britain's pre-eminence in nail-making. Before the introduction of mechanised slitting in the late sixteenth century nail rods had to be split by hand. This was a tedious and costly operation, the need for which was obviated by the adoption of the water-driven methods that had been developed in the Liège region *c.* 1500. Bars of iron were cut into lengths of about one foot each at mechanically powered shears. These lengths were then brought to a red heat in a coal-fired reverberatory furnace and rolled flat. The flattened iron was immediately passed through the slitting rolls whose

[297] David Hey, *The rural metalworkers of the Sheffield region: a study of rural industry before the Industrial Revolution* (Leicester, 1972), p. 34.

[298] GP to Francis Homfray, 6 November 1733.

[299] SA, DD/DN 448.

[300] Walter B. Edgar (ed.), *The letterbook of Robert Pringle. Volume 1: April 2, 1737–September 25, 1742* (Columbia SC, 1972), p. 371.

steel-edged cutters sliced it into long, curling strands, some four to five feet in length. Each length of iron was slit into eight rods, or so an eye-witness who saw the process at Sampson Lloyd's Birmingham mill in 1755 reckoned.[301] As for the thickness of the rods, the gauge of the cutters could be altered to produce the desired effect. At the Crowleys' Winlaton slitting mill, for example, rods were made in thirteen different gauges from $^{3}/_{16}$ of an inch to 1 inch in diameter.[302]

This was a capital-intensive rather than a labour-intensive process. Just four men were required: the master roller, his assistant ('the middleman'), a furnaceman, and a youth ('the drawer') to straighten and bundle the rods. The furnaceman at Lloyd's mill in Birmingham in 1749 was, Samuel Schröder reported, 'a black slave'. He introduced the red-hot iron to the rollers. The middleman fed the iron through the slitting rolls, and the drawer took up and sorted the rods. 'These men are paid 15 to 18 pence a day, apart from the black one, who as a slave receives nothing more than food and clothes.'[303]

Each team of workers was expected to perform five 'heats' in the course of a shift, each of seven hundredweight.[304] Given a good supply of water, a rolling mill working two shifts daily could turn out close on 20 tons of rods a week. But as mill owners knew well, water was not consistently available. Angerstein reckoned that Lloyd's mill processed 17 tons of iron weekly or 600 tons annually, implying that the mill was in use for only 35 weeks in the year.[305] Indeed, slitting mills were often idle during the summer. It was only the coming of autumn rains that allowed the rolls to turn, and sometimes it was only in the depths of winter that storage ponds were sufficiently full to allow round-the-clock working. 'In time to come', the proprietors of Cramond mill told their manager, 'we expect you will cutt 400 Rod Iron & 100 Tons Hoops' per year', distinguishing between the work that could be done by a 'single sett of 4 men' and that done with the 'the assistance of 4 Addition[al] Men [for] 3 months a year.'[306]

---

[301] H.R. Schubert, *History of the British iron and steel industry from c. 450 B.C. to A.D. 1775* (1957), p. 310.

[302] BL, Add. MS 34555, p. 21.

[303] Schröder I, fo. 161.

[304] National Archives of Scotland, GD58/6/1/1, William Cadell to John Lee, 31 March 1760.

[305] *Angerstein*, p. 180.

[306] National Library of Scotland, Cadell of Grange papers, 5381/31, 'RC & G' to Thomas Edington, 3 April 1766.

Slitting mills were few in number in early eighteenth-century England, perhaps no more than thirty, and unevenly distributed.[307] They were clustered along streams that could provide a suitable head of water; they were usually adjacent to coal measures, the essential source of heat energy for the nailer; and they were usually to be found where a large potential workforce was congregated, for nailing, unlike slitting, was profoundly labour intensive. The largest concentration of mills was in the West Midlands, along the river Stour. The south Staffordshire coal measures outcropped just to the east, and by happy chance the social structure and characteristic agrarian practices of the coalfield parishes were conducive to the growth of an industrial workforce. The south Staffordshire plateau was an area of largely poor soils, abounding with unenclosed wastes and heathland. This was not an environment in which arable farming flourished. Indeed, pastoralism was the key feature of the region in the early modern period. It had been colonised rather late in the middle ages, so the communal openfield agriculture that was the hallmark of English manorialism elsewhere was not well-established. The absence of a robust manorialism meant that landholdings were easily splintered and that manorial lords were unable to prevent tenants digging coal for their own use.[308] Similar conditions prevailed in the nailing district of south Yorkshire and north Derbyshire. Coal was present in abundance, whilst woodlands had impeded the spread of arable farming. Mixed farming was practised, with an emphasis on stock-rearing and dairying. Because pastoralism did not make the same insistent demands on labour as arable farming, there were periods of the week, or even seasons of the year, in which farmers and smallholders were free to earn additional income through metalworking.[309]

Nailing hearths were easily built. 'The hearth or fireplace is a *massive* of brick, about 2 ft 6 in. high: the back of the forge is built upright to the ceiling and is enclosed over the fireplace with a hovel which leads

[307] There is no contemporary listing of slitting mills before the 1790s. Forty-eight are mentioned in the 'List of the different Iron Works in England, Wales, Scotland & Ireland to the 1794' (BCA, Boulton & Watt MSS, MII/5/10). Of these, five are definitely post-1750 and several others, whose date of construction is not given, belong in the same category.

[308] Rowlands, *Masters and men*, pp. 4–8.

[309] Hey, *The rural metalworkers*. The region was home to variety of metal working specialisms: see Kathleen M. Battye, 'Scythe makers and other metal workers in the parish of Norton, 1533–1750', *Tools and Trades*, XIV (2005), 46–77.

into a chimney to carry away the smoke'.[310] Simple hand bellows were used to bring the coals to a proper heat, whilst a water trough was kept close by 'to wet the coals in and thereby increase their force; as also to quench the iron in'. None of this represented a massive investment, and nailing hearths could easily be accommodated in lean-to additions to cottages, or in simple purpose-built workshops. The hand tools used in nail manufacture—the hammer, the bickorn, the swage, and the anvil—were commonly valued at just £1 by probate appraisers in the early eighteenth-century Midlands.[311] Just as importantly, the trade was easily learned. Little skill was required, just a willingness to endure the numbing repetitiveness of the task. The nailer took a heated nail rod from the hearth and divided it into appropriate lengths over the sharpened edge of his or her anvil. With a few more blows of the hammer, the nailer achieved the desired cross-section. A few blows more, and the nail had been pointed. The semi-formed nails were then placed one after another in a hole in the anvil. The hammer was brought down smartly to press the protruding iron into a head, and with sufficient force to make the nail jump out of the cavity, leaving it vacant for the next.

All of this was the work of a few seconds. Angerstein reported that a Stourbridge nailer could 'work up two bundles of slit iron per week, with a total weight of 1 cwt which gives him an income of 1s 6d [*per diem*]'.[312] There was a considerable wastage of iron in hammering out nails, especially in the making of smaller nails. A nailer making '4 pundy' nails was expected to return 96lb of nails from the 120lb of rods issued to him. This was equivalent to 28,800 nails over the week, or 4800 a day. In other words, a nailer would spend less than 10 seconds on each nail.[313] As can be imagined, it was a gruelling occupation. 'The handles of some nailers' hammers which survive are evidence of this',

---

[310] Ephraim Chambers, *Cyclopaedia* (1741), quoted in Rowlands, *Masters and men*, p. 27.

[311] Rowlands, *Masters and men*, p. 27.

[312] *Angerstein*, p. 175.

[313] This assumes that the nailer worked for twelve hours daily. In reality, the intensity of work varied across the week. Nailers were notoriously devoted to Saint Monday, implying production at a furious pace on Fridays and Saturdays. Adam Smith gave a rather lower estimate of daily output, albeit for juvenile workers: 'I have seen several boys under twenty years of age who had never exercised any other trade but that of making nails, and who, when they exerted themselves, could make, each of them, upwards of two thousand three hundred nails in a day'. Adam Smith, *An inquiry into the nature and causes of the wealth of nations* (1776), ed. Andrew Skinner (1970), p. 113.

as Marie B. Rowlands has remarked. 'The impression of the fingers and thumb is worn so deep into the handle that barely half an inch of wood remains after a lifetime of use'.[314]

The organisation of the nail trade varied from region to region. In the West Midlands it assumed a classical proto-industrial form. Nail-making was organised on a putting-out basis, with chapmen making use of a dispersed, often part-time workforce. The pattern of production can be glimpsed through the accounts of the Knight partnership, the most powerful iron making concern in the Midland in the 1730s.[315] The Knights ran forges at Wolverley, Cookley, Whittington, and Mitton in the Stour valley that were capable of turning out as much as 2000 tons of bar iron annually. The bar iron produced was processed at independently owned slitting mills.[316] Slitters made a charge for slitting the bars that were put out to them—Sampson Lloyd charged 16*s.* per ton at Birmingham—and returned the rods to the Knights.[317] The Knights then sold the rods on to the wholesale ironmongers who dealt with the thousands of rank-and-file nailers.

Several dozen ironmongers did business with the Knight partnership: between 50 and 75 each year in the 1730s and 1740s. Some were petty chapmen who bought less than five tons in the course of a year. Others were clearly major employers, supplying dozens of nailers at a time, who bought between 100 and 200 tons annually. The smaller chapmen would operate from their own shops or perhaps from rented space at a local inn. The larger dealers would maintain warehouses where a salaried clerk would issue bundles of rods and take in the bags of completed nails. The Crowleys' warehouse at Stourbridge, for example, was a substantial and well-equipped depot. When inventoried in 1728 it contained 131 tons of 'Common tough Rodds', plus two tons of the

---

[314] Rowlands, *Masters and men*, p. 26.

[315] Worcestershire Record Office, Knight MSS, 899:310, Stour works general accounts. For context see Laurence Ince, *The Knight family and the British iron industry, 1695–1902* (Birmingham, 1991), and for a close analysis of production patterns see Göran Rydén, *Production and work in the British iron trade in the eighteenth century: a Swedish perspective* (Uppsala, 1998).

[316] The partnership only acquired its own in-house slitting capacity in 1746 with the building of a mill at Nechells Park. This was part of an important expansion in the Birmingham area, coinciding with the purchase of Aston furnace and the forge at Bromford.

[317] *Angerstein*, p. 180.

more expensive 'Best Tough Rodds' that were used for making rivets, horse shoe nails, and other specialised products.[318]

Nailers took iron on credit. Every Saturday they would collect a bundle or two of nail rods from their chapman's shop. They would return a week later with the bags of nails they had completed during the week. From the chapman's point of view this arrangement had certain advantages. The nailers provided their own tools and procured their own fuel. There were no overheads in the form of plant that the chapman had to cover. There were, though, as in all forms of putting-out, problems of quality control. When the manufacturing process was dispersed across different parishes and chapelries there was no opportunity for supervision. Hence the regularity with which chapmen had to complain about nails that lacked heads or points. From the nailers' perspective, putting-out had something to recommend it. They could take advantage of the chapman's credit rather than advance money of their own. Plus, they retained some control over their work routine, allowing them to seize other earning opportunities as they arose. Yet there were grave disadvantages as well. The independence of the nailer was more formal than real. 'In Staffordshire and other Nailing Countreys it is Usual With all Buyers of Nailes to Oblige the Workmen to take a Certain Quantity of Iron thereby to prevent their buying of Iron or Working other Iron than they received of the Master that Employed them'.[319] This subordination was compounded by a tendency to indebtedness. All nailers took advances in the form of nail rods. By returning an appropriate weight of nails they could redeem their debt, but if they exceeded the permitted wastage or returned defective nails they were penalised. Simple human fallibility meant that such penalties were easily incurred and that a nailer's debt could not always be discharged in full. Because nailers were so often men and women of meagre resources they had few ways of cancelling their debts and so indebtedness became a chronic condition.

In the North East of England an entirely different organisational pattern prevailed. Here, the centralised manufactory was predominant. The symbiotic linkage between domestic manufacturing and part-time agriculture that had stimulated proto-industrial nailmaking in the West

---

[318] Suffolk Record Office (Ipswich), HA1/GD/5/11, 'An Inventory of goods in the warehouse at Stourbridge'.

[319] BL, Add. MS 34555, p. 174.

Midlands and south Yorkshire was absent. The North East's coastal plain, far from being a zone of splintered landholding and weak manorial control, was characterised by large estates and intensive agriculture. The North East did have coal in abundance, however, and well-established maritime links with London. It was these features that attracted the attention of Ambrose Crowley, the effective founder of the North East's nail trade, in the 1680s. Crowley, the son of a Stourbridge hardware manufacturer, was apprenticed to a London ironmonger and lost little time in establishing himself as a major dealer in nails and other ironwares in the capital. Crowley, a man of unquenchable energy and ambition, depended upon the ironmasters and ironmongers of his native region for supplies. Inevitably, given Crowley's congenital impatience, their performance in this area did not meet with his approval, so he resolved to take matters into his own hands and manufacture his own nails. He did so in dramatic fashion, transferring the best part of his business to the North East. He built a factory at Sunderland in 1682, where over 100 workers were soon employed. He switched to a larger site at Winlaton in the Tyne valley in 1691, and founded an additional, far larger Tyneside plant at Swalwell in 1707.

The institutional and agrarian preconditions for proto-industrial nailmaking were not to be found in the North East, but Crowley could console himself with the thought that living costs in the North East were substantially lower than in the West Midlands ('Vitalls is above 1/3 cheaper than in the present naill cuntry'), enabling his workmen to compete with the low-waged, semi-rural manufacturers of south Staffordshire. Iron could be imported from Sweden or the Low Countries, or taken from domestic producers in Yorkshire, Nottinghamshire, and Derbyshire whose works were accessible via the Humber and the Trent.[320] More telling, perhaps, was the opportunity for monitoring and disciplining sloth and workplace fraud that the gathering together of workers under the surveillant eye of the nailmaster afforded. Indeed, Crowley was such a martinet that he wrote his own 'law book', prescribing in minute detail the procedures that were to be followed in his works.[321] The slitting mill that Crowley built in the last years of the seventeenth century at Winlaton Mill, a riverside site below the hilltop

[320] Flinn, *Men of iron*, pp. 34–42.

[321] M.W. Flinn (ed.), *The law book of the Crowley ironworks* (Publications of the Surtees Society, CLXVII, 1957). This is a partial transcription of the original document, which is BL, Add. MS 34555.

factory, had the capacity to make 500 tons of nail rods annually. 'Large quantities of these rods are consumed at Winlaton and Winlaton Mill', Angerstein noted, 'where more than 300 workers always are at work'.[322] The mill at Swalwell was 'used partly for rolling heavy sheets and partly for slitting nailrods'. Most of the latter were worked up in on-site workshops, two for large ship nails, and 'ten for smaller nails'.[323]

The Crowley works provided an organisational template from which other nailing enterprises in the North East were struck. An additional slitting mill was built in the Tyne valley *c.*1719—at Teams, just downstream from Swalwell—then another at Bebside in the Blyth valley in 1736. The Bebside site is comparatively well-documented, so the influence of the Crowley model of centralised manufacture can be readily seen. Founded by William Thomlinson, a Newcastle merchant, the Bebside mill soon passed into the hands of Harrison, Bannister & Hallett, the London-based firm in which Josias Wordsworth, the noted Baltic merchant, was a partner.[324] It comprised 'a Slitting Mill, and Nailors Shops, with several Houses, Warehouses, and other buildings'.[325] A later survey clarifies the layout of the site. The slitting mill was, naturally enough, at the riverside. The nailing shops, which overlooked the deeply incised river valley, were in three parallel terraces: North Row, Middle Row, South Row. Since each terrace comprised 'Six rooms below, and Six above', there was room for 36 nailers to be at work simultaneously. The finished nails were sent by lighter to the port of Blyth, two miles downstream in tidal waters.[326]

From Blyth, Bebside nails entered the coils of the international market. They were shipped to Harrison, Bannister & Hallett's London warehouse. Or rather warehouses, for the firm had goods secreted at a variety of sites in the City of London and, far more extensively, at Deptford. The 'nail warehouse' at Deptford contained goods to the value of £1,026 when inventoried in 1751, and it was just one part

---

322 *Angerstein*, p. 265.

323 *Angerstein*, pp. 260–61.

324 Northumberland Record Office, ZAN M13/C7, Revd R. Thomlinson to J. Thomlinson, 13 March 1739, and to W. Thomlinson, 8 November 1740; TNA: PRO C11/822/3. The other partners included William Harrison (d. 1745), the Wealden ironmaster and gunfounder, (for whom see Henry Cleere and David Crossley, *The iron industry of the Weald* (Cardiff, 1995), pp. 200–13), and John Bannister (d. 1743) the former general manager of the Crowley empire.

325 *Newcastle Courant*, 10 March 1750.

326 Northumberland Record Office, ZMD 66/46, terrier for survey of Bebside, 1771.

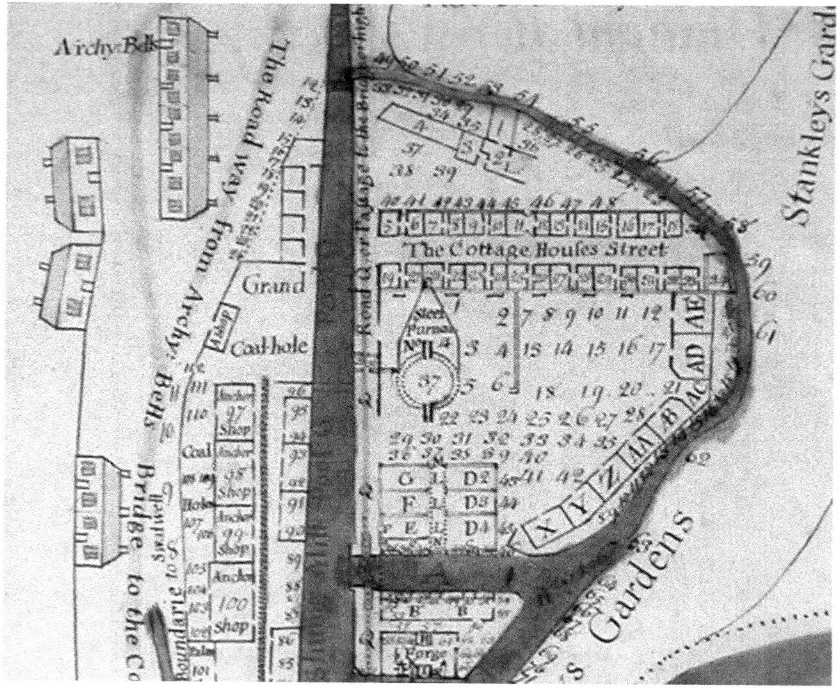

Illustration 2.26. Swalwell works.

Courtesy of Tyne and Wear Archives Service (DX104/1).
Caption: When inventoried in 1728, which may have been the occasion for the drawing up of this map, the plant included two steel furnaces (although only one, 'No. 4', with its conical chimney, is shown here), associated forge hammers, a slitting mill, a blade mill, four anchor shops, air furnaces valued at £100, three warehouses, five hoe makers' shops, and shops for the making

of a complex of riverside storehouses, shops, and garrets used by Crowley Hallett & Company, the successor firm to Harrison, Bannister & Hallett.[327] The great southward loop of the Thames as it rounded the Isle of Dogs was the point of departure of much of the hardware manufactured in the English provinces. Samuel Tossick, the London merchant who acted as agent for the nail business of William Spencer, the Yorkshire ironmaster, also kept a warehouse at Deptford.[328] Just a few hundred yards downstream was the Crowley depot at Greenwich. This was a colossal facility. When inventoried in 1728 it required 138 densely packed pages to itemise the stock, which was valued at over £48,000.[329] A little further downstream, on the opposite bank of the river, lay the East India Company's shipyard at Blackwall from where Swedish iron and English steel were dispatched to Bengal and the Coromandel Coast.

The warehouses that lined the lower reaches of the Thames contained goods from both the North East and the West Midlands, for the major export houses did not restrict themselves to one source of supply. Harrison, Bannister & Hallett's Deptford stores contained casks of nails from Midland ironmongers like the Homfrays of Stourbridge, the Finches of Dudley, and the Molineuxs of Wolverhampton, as well as material from their Bebside factory.[330] Similarly, the Crowleys drew upon outworkers in the West Midlands, maintaining their own warehouses at Wolverhampton, Walsall, and Stourbridge. The product was identical whether it was made in Wolverhampton or Winlaton. What divided the nail trade was the form of discipline imposed upon the workforce. In the North East an authoritarian model prevailed in which centralised factories were the norm. The nail factory had no technical advantages over the simple nailing hearth of the out-worker. It was, after all, nothing more than an agglomeration of such hearths; nailmaking remained non-mechanised long into the nineteenth century. Nor did it save on the costs of distributing and collecting materials, for these costs were

---

[327] BL, OIOC, MSS Eur F 218/115.

[328] Hey, *Rural metalworkers*, p. 43.

[329] Suffolk Record Office (Ipswich), HA1/GD/5/1, 'An Inventory of the Goods which were at Greenwich at the Decease of John Crowley Esq Jany 2 1727/8'. The Crowleys also had goods to the value of £10,924 stored at six different locations in the City.

[330] The Homfrays, Finches, and Molineuxs all bought rod iron from the Knight partnership in the 1730s and 1740s. John Finch and John Finch junior also bought Russian iron from Graffin Prankard, as did (on a far larger scale) Francis Homfray and his widow Mary Homfray.

usually borne by the nailer who made the weekly journey to and from the chapman's shop. The factory did, however, present its proprietor with the possibility of exercising a more exact supervision of the work process. It afforded an opportunity to monitor workplace malpractice more closely, and to regularise working patterns.

In the West Midlands it was debt bondage that underwrote the authority of the nailmaster. Nailers could enjoy little freedom of action when they were inescapably beholden to their chapmen.[331] There was but one way in which nailers could reduce the burden of debt that oppressed them. That was to demand a higher price for their product, and to do so collectively and in the most direct of ways. Nailers, the House of Commons was told in 1738, were 'continuously rising in a tumultuous manner' in the West Midlands, mobbing the houses and warehouses of their employers. *Aris's Birmingham Gazette* gave an account of one such tumult in 1745. The nailers 'rose up in a very considerable number and went to the masters hereabouts and obliged them to give them money and to sign an article to raise the price of nails.'[332] But nailers were also aware that a greater power stood behind the nail chapmen: the ironmasters who furnished the rod iron.

The Midland ironmasters were a cohesive body, bound together in a sequence of interlinked partnerships and animated by a certain *esprit de corps*.[333] They had a long tradition of collective organisation, one made visible in their regular meetings at Stourbridge. Here, said the Swedish traveller Kahlmeter in 1725, ironmasters met every month 'to confer on their business affairs and interests, and to agree upon the division of the market for their iron'.[334] They set price schedules not just for bar iron but for rods as well, even though the slitting of rods

[331] Admittedly, many nailers were so impoverished and the costs of recovering debts so burdensome that chapmen often had to write off debts. 'Robert Chambers of Smethwick', we are told, 'was in a small way of business, described as a petty chapman. In the year before his death, 1727–8, he bought two tons of rod from the Stour mills. At his death his neighbours listed the names of twenty-one nailers who owed him small sums. The total was £16 19s, that is to say, just about half his year's bill for rod iron. The praisers wrote off the whole of these debts, amounting to a considerable proportion of his trading turnover, as "mostly desperate".' Rowlands, *Masters and men*, pp. 81–82.

[332] Quoted in Rowlands, *Masters and men*, p. 83.

[333] Chris Evans, 'The corporate culture of the British iron industry 1650–1830', in Göran Rydén (ed.), *The social organization of the European iron industry 1600–1900* (Stockholm, 1997), pp. 121–46.

[334] Quoted in Hildebrand, 'Foreign markets for Swedish iron in the eighteenth century', 28.

was often undertaken by independent mill owners. Edward Knight, when announcing an increase of 20 shillings per ton in the price of Midland iron in 1731, stipulated the new market rate for five grades of bar iron *and* for the rods slit from them:[335]

Table 2.5. Bar and rod iron prices in the West Midlands, 1731.

	Bars	Rods
Best Bars	18.16.0	20.10.0
Forrest Bars	17.16.0	19.10.0
Ord[inar]y Bars	17.16.0	19.10.0
Blend Bars	17.07.6	19.00.0
Coldshort	16.17.6	18.10.0

Source: Religious Society of Friends Library, Lloyd MSS, 210/1/86, Edward Knight to unidentified correspondent, 2 November 1731.

Control over the commodity rested with the ironmaster, to whom the slitters were no more than subordinate contractors. The ironmongers who bought the rods were thus confronted with a powerful price-fixing cartel, and so, in their turn, were the nailers who worked up the rods. This was something the nailers appreciated all too well, for their periodic protests were directed not just against their immediate employers but against the ironmasters as well. Nail rods, Angerstein explained in the early 1750s, were put out to West Midland nailers at £22 per ton.

> This has recently been increased by £1 10s. per ton by Messrs Knight and Spooner who in this country increase the price of iron as they please, which gives the workers, who do not get more for their labour, reason to groan and be angry. They recently sent a 'fiery cross' to Mr Knight at Wolverley, threatening to pull down the house that he has recently built, which cost him £5000, unless he agreed to sell iron at the old price. Due to this message, Mr Knight was compelled to have a guard around his house with loaded guns and cannons for two weeks, until the excitement cooled down but, in spite of this, the price remained the same.[336]

The ability of Midland ironmasters to maintain a high price level for nail rods was secure so long as they were the sole suppliers of bar iron in their region. That control was, of course, compromised by the

[335] Religious Society of Friends Library, Lloyd MSS, 210/1/86, Edward Knight to unidentified correspondent, 2 November 1731.
[336] *Angerstein*, pp. 175–76.

availability of iron from the Baltic, but until the 1720s the inflow of Swedish iron was not great and was largely directed to special uses such as steel manufacture, not to nail making. It was this that accounted for Edward Knight's confidence when announcing an increase in bar iron prices in the autumn of 1731: 'The Ironma[ste]rs do not seem to doubt of keeping up the Advance of 20s per tun in the Rod Iron provided there do's not come a greater Quantity of Forreign Iron into the Markett than we are yet appriz'd of'.[337] The influx of Russian iron in the 1730s, however, threatened a radical upheaval, for Russian iron, unlike Swedish, was eminently suitable for nailing. Iron from St Petersburg could therefore corrode the easy complicity that allowed English ironmasters to govern their market.

The actions of Graffin Prankard in introducing Russian iron into the Severn valley had large consequences. He started in a small way, buying up small odd parcels on the Dutch market in the late 1720s, but from 1730 he began to import directly from St Petersburg. Thereafter his sales of Russian iron saw a dramatic if discontinuous rise from little more than 100 tons in 1732 to over 600 tons in 1738. From the outset Prankard targeted Midland slitting mill proprietors like Francis Homfray and Sampson Lloyd. 'I presume thee art Sensible', he told Homfray in 1732, 'that I sold Sampson Lloyd a Large Parcel of Muller Fabrick Russia wch...is Deemed ye Mildest Collshire [coldshort] of all'.[338] Russian iron was not only cheap, but Prankard could offer the additional inducement of having it slit at Congresbury mill, just west of Bristol, before it was shipped into the water-short Midlands. The strategy bore fruit. Sampson Lloyd became the mainstay of Prankard's trade in Russian bars. He accounted for 64.2 per cent of Prankard's sales of Russian iron in the Midlands in the 1730s. The Homfrays took another 16.5 per cent.

As the flow of Russian iron gathered force the Midland ironmasters began to stir. They were, so it seemed, on the brink of losing their most valued market, that for nail rods. Prankard anticipated a cut in the price of 'ye Slitting English Iron' in 1733: 'they [the ironmasters] groans very much under ye Load of fforeign Iron & I am afraid will fall

[337] Religious Society of Friends Library, Lloyd MSS, 210/1/86, Edward Knight to unidentified correspondent, 2 November 1731.

[338] GP to Francis Homfray, 14 November 1732. See also GP to Sampson Lloyd, 19 March 1734.

it'.[339] Price warfare duly ensued. Yet price cutting was not an effectual response to imports that might be priced at under £13 per ton, so other strategems to nullify the effect of Russian iron were set in motion. In 1736 Prankard claimed that a 'Set of Men yt Envey mee' had launched a scheme for 'Ingrossing all ye Comm[on] Russia Iron in Muscovy for a long term of years wch if they Succeed in must be prejudicial to private traders'. The authors of this (abortive) initiative were, he claimed, 'the Iron Masters in Worcester and Stafford Shire and thereabout', intent upon preventing Russian iron arriving on the British market at a rate which undercut their own product.[340] This audacious manoeuvre came to naught, but it signalled the opening of an era of controversy and confusion in the international iron trade. The emergence of Russian iron onto the world market posed serious problems not just for British ironmasters; it threatened Sweden's hegemony in northern Europe and beyond. As result, in the 1730s and 1740s ironmasters, merchants, and policy makers in both Britain and Sweden pondered new strategic directions for their trade. At stake was command of the British market and with it the best part of the Atlantic basin.

Charleston

When the *Baltick Merchant* sailed for Charleston in 1735 she was carrying 70 casks of nails, containing more than two million nails of various sorts. Nearly 500 bars of Swedish iron had also been lowered into her hold, together with bars of German steel and faggots of English steel. Whip saws, saw files, ploughshare moulds, hoes and gunpowder completed the cargo.[341] This was an extraordinarily utilitarian consignment. There was nothing modish or ornamental: no ceramic wares, no fine furniture, no glassware, no millinery, and no fabrics; none, in fact, of the consumer goods that were routinely despatched to the Chesapeake or the Delaware. The goods listed on the *Baltick Merchant*'s manifest marked Charleston out as a place apart.

---

[339] GP to William Vigor, 5 March 1733.

[340] GP to 'Respected Friend', 3 January 1736, and to William Vigor, 10 January 1736.

[341] SA, DD/DN 448.

Charleston—or Charles Town as it was known to its colonial inhabitants—in the early 1730s was a town of some 4,500 inhabitants. Situated on a tongue of land at the confluence of the Cooper and Ashley rivers, it was the commercial centre of South Carolina. It was British North America's fifth largest city, some way behind Boston (13,000 inhabitants), Philadelphia (11,500) and New York (8,600), but neck-and-neck with Newport, Rhode Island. The picture that the colony's propagandists painted of Charleston was one of order, godliness and prosperity: 'There are between 5 and 600 Houses in Charles Town, the most of which are very costly; besides 5 handsome Churches, viz. one for those of the Church of England, one for the Presbyterians, one for the Anabaptists, one for the Quakers, and one for the French.'[342] 'The Inhabitants', another booster trumpeted, 'by their wise Management and Industry, have much improv'd the Country, which is in as thriving Circumstances at this Time, as any Colony on the Continent of English America'.[343] Wealth there was, but it had been born of violence and ruthless expropriation, not order.

The years following the foundation of South Carolina in 1670 were years of carnage. The earliest English settlers had come to the area from Barbados. Conscious of the spread of a sugar monoculture in the West Indies and the demand that it generated for labour, the English were soon encouraging the Native Americans with whom they traded to raid neighbouring communities for slaves. This triggered a long series of Indian wars that furnished a steady supply of captives for the plantations of the Caribbean and resulted in a massive depletion of the indigenous population.[344] Intertwined with these bloody developments was a growing trade in deerskins, supplied by Native American hunters and eagerly awaited by European leather workers. The process reached its savage apogee in the Yamasee War of 1715–16 that left thousands of acres denuded of human inhabitants.[345]

---

[342] 'A description of the Province of South Carolina, drawn up at Charles-Town in Sept. 1731', *The Gentleman's Magazine*, XX (August 1732), p. 896.

[343] John Lawson, *A new voyage to Carolina* (1709), p. 2.

[344] Alan Gallay, *The Indian slave trade: the rise of the English empire in the American south, 1670–1717* (New Haven CT and London, 2003).

[345] Daniel K. Richter, 'Native peoples of North America and the eighteenth-century British empire', in Marshall, *British Empire*, pp. 352, 360. See also Richard L. Haan, 'The "trade do's not flourish as formerly": the ecological origins of the Yamassee War of 1715', *Ethnohistory*, XXVIII, 4 (1982), 341–58.

As the coastal lowcountry was emptied of its native residents it was re-populated with a new racial group and dedicated to the production of a new commodity for international markets. The commodity was rice, cultivated by African slaves. Experiments in the growing of rice had begun in the 1690s as local planters sought a staple crop that would bring them the fabulous wealth that sugar had brought to their counterparts in the West Indies. Climatic and environmental conditions were not so favourable that Carolina growers could raise cane to compete with that of the sugar islands (nor cultivate tobacco to match that of the Chesapeake), but the lowcountry's abundant swamps lent themselves to the planting of rice. By the 1710s the crop was the critical element in the local economy. Rice never achieved the importance of sugar or tobacco in the wider Atlantic economy, but it revolutionised life in Carolina, making the province the richest in British North America.[346] South Carolina also became home to the American mainland's most brutal slave regime. It was not coincidence.

Rice cultivation was enormously labour intensive. The conversion of marshes into rice fields could only be accomplished through an injection of African labour, for white servants were in short supply. English migrants found life in the Chesapeake, harsh though it often was, much preferable to the exhausting routine of planting, harvesting and processing that rice imposed on its growers in the Carolinas. Field hands were condemned to endless labour with the hoe, breaking up the soil and clearing weeds.[347] The work was 'peculiarly unwholesome, and even fatal to health'. Slaves had to stand 'ancle, and even mid-leg deep in water... exposed all the while to a burning sun, which makes the very air they breathe hotter than the human blood; these poor

[346] Coclanis, *The shadow of a dream*, especially chapter 3; Ira Berlin, *Many thousands gone: the first two centuries of slavery in North America* (Cambridge MA, 1998), chapter 6; Joyce E. Chaplin, *An anxious pursuit: agricultural innovation and modernity in the Lower South, 1730–1815* (Chapel Hill, 1993).

[347] Africans also had a greater immunity to the malarial disorders that were endemic in the lowcountry. See Peter H. Wood, *Black majority: negroes in colonial South Carolina from 1670 through the Stono Rebellion* (New York, 1975), chapters 2 and 3. It should also be stressed that many Africans were experienced farmers of rice, which was a staple food in West Africa. The crop was not grown in northern Europe. See Judith A. Carney, *Black rice: the African origins of rice cultivation in the Americas* (Cambridge MA, 2001).

wretches are then in a furnace of stinking putrid effluvia'.[348] Coercion, and nothing less, was the basis of planters' fortunes.

The province that had once been an exporter of Amerindian captives now bought in African slaves on a massive scale.[349] At first, Africans were obtained through Caribbean slave marts, but by 1714 a direct trade with the Guinea coast was underway. Imports remained modest until the mid-1720s, but then an upward surge began, culminating in 1738 when 3,658 slaves were disembarked in the Carolinas in a single year.[350] Rice brought about an 'Africanization' of South Carolina.[351] Blacks had formed a minor part of the province's non-indigenous population in its early days, just 200 individuals out of 1200 in 1680. Yet by 1700, as rice exports began to climb, blacks made up 43 per cent of South Carolina's inhabitants. By 1720 the figure was 70 percent. Carolina, as a Swiss migrant remarked in 1737, 'looks more like a negro country than a country settled by white people'.[352] In the rice-growing lowcountry the dominant language was a pidgin that drew on the linguistic heritage of West Africa as much as it did on English.[353] The Europeans clustered in and around Charleston. In part, this was a legacy from the Indian wars, one dictated by a basic need for security during the many periods of mayhem. It was also a response to the conditions of rice cultivation. Planters were fearful of the numbers and the disturbingly alien culture of their chattel labourers. Such fears were amply borne out by the disclosure of planned slave insurrections: 'a very wicked and barbarous

---

[348] *American husbandry; containing an account of the soil, climate, production and agriculture, of the British colonies in North-America and the West-Indies* (1775), pp. 393–94.

[349] Wood, *Black majority*; Daniel C. Littlefield, *Rice and slaves: ethnicity and the slave trade in colonial South Carolina* (Baton Rouge, 1981).

[350] David Richardson, 'The British slave trade to colonial South Carolina', *Slavery and Abolition*, XII, 3 (1991), 125–72; Kenneth Morgan, 'Slave sales in colonial Charleston', *English Historical* Review, CXIII (1998), 905–27; Eltis *et al., The transatlantic slave trade.* The influx of slaves gained extra momentum from 1731 when rice was removed from the list of enumerated articles that had to be routed through a British port before re-export to European markets. Carolina rice could now be sent direct to Iberian and Mediterranean consumers.

[351] For the numbers see Russell R. Menard, 'The Africanization of the Lowcountry labor force, 1670–1730', in Winthrop D. Jordan and Sheila L. Kemp (eds), *Race and family in the colonial* South (Jackson MI, 1987), pp. 81–108. For the cultural impact see Leland Ferguson, *Uncommon ground: archaeology and early African America* (Washington DC, 1992), and Philip D. Morgan, *Slave counterpoint: Black culture in the eighteenth-century Chesapeake and Lowcountry* (Chapel Hill NC, 1998).

[352] Quoted in Wood, *Black majority*, p. 132.

[353] This was the origin of 'Gullah', the Black dialect spoken on the Sea Islands south of Charleston into the twentieth century.

plott', was uncovered in 1720, for example, 'of the Negroes rising with a designe to destroy all the white people in the country'.[354]

These factors—the growing dependence of the colony upon rice exports, and the dependence of rice exports upon slave imports—determined South Carolina's articulation with the wider Atlantic economy. Rice had to be carried to European markets, yet there was a restricted local market for European manufactured goods. The white settlers who spread up the Delaware and Hudson valleys, thereby populating the hinterlands of Philadelphia and New York, had no counterparts on the banks of the Cooper or Santee rivers.[355] The appetite for European consumer goods was therefore far lower among Carolina's colonists. African slaves, after all, exercised little in the way of consumer choice. It is significant, in this respect, that Charleston was slow to develop an autonomous merchant class of the sort found in more rounded entrepôts like Philadelphia or Boston. Before 1750 her merchant houses were essentially offshots of London or Bristol-based partnerships that were concerned with rice exports and with little else.[356]

Carolina was nonetheless a growing market, even if that market lacked the multi-dimensionality of the Middle or Northern colonies. Quite apart from anything else, its population grew from 5,704 in 1700 to 45,000 in 1740. The Indian trade flourished, despite the devastations of the Yamasee War, not least because bovine epidemics in Europe cut the supply of cowhides and drove up the demand for deerskins. And for European traders to obtain the skins, trade goods had to be offered

---

[354] Quoted in D.D. Wax, '"The great risque we run": the aftermath of the slave rebellion at Stono, South Carolina, 1739–1745', *The Journal of Negro History*, LXVII, 2 (1982), p. 137.

[355] T.H. Breen, 'An empire of goods: the anglicization of colonial America, 1690–1776', *Journal of British Studies*, XXV (1986), 467–99; Nuala Zahedieh, 'London and the colonial consumer in the late seventeenth century', *Economic History Review*, XLVII, 2 (1994), 239–61.

[356] Jacob M. Price, 'Economic function and the growth of American port towns in the eighteenth century', in *idem*, *The Atlantic frontier of the Thirteen Colonies and States: essays in eighteenth-century commercial and social history* (1996), pp. 162–63 (first published in 1974), stresses the under-development of Charleston's merchant class. R.C. Nash, 'Urbanization in the colonial South: Charleston, South Carolina, as a case study', *Journal of Urban History*, XIX, 1 (1992), 3–29, and Peter A. Coclanis, 'The hydra head of merchant capital: markets and merchants in early South Carolina', in David R. Chesnutt and Clyde N. Wilson (eds), *The meaning of South Carolina history: essays in honor of George C. Rogers, Jr.* (Columbia SC, 1991), pp. 1–18, emphasise the wealth and diversity of the port's merchant class.

Illustration 2.27. Charleston in the mid-eighteenth century.

An Exact Prospect of Charlestown, the Metropolis of the Province of South Carolina (1762). © American Antiquarian Society. Charleston's quayside is seen from across the Cooper river. On the left, a battery defends the town against attack from the sea. Upriver, piers jut out into the stream to allow ocean-going ships to load up with rice. Back from the wharves and warehouses were the principal streets of the 'Metropolis of the Province', where the great planters resided for most of the year. 'An European at his first arrival must be greatly surprised when he sees the elegance of their houses, their sumptuous furniture, as well as the magnificence of their tables; can he imagine himself in a country, the establishment of which is so recent?'[357]

[357] J. Hector St. John [de Crèvecoeur], *Letters from an American farmer* (1782), p. 215.

in exchange: 'Lead, Powder, coarse Cloth, Vermillion, Iron Ware, and some other Goods, by which they have a very considerable Profit'.[358]

Above all, the extension of rice cultivation along the coast called for a wholesale reshaping of the landscape. This, in turn, rested upon an infusion of European-made *matériel*: axes, hoes, spades, ploughshares, ox chains and the like. It was this requirement that attracted the attention of metalware manufacturers in Britain.

John Crowley was exporting sizeable quantities of iron goods to South Carolina in the 1720s. This was understandable. His firm manufactured an array of goods expressly for plantation agriculture. The inventory made after Crowley's death in 1727 revealed that both 'Barbados' and 'Virginia' hoes were manufactured at Swalwell, each in eight different gauges.[359] What was less predictable was that the firm should acquire its own fleet after the Peace of Utrecht (the *Crowley* in 1715, the *Ambrose* in 1716, the *Theodosia* in 1718, and the *John* in 1721) and engage directly in the export trade. These vessels spent part of the year ferrying materials back and forth between the Crowleys' Tyneside works and their Greenwich depot, but they would then sail for the American colonies.[360] Charleston provided a ready market for the sort of metalwares that the *Crowley* or the *Ambrose* might carry.[361] That much is evident from the scale of the debts incurred by the town's merchants. Several of them owed John Crowley sums in excess of £1,000 at the time of his death.[362] The prominent Huguenot partnership of John Guerard, Benjamin Godin and Benjamin de la Conseillere—merchants, planters, Indian traders, and slavers—stood in this position. But the biggest debtor of all was Joseph Wragg, another merchant-planter, whose brother Samuel was

358 'A description of the Province of South Carolina', p. 896. See also Kathryn E. Holland Braund, *Deerskins and duffels: the Creek Indian trade with Anglo-America, 1685–1815* (Lincoln NE, 1993), pp. 121–25.

359 Suffolk Record Office (Ipswich), HAI/GD/5/15, 'Goods in Robt Armstrongs hands'. The Crowleys also specialised in producing the hatchets that were an important commodity in South Carolina's Indian trade: see *Angerstein*, p. 264.

360 TNA: PRO, CO 5/508–509, South Carolina shipping lists, 1716–1719, 1721–1735.

361 Among the goods demanded of a London merchant by his Charleston correspondent in 1738 were 'Crowley best Broad Hoes & 12 Dozen Narrow, assortment of good stock Locks, Plated assortment of Pad Locks, & Carpenters Hammers, assortment of Brass Garnet Hinges, A Dozen of Chimney Backs sorted. Iron Pots sorted & of all sizes[.] 4d, 6d, 10d, & 20d Clasp Nails in small Casks a pretty Large Quantity...' Robert Pringle to Christopher Bradgate, in Edgar, *The letterbook of Robert Pringle*, p. 50.

362 Suffolk Record Office (Ipswich), HAI/GD/5/2, 'Credit Ledger A'.

Illustration 2.28. 'Indiens going a hunting', 1736.

Courtesy of the Royal Library of Denmark (Manuscript Department, Ny kgl. Saml. 565, 4°).
Caption: This watercolour by Philip Georg Friedrich von Reck shows native hunters of the American southeast with a variety of European goods. The figure on the left wears a traditional leather matchcoat, but his companion, musket in hand, is dressed

the colony's agent in London and a major metropolitan slave merchant. He owed £3,914.[363]

John Crowley's ships would sail for Carolina with a cargo of iron-wares, swinging south to Madeira to pick up some pipes of the local wine. On their return they would carry rice, deerskins, and timber products. When the *Crowley* cleared Charleston in November 1723, for example, she was loaded with 222 barrels of rice, 457 barrels of pitch, 267 barrels of tar, and 5 chests of deerskins.[364] Graffin Prankard pursued the same course. The *Parham*, launched in 1722, sailed for Charleston every winter. Her cargo would include metalwares such as hoes and chains, Swedish bar iron, English steel, and nails by the hundred thousand. Lead shot and gunpowder, staples of the Indian trade, also featured prominently. Salt or coal served as ballast.[365] The return cargo from Charleston was of course rice, augmented by dyestuffs such as indigo and logwood. This was a flourishing trade, for Prankard soon built a new, far larger ship to join the 100-ton *Parham*. The 226-ton *Baltick Merchant*, registered at Bristol in 1732, was capable of carrying over 1300 barrels of rice.

There was no paradox in a ship named the *Baltick Merchant* engaging in transatlantic trade, for Graffin Prankard was seeking to capitalise on a potential symmetry between Baltic commerce and the passage of goods to and from Charleston. There was a complementarity between Swedish iron and Carolina rice that would allow Prankard to employ his shipping in a year-round circuit. In May, just as Prankard's ships were entering the Baltic, thousands of Africans were spreading out across the rice fields of Carolina to plant the new crop. During the summer, as the *Baltick Merchant* made her way back across the North Sea, African field hands were occupied with irrigating, hoeing and weeding. The rice harvest, which began in late August and lasted through to October, coincided with the fitting out of Prankard's ships for the transatlantic phase of their circuit. During November and December, as the *Baltick Merchant* struggled across a stormy Atlantic, slaves were engaged in laboriously 'pounding out' the rice in order to separate the husk from the grain.

363 Stuart O. Stumpf, 'The merchants of colonial Charleston, 1680–1756', (Ph.D. thesis, University of Michigan, 1971), pp. 75, 112.

364 TNA: PRO, CO 5/509.

365 See the invoices copied into Graffin Prankard's letterbooks for 30 October 1729 (the *Parham*), 13 July 1730 (the *Lyon*), 25 August 1731 (the *Whatley*), 20 November 1731 (the *Parham*). See also the accounts of sales for the *Baltick Merchant* in SA, DD/DN 448.

At the year's end, when the *Baltick Merchant* tied up at Charleston, hundreds of barrels of rice were ready to be stowed on board. This rice would be delivered to Hamburg or Amsterdam in April or May. Then the *Baltick Merchant* would pass eastward through the Sound once more, ready for another loading of Swedish bar iron.

This pattern of trade throve through the 1730s. The *Baltick Merchant* made the trip to Charleston every year. So too did vessels chartered by Graffin Prankard, such as the *Charming Molly* and the *Whitfield*, both of which sailed from Bristol in 1734. But the headlong development of South Carolina's rice economy was about to undergo a sharp deceleration. The outbreak of war between Britain and Spain in 1739 brought a general disruption to Atlantic traffic, whilst the slave rebellion at Stono, near Charleston, delivered an abrupt check to the Carolina trade in particular. The Stono uprising was, in fact, facilitated by Anglo-Spanish antagonism. The armed slaves who gathered at Stono on 9 September 1739 had heard of an edict issued by the Spanish governor of Florida promising freedom to refugee English slaves.[366] Those who marched south, killing many of the Europeans they encountered *en route*, were intent on reaching the Spanish stronghold at St Augustine. The rebels were surrounded by militia forces before the day was out and subjected to merciless reprisals, but the brevity of the rebellion could not disguise its seriousness. Nearly two dozen whites had died in an enterprise that spoke of concerted planning among its participants. The colony's rulers were seized by panic.

South Carolina's General Assembly devoted the winter of 1739–1740 to upgrading the repressive mechanisms needed to counter future outbreaks. The legislators met in an atmosphere of dread. The 1730s was a time of mounting slave resistance in the Caribbean islands with which Carolina had so marked a typological affinity. The British authorities in Jamaica were engaged in a bitter war of suppression against the

[366] The rebels, it was suggested at the time, originated in Angola where 'Thousands of the Negroes profess the Roman Catholic Religion' and where Portuguese, which was 'as near Spanish as Scotch is to English', was widely spoken. 'An Account of the Negroe Insurrection in South Carolina', (*c.* 1740), quoted in John K. Thornton, 'African dimensions of the Stono Rebellion', *American Historical Review*, XCVI (1991), 1102. See also Edward A. Pearson, '"A Countryside Full of Flames": a reconsideration of the Stono Rebellion and slave rebelliousness in the early eighteenth-century South Carolina Lowcountry', *Slavery and Abolition*, XVII (1996), 22–50, and Mark M. Smith, 'Remembering Mary, shaping revolt: reconsidering the Stono Rebellion', *Journal of Southern History*, LXVII, 3 (2001), 513–34, for the importance of African notions of masculinity and religiosity respectively in explaining the rebellion.

Illustration 2.29. *View of Mulberry* by Thomas Coram.

Courtesy of Gibbes Museum of Art/Carolina Art Association.
Caption: The pastoral calm suggested in this painting of Mulberry plantation in Berkeley County belies the violence and cultural tension of the Carolina lowcountry. The brick-built mansion, which is rather grander than most planters' houses of the early eighteenth century, borrows from the European baroque tradition. It was built at the time of the Yamasee War by Thomas Broughton, a rice planter, Indian trader and eventually Carolina's Governor. The solidity of the structure did not merely emulate European style, it had defensive advantages too—firing slits pierced the cellar walls. The slave quarter speaks of a more ambivalent architectural heritage. The arrangement of workers' housing along an avenue leading to the big house is redolent of a European estate—several Swedish *bruk* were laid out similarly—but the high-pitched thatched roofs of the huts betray an African influence. Slaves are shown strolling about with hoes, their badge of office, slung over their shoulders.

'Maroons', the runaway slaves who defied their erstwhile masters from mountain fastnesses in the interior of the island, whilst a major revolt was only just thwarted in Antigua in 1736. Rebellious outbreaks sprouted across the Caribbean whether the islands were claimed by the English, the Spanish, the French, the Dutch, or the Danish.[367] These insurrections were echoed in the Carolinas. Slave conspiracies were detected in 1730, 1733, 1734, 1737, and 1738. Perhaps the assemblymen also felt premonitory tremors of the insurgence that was shortly to flare up in other parts of continental North America, most notably in New York in 1741.[368] Amid such tensions South Carolina's rulers were inescapably drawn to the question of the province's racial imbalance. Steps were needed, it was decided, to curb the continuing inflow of African labour. Unless this was done, blacks would reach such a numerical preponderance that the Europeans would lose the coercive critical mass upon which their security rested. Moreover, it was felt necessary to reduce the ratio of African-born slaves in the unfree population. Africans, it was thought, were intransigently wedded to memories of their former freedom, whereas American-born blacks, knowing nothing but servitude, were more biddable. Accordingly, the 'Negro duty bill', enacted in April 1740, placed a prohibitively high tax on the importation of slaves.[369] The effect was instantaneous. Slave sales collapsed: 22,215 slaves had been landed in the Carolinas in the 1730s, but just 2,841 were disembarked in the 1740s.[370] Nearly twenty years would pass before slave imports returned to their former level, and so the Carolina economy lost the ebullience that had attracted first the Crowleys, then Graffin Prankard in the aftermath of Queen Anne's War.[371]

1740 was a sombre year in South Carolina. The previous year had seen the rising at Stono and a yellow fever epidemic that carried off hundreds. Now, the central part of Charleston was destroyed by fire.

---

[367] See Richard B. Sheridan, 'The formation of Caribbean plantation society, 1689–1748', in Marshall, *British Empire*, p. 406.

[368] Peter Linebaugh and Marcus Rediker, *The many-headed hydra: sailors, slaves, commoners, and the hidden history of the revolutionary Atlantic* (Boston MA, 2000), pp. 174–210.

[369] Wax, '"The great risque we run"'; Wood, *Black majority*, pp. 323–26.

[370] Eltis *et al., The transatlantic slave trade.*

[371] See Stephen G. Hardy, 'Colonial South Carolina's rice industry and the Atlantic economy: patterns of trade, shipping, and growth, 1715–1775', in Greene, Brana-Shute and Sparks, *Money, trade, and power*, pp. 108–140, especially pp. 111–12. Between 1722 and 1738 the value of rice exports grew at an annual rate of 13.9 per cent. Between 1739 and 1763 the growth rate shrank to just 1.3 per cent.

The conflagration of 18 November 1740 consumed over 300 homes, the city's rice warehouses and numerous stores on the waterfront. The total loss was estimated at £250,000. 'From one of the most flourishing towns in America', the *Gentleman's Magazine* reported, 'Charlestown is at once, in five hours time, reduced to ashes.'[372] 1740 was also a year of catastrophe for Graffin Prankard, a man who had prospered mightily during the boom years of the rice trade.

The *Baltick Merchant* sailed from Charleston in May 1740 with her usual cargo of rice and logwood. All was well until the ship was within sight of the Scilly Isles; then she encountered a Spanish privateer. Being so close to home, the men of the *Baltick Merchant* resolved to make a fight of it. A four-hour pursuit ensued. The two ships were well matched in terms of cannon, but the Spanish vessel was far more heavily crewed, and as soon as the ships came within musket range this numerical superiority began to tell: 'we had', said Nathaniel Alloway, the *Baltick Merchant*'s master, 'no hands to stand by our small arms whilst others fought the guns'. The deck of the *Baltick Merchant* was swept by unanswered musket fire. Now, Alloway continued, 'I had... the mortification of seeing one of my sailors drop down dead on the spot just by my side'. At any other time, Alloway reflected, this 'would have been a very shocking sight, but at this time had no effect on any body as I could perceive, so much had the noise of guns and the heat of action altered our natures. They were now along our side fireing volleys of small arms on us from upwards of 100 men, so that we were obliged to quit the deck'.[373] With two crewmen and a passenger dead, and three others wounded, Alloway asked for quarter.

The loss of the *Baltick Merchant*, Graffin Prankard's pride and joy, was a grievous blow to his business, one compounded by the near simultaneous wreck of the *Seaflower*, a chartered vessel, in the Gulf of Finland. In Bristol his creditors scented danger and descended on his house on St Augustine's Back. Although Prankard was saved from bankruptcy by the intervention of his wealthy son-in-law Caleb Dickinson, his years as a front-rank merchant were at an end. It was a disappointing conclusion to his commercial career, but in his time Graffin Prankard

[372] Quoted in Matthew Mulcahy, '"Melancholy and Fatal Calamities": disaster and society in eighteenth-century South Carolina', in Greene, Brana-Shute and Sparks, *Money, trade, and power*, p. 282.

[373] Quoted in J.H. Bettey, 'The capture of the *Baltick Merchant* 1740', *Mariner's Mirror*, LXXVI, 1 (1990), 37.

had embodied the most important trends in Anglo-Baltic trade. More than that, he had pioneered forms of commerce that brought together the Baltic and Atlantic worlds.

The first decades of the eighteenth century saw British merchants consolidate their hold over the iron trade in northern Europe. With mounting demand for malleable iron on British markets, London's established Baltic merchants strengthened their ties with Stockholm and Gothenberg, and new actors, Graffin Prankard among them, entered the Baltic trade, shipping Swedish iron around Lands End to western markets that had previously stood proof against it. The development of hardware manufacturing in the British Isles made for a larger and more variegated market, with important consequences for iron producers to the east of the Sound. A greatly increased demand for steel pushed British merchants into seizing control of the supply of 'Orground' iron. The *Leufstawerken* forges became adjuncts to the English steel industry; and so Walloon forgemen were pressurised to abandon the working practices to which they were long accustomed. Similarly, as the thirst for coldshort nail rods in the West Midlands led merchants to St Petersburg, the outlet for Siberian iron, so more and more enserfed workers found themselves corralled on distant Ural estates.

The growth of British hardware manufacturing did not just affect communities in *Bergslagen* or the Urals; it impacted upon peoples of the long Atlantic littoral. The products of Birmingham and Swalwell were landed along the surf-lashed beaches of west Africa and swung aloft from ships' holds onto the piers of Charleston harbour. This Atlantic demand was critical for the entire British economy, for it was here, not in the traditional markets of continental Europe, that exports of British manufactured goods found new outlets. Yet the turn westward, far from loosening Britain's ties with northern Europe, actually deepened British dependence upon Baltic resources. Atlantic commerce could not progress without 'Orground' iron, Riga hemp and planking from Danzig.

The advance of transatlantic exchange was no smooth process, however. It was wracked by violence and instability. It could hardly have been otherwise. Much of New World agriculture relied upon a brutally coercive system of labour. And the plantation economies practised a cash-crop monoculture that was singularly vulnerable to the convulsions of a global market. That Britain's colonies were open to attack by rival imperialisms exacerbated the problems. As the example of South Carolina demonstrated, feverish expansion could be followed by a shuddering halt. After the Stono rebellion one sub-route of the commodity chain that extended from the hardware manufacturing zones of Britain to the Atlantic colonies fell into abeyance. Nevertheless, the Atlantic economy was not given to stasis. Quiesence in one sector rarely extended far.

And so it was in this instance. The occlusion of the Carolina market coincided with the opening up of new possibilities in the Middle Colonies. There was more to the Chesapeake than tobacco farming; Virginia and Maryland were also endowed with vast stands of timber and beds of iron ore. To ingenious minds in Britain this suggested a further use for the Middle Colonies. They could produce pig iron for the mother country. If smelting were to be promoted in Virginia and her neighbours, Britain could be furnished with iron from within her own empire—a key mercantilist desideratum. If that was the case, the huge importation of Swedish and Russian iron, an affront to mercantilist sensibilities, would cease to be a necessity and Britain's relationship with the Baltic would be transformed.

CHAPTER THREE

THE INTERNATIONAL IRON TRADE AT A CROSSROADS: SWEDISH AND BRITISH DEBATES, 1730–1760

The international market for bar iron was heavy with change in the 1730s and 1740s. Demand on the British market, the pivot about which all other markets in northern and western Europe revolved, continued to rise. This, on the face of things, was greatly to the advantage of the Swedish exporters who had exercised hegemony over the international iron trade for a century. As long as British ironmasters were hamstrung by energy shortages the Swedes could look forward to tightening their grip on the British market. Yet Swedish policy makers were becoming aware that they faced constraints of their own. There were impediments, both political and ecological, to a further expansion of iron production in *Bergslagen*. Indeed, influential voices within Sweden began to ask whether a cap on output might be appropriate. The Swedes were also conscious of a growing threat to their hegemony from Russia. Siberian iron was now beginning to appear on western markets in appreciable quantities, subjecting the common sorts of Swedish iron to stiff competition. There were other troubling signs. English ironmasters were growing restive over the volume of Baltic iron entering their home market and began to press for more effective barriers to further import penetration. British hardware manufacturers had apprehensions of their own, but from a very different perspective. They were concerned at the inelasticity of supplies from the Baltic, not their excess. Trade embargoes such as that of 1717–1719 had thrown Britain's dependence on Swedish bar iron into sharp relief. Renewed tensions between Britain and Sweden in the early 1730s led many hardware manufacturers in Britain to look for a substitute for Baltic iron. They looked to America. In the 1730s and 1740s they mooted the creation of an imperial iron industry that would span the Atlantic, one that would dissolve British reliance upon Swedish or Russian bar iron. In Britain as in Sweden, then, the second quarter of the eighteenth century was a time of intense debate. Ministers, merchants and manufacturers in both countries sought to re-jig the mercantilist frameworks within which iron was made, traded and consumed.

Sweden and Britain: rival mercantilisms

For Swedes, as for Britons in the mid-eighteenth century, mercantilism was an article of faith. *Inledning til almänna hushålningen* (1747), one of the most important economic texts of the age in Swedish, extolled the virtues of a positive balance of trade. 'Foreign trade', wrote its author, Anders Berch, professor of economics at Uppsala University (and the father of our guide to Leufsta, Christer Berch), 'is that which makes a country either prosperous or poor and deserves therefore the greatest attention.'[1] This was a well-worn orthodoxy, which has led some to think of Berch as a traditional mercantilist, whose views were derivative and whose importance lay in policy-making rather than theoretical insight.[2] Yet mercantilism, as we have seen, was not a fixed mode of thought; it developed over time. Indeed, a close reading of Berch reveals a set of views on the nature of production that aligned him with British contemporaries such as Defoe and Postlethwayt, not the older generation of mercantilist thinkers. Berch's was an 'enlightened' mercantilism. He divided the economy into four parts—agriculture, mining-metalworking, craft manufacturing, and trade ('*Landtskötsel, Bergwärk, Handaslögder och Handel*')—that were co-equal and interdependent. In a passage that anticipated Joseph Massey's depiction of the economy as 'one vast piece of machinery', Berch wrote the following:

> Between these parts [*näringsmedel*] is a fairly strong connection... as one sometimes increases or decreases, so the others feel augmented or diminished: Agriculture and Mining-Metalmaking are the foundation to the other parts; Craft is totally dependent upon them; [and] Trade demands help from all three. [Yet] Agriculture and Mining-Metalmaking without Craft and Trade become impotent; Craft without Trade can never increase

[1] Anders Berch, *Inledningen til den almänna hushålningen. innefattande grunden til politie, oeconomie och cameral wetenskaperne* (Stockholm, 1747), p. 272.

[2] Heckscher maintained that Berch added little to economic discourse during his long academic career: Eli F. Heckscher, *Sveriges ekonomiska historia från Gustav Vasa* (Stockholm, 1949), pp. 828–33. See also Karl Petander, *De nationalekonomiska åskådningarna i Sverige. Sådane de framträda i litteraturen* (Stockholm, 1912), pp. 51–76, where Berch is appraised more positively, although still identified as belonging to 'official mercantilism' (p. 53). Recent writers, like Sven-Eric Liedman and Lars Magnusson, are somewhat more nuanced, but still adhere to the view that Berch promoted trade as the only way for a country to generate wealth: Sven-Eric Liedman, *Den synliga handen. Anders Berch och ekonomiämnena vid 1700-talets svenska universitet* (Värnamo, 1986); Lars Magnusson, *Merkantilism. Ett ekonomiskt tänkande formuleras* (Stockholm, 1999), pp. 249ff.; Lars Magnusson, *Äran korruptionen och den borgerliga ordningen. Essäer från svensk ekonomihistoria* (Stockholm, 2001), pp. 36ff.

> a country's wealth; [and] Trade without Craft is always harmful or at least less profitable...so that one link within this chain leads towards the same purpose, and the one sets the others in perfect motion.[3]

Production and trade, Berch continued, formed a common foundation 'upon which the public good of society will rear'.[4] Wealth creation depended upon productive human labour. Indeed, Berch's writings betray an attention to labour worthy of the *encyclopédists*. 'The means by which all created things are employed and cultivated', he announced, 'is diligent labour.'[5]

Berch was not unique in drawing attention to the importance of craft production. Indeed, Swedish thinkers had never been wholly neglectful of manufacturing. Policy makers and theorists during Sweden's Age of Greatness had been fixated on exploiting the country's natural resources to the full, aiming for national autarky in manufactured goods and thereby achieving an impregnably positive balance of trade.[6] Indeed, they employed an expression for trade (*näring*) that united conceptually both production and exchange, the phenomena that Defoe would much later yoke together in his phrase 'dealing and manufacturing'. Anders Berch, in other words, was the legatee of an intellectual tradition whose attention to productive industry was in advance of rival British schools of thought.

Where Berch departed from his British contemporaries was in his attitude to the state's regulation of the economy. (It is here that later critics of Berch, who have dismissed him as a 'conservative mercantilist', are on firmer ground.) Whereas the tendency in mid-eighteenth-century British economic discourse was towards economic liberalism, Berch remained an adherent of the German cameralist school.[7] As the subtitle of his *magnum opus* declared, the 'National Economy' (*Almänna hushållningen*) had three inter-related aspects to it: the polity, the economy and 'cameral science' (*Politie, Oeconomie och Cameral Wetenskaperne*). By cameral science, Berch meant the ways whereby all productive activity was put at the service of the state. Cameralism dictated that economic life should be governed by formal rules; that the state should allocate

---

[3] Berch, *Almänna hushålningen*, pp. 10f. For Massey, see above, pp. 9–10.

[4] Berch, *Almänna hushålningen*, unpaginated foreword.

[5] Berch, *Almänna hushålningen*, p. 6.

[6] Leif Runefelt, *Hushållningens dygder. Affektlära, hushållningslära och ekonomiskt tänkande under svensk stormaktstid* (Stockholm, 2001), pp. 128–151.

[7] Heckscher, *Sveriges ekonomiska historia*, pp. 826ff.

particular tasks to particular social groups; and that the state should intervene to ensure that production and exchange were conducted harmoniously. The objective was the preservation of social order and the effective raising of revenue for the state.[8] A self-regulating economy of the sort advocated, albeit partially and cautiously, by many British writers of the time was explicitly ruled out.

These theoretical differences reflected differing practices and policies in Sweden and Britain. In Sweden, the central state had a pervasive presence in economic life. The production and marketing of goods was heavily regulated. In Britain, the role of the state was more muted. Or more accurately, the role of the state was far more uneven. Many goods and services circulated on unregulated markets, but when it came to foreign trade the state loomed extraordinarily large, brimming with mercantilist ambition.

The cornerstone of British mercantilism was the Navigation Act of 1651, which stipulated that foreign goods could only be brought into British ports in ships that were British-owned and British-crewed or in vessels from the exporting country. The intention was to exclude the Dutch, who at that time dominated the carrying trade of Europe. In that, the Act of 1651 was conspicuously successful. The British pegged back the Dutch merchant fleet, first in home waters, then in more distant seas. Indeed, the Navigation Acts (for there were several) were instrumental in consolidating Britain's seaborne empire. Colonial trade was to be conducted in British or colonial-built vessels, and key commodities such as sugar, tobacco and rice had to be landed at an English port before they could be forwarded to European destinations.[9]

Mercantilist writers exulted in this ruthless economic nationalism. The Navigation Act, wrote Josiah Child in 1693, was 'one of the choicest and most prudent *Act* that ever was made in *England*, and without which we had not now been *Owners* of one half of the *Shipping*, nor Trade, nor employed one half of the *Sea-men* which we do at present'.[10] The Navigation Act of 1651 inaugurated a new phase of aggressive imperialism, for the quest for maritime hegemony was military as well

---

[8] Berch, *Almänna hushålningen*, unpaginated foreword, and pp. 361ff.

[9] Stanley Engerman, 'Mercantilism and overseas trade, 1700–1800', in Roderick Floud and Donald McCloskey (eds), *The economic history of Britain since 1700. Volume 1: 1700–1860* (2nd edn, Cambridge, 1994), pp. 196ff.

[10] Sir Josiah Child, *A new discourse of trade* (1693) in Lars Magnusson (ed.), *Mercantilism. Volume III: a science of trade* (London, 1995), p. 72.

as commercial. Between the 1650s and the 1750s the British waged a series of wars, first against the Dutch, then against the French, for the control of trade and colonial wealth. An expensively maintained fleet was deployed to police sea lanes and protect colonial harbours.[11] Escalating military expenditure was not seen as a drain on commercial wealth. On the contrary, naval prowess, whether used in the reduction of rival fleets or the extirpation of pirates, was identified as the essential underpinning of national prosperity. It was clear, wrote Thomas Lediard, author of *The naval history of England* (1735), 'That our trade is the Mother and Nurse of our Seamen; Our Seamen the Life of our Fleet; And our Fleet the Security and Protection of our Trade: And that both together are the WEALTH, STRENGTH, and GLORY of GREAT BRITAIN'.[12] Christer Berch, visiting Britain at the end of the 1750s, agreed: 'when the question is about the defence of increased trade and shipping the English Nation is among the first to shout for war, as it is therein that the source of [their] wealth can be found'.[13] At first, the drive for naval supremacy was paid for on a hand-to-mouth basis, but after 1688 the British state was re-launched on a more secure fiscal foundation. Military expenditure was paid for by long-term borrowing, secured against future tax receipts. The foundation of the Bank of England in 1694 symbolised the emergence of a money market in London that was able to fund the National Debt and eager to do so. Financial innovation gave rise to fiscal stability; together they enabled Britain to stand forth as the 'military *Wunderkind* of the age'.[14]

To foreign observers, eighteenth-century Britain was a paradoxical place. The British state was both potent and self-denying. The Royal Dockyards spoke of an enormously powerful military state, and the Excise and Customs services were highly professional bureaucracies that went about their business with unexampled fiscal efficiency. Yet in other respects the state was unobtrusive, leaving unchecked commodity chains and credit flows that in other parts of Europe were channelled by government agency. A martial carapace sheltered an internal regime

[11] Daniel Baugh, 'Maritime strength and Atlantic commerce: the uses of "a grand marine empire"', in Lawrence Stone (ed.), *An imperial state at war: Britain from 1689 to 1815* (London, 1994), pp. 185–223.

[12] Quoted in Baugh, 'Maritime Strength and Atlantic Commerce', p. 195.

[13] Uppsala Universitetsbibliotek, Handskriftsavdelningen X407b: 'Christer Berch relation öfver sin resa, 1757–1761'.

[14] John Brewer, *The sinews of power: war, money and the English state, 1688–1783* (1989), p. xiii.

of economic liberalism. The contrast was marvelled at by Swedish commentators who were accustomed to the cameral rigour of their native land. Swedes could also reflect ruefully on Britain's post-1688 colonial success, for it coincided with the collapse of Sweden's own imperial pretensions. Whereas the Treaty of Utrecht (1713) that ended the War of Spanish Succession announced Britain's elevation to great power status, the Treaty of Nystad (1721) that terminated the Great Northern War brought Sweden's 'Age of Greatness' to a close.

Sweden's military and diplomatic eminence in the seventeenth century had been achieved through an influx of foreign capital and the effective exploitation of the mineral resources of *bergslagen*. The process had been overseen by a robust state apparatus, which marshalled resources with such care that Sweden was able to compete militarily far beyond her demographic means. Military failure, when it finally came, pushed Sweden back into the diplomatic periphery from which she had erupted with such startling force in the 1620s. Economically, however, Sweden remained a force. Indeed, all important indicators show continued expansion in the eighteenth century. Agricultural development was marked, the population grew, and so did the proportion of the population that dwelt in towns.[15] Moreover, the Stockholm-based international merchants who had played such a critical role in establishing Sweden as a major trading power continued to provide capital and commercial expertise, enabling their host country to thrive as an exporter of metals and timber products.

The question confronting Swedes after 1721 was whether the state could make use of the country's continued economic success to restore her lost political fortunes. Opinion was divided in the politically fractured decades that followed the Peace of Nystad. The nobility had seized the opportunity afforded by the death of Charles XII in 1718 to curtail royal absolutism. In the new 'Age of Liberty' (*Frihetstiden*) Sweden was no longer subject to the personal rule of a monarch but to the direction of a council of state headed by the king. The council of state, moreover, was responsible to an assembly (*Riksdag*) of the four estates: the nobility, the clergy, the bourgeoisie, and the peasantry. Rival factions competed for control of the *Riksdag*. The 'Hats' brought together the

---

[15] C.-J. Gadd, *Den Agrara Revolutionen 1700–1870* (Stockholm, 2000); J. Myrdal, *Jordbruket under feodalismen 1000–1700* (Stockholm, 1999); S. Lilja, *Tjuvehål och stolta städer. Urbaniseringens kronologi och geografi i Sverige (med Finland) ca 1570-tal til 1810-tal* (Stockholm, 2000).

mightiest noble families, the merchant elite of Stockholm, and powerful state functionaries. They yearned for a bellicose foreign policy, aimed at recovering the provinces that had been lost in 1721. The disastrous war with Russia of 1741–1743 and Sweden's participation in the Seven Years' War came at their behest. The Hats' political adversaries, an anti-aristocratic grouping known as the 'Caps', were far more inclined to accept the new balance of power in the Baltic.[16]

War or no war, economic development was seen as an integral part of national renewal, and for that reason the Swedish state's mercantilist bent became still more pronounced during *Frihetstiden*.[17] Enrichment would be the prelude to reconquest. State policy took a new turn in the 1720s, with twin initiatives to bolster the economy. The first of these, announced in the *Produktplakat* of 1724, echoed the British Navigation Acts. Henceforth non-Swedish vessels could only bring in goods from their country of origin. This was a blow to the Dutch, who specialised in the carrying trade, but it also struck hard at the British who usually shipped salt from the Mediterranean or French wines to help pay for Swedish iron. The intention was to promote Swedish shipping and shipbuilding; it succeeded on both counts.[18] The second plank of the new policy, *Manufakturpolitiken*, involved state promotion of new industrial enterprises, especially those that could contribute to import substitution. New workshops were given special legal status and furnished with credit by the state. The effects were most noticeable in the woollen textiles sector, which underwent rapid expansion. Mining and metalworking, the well-established core industries, were not neglected, however. The legal framework within which ore extraction and smelting took place was also overhauled during the Age of Liberty.

The *Bergscollegium* (Board of Mines), which had overseen mining and metal processing since its inception in 1649, was no minor body. It was one of the principal agencies of the Swedish state.[19] Its first

---

[16] Göran Behre, Lars-Olof Larsson and Eva Österberg, *Sveriges historia 1521–1809. Stormaktsdröm och småstadsrealitet* (Stockholm, 2001), pp. 238–43, 268–81; Lars Magnusson, *Sveriges ekonomiska historia* (Stockholm, 1996), pp. 273–81.

[17] Magnusson *Sveriges ekonomiska historia*, pp. 246ff. See also Magnusson, *Äran Korruptionen*, pp. 25ff.

[18] Staffan Högberg, *Utrikeshandel och sjöfart på 1700-talet. Stapelvaror i svensk export och import 1738–1808* (Stockholm, 1969), pp. 13–33.

[19] This and the following paragraphs are based on Bertil Boëthius and Åke Kromnow, *Jernkontorets historia. De I. Grundläggningstiden*, Stockholm 1947, pp. 1–27; Birgitta Ericsson, Karl-Gustaf Andersson and Per-Arne Karlsson, 'Privilegiegivningen till järnbruk och järnmanufaktur i Sverige under frihetstiden', in *Industri og Bjergværksdrift. Privilegier*

President, Carl Bonde, was a cousin of Axel Oxenstierna, Gustavus Adolphus's chancellor and the *de facto* ruler of Sweden after his master's death in battle in 1632. Bonde's successors were men of equal weight, who presided over formal meetings of the *Bergscollegium* amid baroque pomp, enthroned under a canopy of blue velvet.[20] This 'aristocratic senate', served by a twenty-strong office staff, directed the affairs of twelve regional bureaux.[21] The chief officer at regional level—the *Bergmästare*—enforced the social and spatial division of labour that had been settled upon in the mid-seventeenth century in a classical piece of mercantilist (or better still, cameralist) policy formulation. *Bergsmän* were to take charge of mining and smelting, whilst *brukspatroner* in adjacent districts were entrusted with the refining of pig iron into bars. Each *bruk* was permitted a set output that was related to the charcoal resources at the disposal of the *brukspatron*. It was in this way that the natural riches of the country were to be most effectively harvested for the benefit of the state.

In the seventeenth century the *Bergscollegium* had been directly subordinate to the Crown. In the Age of Liberty, with the *Riksdag* playing a more prominent part in public affairs, matters became more complex. It became customary for the *Riksdag* to devolve discussion of much regular business to standing committees (*deputationer*). The parliamentary session of 1723 saw the establishment of a *bergsdeputation* to consider matters of policy concerning the iron trade. The *bergsdeputation* was composed of deputies drawn from each of the four estates, selected for their familiarity with iron making. It was expected that the deputies would liaise with the *Bergscollegium*—a prospect strengthened by the *bergsdeputation*'s habit of meeting at the 'Old Mint' (*Gamla Myntet*) where the *Bergscollegium* was housed.

---

i Norden i det 18. århundrade (Oslo, 1985), pp. 148–165; Maths Isacson, 'Bergskollegium och den tidigindustriella järnhanteringen' *Dædalus 1998. Människa, teknik, industri*, pp. 43–58; Johan Axel Almquist, *Bergskollegium och bergslagsstaterna, 1637–1857: administrativa och biografiska anteckningar* (Stockholm, 1909).

[20] Bertil Boëthius, 'Hammarkommissionerna på 1680-och 1720-talet. En studie över deras ställning i bergskollegiets brukspolitiska system', in *En Bergsbok till Carl Sahlin* (Stockholm, 1921), pp. 193f.

[21] For an analysis of these local mining courts see Per-Arne Karlsson, *Järnbruken och ståndssamhället. Institutionell och attitydmässig konflikt under Sveriges tidiga industrialisering 1700–1770* (Stockholm, 1990); Anders Florén, 'The making of the forgeman. Social relations and bar iron production in Sweden c. 1650–c. 1750', in Hans-Jürgen Gerhard, Karl Heinrich Kaufhold and Ekkehard Westermann (eds.), *Europäische Montanregion Harz* (Bochum, 2001), pp. 193–211.

Swedish dilemmas

The deputies and officials who gathered at *Gamla Myntet* had much to discuss, for by the 1730s conditions on the international market were shifting. Although the demand for iron in Britain continued to grow, the first intimations of cut-price Russian competition could be felt. In response, Swedish policy makers began to divide between those who felt that their iron exports could only be maintained by lowering prices and those who believed that the superior quality of their iron gave Swedes an effective monopoly, not least in the critical English market. For the 'monopoly' school of thought, which was strongest amongst the Hats, the price of Swedish iron could be safely ratcheted up on the international market, perhaps by restricting output. For the opposing camp, a regime of high prices would merely stimulate production in rival centres of iron making, most notably in Russia but also in British North America.[22] Swedish officials made several investigative trips to Russia in the 1730s and 1740s in order to assess the Russian threat, but the 'monopoly' party, then in the ascendant, was sceptical about its magnitude.[23] 'The notion that England may be furnished with Iron elsewhere, (Muscovy and America are named) is treated as a wild thought', Edward Finch, the British resident in Stockholm, told his superiors in 1733.[24]

Such hubris was to cost the Swedes dear. Muscovite iron was a very real threat. Russian bar iron achieved visibility on the British market for the first time in the 1730s. Imports from St Petersburg had averaged a meagre 712 tons per annum in the 1720s, but in the 1730s they nearly quadrupled to an annual average of 2,612 tons.[25] This was but little when compared with the surge in Russian imports that was to come in the 1750s and 1760s, but it was a clear indication of the direction of change, as Graffin Prankard, from his vantage point in Bristol, could sense. The Russians 'having fallen into the Making of

[22] For a recent introduction to the debate see Anders Florén and Göran Rydén, 'A journey to the market society: a Swedish pre-industrial spy in the middle of the eighteenth century', in Ragnar Björk and Karl Molin (eds), *Societies made up of history: essays in historiography, intellectual history, professionalisation, historical social theory and proto-industrialisation* (Uppsala, 1996), pp. 265–307.

[23] Ågren, *Iron-making societies*, pp. 1–4.

[24] TNA: PRO, CO 388/32, Edward Finch to Lord Harrington, 31 January 1733.

[25] S.J. Newman, 'Russian foreign trade, 1680–1780: the British contribution', (University of Edinburgh Ph.D thesis, 1985), p. 287.

Iron & Vending it so low', he informed Francis Jennings in 1735, 'will be a means to keep Down the Sale of so much Sweeds'.[26] This was no chance observation, for Prankard was an active buyer of the Russian product. His trade with Russia had begun in 1730 with a single tentative voyage, but in 1738 Prankard sent eight ships to St Petersburg, returning with 560 tons of iron, plus 260 tons of hemp and 40 tons of linen.[27] There was much here for Swedish policy makers to ponder, but for many of them at the start of the 1730s it appeared that the difficulty lay with the mismanagement of Sweden's own industry, not the exports of a rival.

Addressing the *Riksdag* in 1734, Fredrik Gyllenborg, ironmaster and future President of the *Bergscollegium*, diagnosed the problem facing the Swedish iron industry as one of over-production. Too many ironmasters had increased their make during the 1720s when iron had fetched a good price on international markets. In doing so, they had driven their industry into recession. Gyllenborg's remedy was a return to the output levels set out in the tax assessment of 1695 (*hammarskattetaxeringen*). No drastic action was taken, however. The authorities contented themselves with a restatement of the golden rule of Swedish iron making: that production at a *bruk* should never outpace the charcoal resources allotted to it.[28] The curb that Gyllenborg suggested ran counter to the interests of too many *brukspatroner*. Those in the western county of Värmland, where the iron industry was not so well established, argued that they still had under-exploited timber at their disposal, which would be wasted if the restrictive *hammarskattetaxeringen* of 1695 was to be their benchmark. Opponents of capping, some of them very powerful, could also be found in the older districts of *Bergslagen*. As we have seen, the De Geers, anxious to respond to the British demand for 'Orground' iron, were committed to expanding their Uppland *bruk* throughout the 1720s and 1730s.[29]

---

[26] GP to FJ, 10 February 1735.

[27] GP to William Vigor, 31 May 1738.

[28] This and the following paragraphs see Ericsson, Andersson and Karlsson, 'Privilegiegivningen till järnbruk och järnmanufaktur', pp. 278ff.; Boëthius and Kromnow, *Jernkontorets historia*, I, pp. 40–60.

[29] Bergmästarämbetet i Gävleborgs, Uppsala och Stockholms län, Bergmästarens tjänsteberättelser Uppland, Gästrikland, Hälsningland, Västernorrland 1737, B II:5, ULA. It was not until the 1740s that Louis De Geer was prepared to reduce capacity at a few *bruk*, such as Österby and Forsmark. See LDG Nordencrantz, 23 February 1744, Börstorpssamlingen, vol 82, signum E 3027, RA.

The debate on output levels was fluid and inconclusive in the mid 1730s.[30] Anders Nordencrantz, one of the key participants in the controversy, spoke exasperatedly of 'many years of accumulated contradictions and disputes...some exclaiming that we forge too much, others too little...some saying the price is too high, some too low; certain people that we are threatened by foreign works, others that this is no problem at all'.[31] For Nordencrantz, a parliamentarian and a prominent political economist, squabbles about the volume of iron that should be produced were secondary. The fundamental issue was the structure of the market, not the quantity of iron that circulated upon it. *Brukspatroner* were vulnerable because they were many whilst their customers, the merchants of Stockholm and Gothenberg, were few. It was a buyers' market, and therefore one characterised by low prices. There was an allied difficulty. Because the production-and-marketing cycle of iron was so protracted, *brukspatroner* depended upon the capital-rich export merchants for credit.[32] The ironmasters were subordinate to the merchant elite, who were themselves the agents of overseas buyers. It was an arrangement that offended every mercantilist precept.

This was a message driven home by Lars Salvius, journalist, publisher and ardent Hat. During the parliamentary session of 1738, which saw the Hats installed in power, Salvius propounded his 'reform-mercantilist' views in a weekly journal entitled *Tanckar öfwer den Swenska Oeconomien igenom Samtal yttrade* ('Thoughts on the Swedish Economy uttered in dialogue'). The principal characters in the dialogue were *Fru Swea* ('Madame Sweden') and *Herr Mentor*, her guide to a proper—that is to say Hat—mode of government. They were joined at appropriate points by *Fru Oeconomia*, the embodiment of British interests.[33]

In issue 19 of *Tanckar öfwer den Swenska Oeconomien* questions were raised about Sweden's iron trade. A fictional correspondent asked *Herr Mentor* how it could be that bar iron was exported in its raw state, without

[30] As Bertil Boëthius and Åke Kromnow remarked, it is well to 'remember that the principles of free trade were asserted with such an emphasis...so soon before the victory of the politics of monopoly'. Boëthius and Kromnow, *Jernkontorets historia*, I, p. 44.

[31] Anders Nordencrantz, Ödmjukt memorial, manuscript from February 1743, Börstorpssamlingen, volym 82, signum E 3027, RA. For brief details of Nordencrantz's career see Sten Lindroth, *Svensk lärdomshistoria. Frihetstiden* (Stockholm, 1978), pp. 102–09; Magnusson, *Äran korruptionen och den borgerliga ordningen*, pp. 51–79.

[32] Boëthius and Kromnow, *Jernkontorets historia*, I, pp. 100ff.

[33] [Lars Salvius], *Tanckar öfwer den Swenska Oeconomien igenom Samtal yttrade* (Stockholm, 1738). The following paragraphs are based upon this unpaginated text.

further manufacturing. Surely, the writer continued, this resulted in Sweden gaining only a third of the iron's 'true value'? At first, *Mentor* appears reluctant to answer: 'it would take a lot of reflection and the answer is so dry and unpleasant that no-one can be bothered to hear it'. Instead, he launches into a jocular song about the spoliation of Swedish mineral resources.

Mentor is cut short by *Fru Swea*, who, ordering him to be serious, invites *Fru Oeconomia* to join them. *Herr Mentor* is a little coy about speaking in front of *Fru Oeconomia*—'what I will say should not be for her'—but it is not long before he warms to his theme. *Mentor* descants upon the necessity for a 'Swedish Iron-Company' that could supply *brukspatroner* with short-term credits. Such an enterprise would loosen the grip that the great merchant houses of Stockholm, which were beholden to foreign capital, had over the nation's iron industry. *Fru Oeconomia* was scornful; the foreign interests for whom she stood proxy were too powerful. They had the resources to stockpile iron until the Swedish company broke: 'my people', she reminded *Mentor*, 'have the advantage of possessing large iron warehouses'. (*Fru Oeconomia* was correct in thinking that foreign merchants held significant stocks. Samuel Shore and Graffin Prankard's pursuit of monopolistic mastery in the 1730s, described above, had been compromised by the presence of *Leufstawerken* iron in the hands of the Grills in Amsterdam and Henry Norris of London. Eric Touscher, the *directeur* at Leufsta, believed that the Grills held substantial stocks 'both within and without the country' and was awestruck at the Dutch merchant house's capacity 'to endure so long with so much capital standing idle'.)[34] *Fru Oeconomia* also boasted that her people had the opportunity of importing iron from America 'without fetching it from here'. Now it was *Herr Mentor*'s turn to be scornful. If the 'Iron-Company' could amass the capital needed to fund the industry for three years, he insisted, the British stranglehold would be broken. He was also sceptical about the practicality of bringing iron across the Atlantic. 'Between having and getting', he chided *Fru Oeconomia*, 'is a big difference.' Finally, he wondered aloud about the strategic advisability of relying on the American colonies for iron. Could it be wise, added *Fru Swea*, to give colonists the means of making guns?

Tanckar öfwer den Swenska Oeconomien returned to the international iron market in issue 34 when *Mercurius*, an authority on 'iron making around

[34] ET to Anders von Drake, 19 August 1738, Leufsta Arkivet, vol. 167, RA.

the World', joined the usual cast of characters. The discussion turned once more to America, and once again the impolicy of encouraging colonial iron production was given a full airing. The position of Russian iron was then reviewed. The export from St Petersburg had risen markedly in the previous decade, but *Herr Mentor* dismissed this as an entirely artificial phenomenon. Russian iron could only survive because of the politically inspired support of the British. And this could not endure, not when the quality of the iron was so poor. *Mercurius* then took the opportunity to make a still more sweeping assertion: neither the Russians, nor the Americans, nor any other set of people could take the place of the Swedes. Nowhere else other than Sweden was iron made in sufficient quantity or of suitable quality to meet British demand. Swedish iron was indispensable.

This reassuring conclusion formed the basis of Swedish policy in the 1740s after the Hats' accession to power. There was a further thickening of the institutional framework that governed iron making. To the *Bergscollegium*, which regulated mining and processing, and the *bergsdeputation*, in which members of the *Riksdag* debated the strategic direction of the national iron industry, was added the *Bruksriksdag*, an assembly of ironmasters that met during parliamentary sessions. These bodies, all Hat-dominated, represented a conflux of bureaucratic and capitalist interests, devoted to the exploitation of Sweden's perceived monopoly.[35] (For this reason it is inappropriate to see the policy shifts of the 1740s as being imposed on *brukspatroner* by an alien state apparatus; ironmasters were comfortably embedded within a corporate structure that took in both industry and state.) Another organisation came into being in 1747: the *Jernkontor* ('Iron Bureau'). Its purpose was to furnish credit to Sweden's under-capitalised *brukspatroner*, the very function that *Herr Mentor* had proposed for his putative 'Swedish Iron-Company'.[36] The *Jernkontor* was also intended as a buyer-of-last-resort of iron for which there was no immediate demand. It would, as Eric

---

[35] Ericsson, Andersson and Karlsson, 'Privilegiegivningen till järnbruk och järnmanufaktur', p. 268.

[36] The *Jernkontor* was an ambiguous institution that straddled the divide between the public and the private. In that, it was characteristic of the age, or so it is argued in Martin Melkersson, *Staten, ordningen och friheten. En studie av den styrande elitens syn på statens roll mellan stormaktstiden och 1800-talet* (Uppsala, 1997), pp. 210f., which asserts that the boundary between state organisations and other types of institution in eighteenth-century Sweden was blurred.

von Stockenström, one of its founders, stated, 'support bar iron to a price appropriate to its value'.[37]

The way was now clear for the Hats to enforce the capping of production that was so dear to them. Decrees issued during the parliamentary session of 1746–47 prohibited the creation of new *bruk* and warned against the breaching of output limitations at existing forges. 'Excess forging' (*översmidet*) was identified as the besetting sin of the industry, with the main culprits being the *bergsmän*, who routinely contravened *Bergscollegium* regulations.[38] New production quotas were drawn up, superseding those of 1695. The new tax regime, embodied in *1748 års Hammarskattelängd*, brought half a century of gathering debate to a close. *Översmidet* was outlawed; it was 1803 before any fresh increase in bar iron output was authorised.[39]

The events of the 1740s marked a fresh phase in a long controversy about the role of the state in Swedish iron making. The foundation of *Jernkontoret* echoed earlier attempts to unify the marketing of Swedish iron under a single authority. That had been a project of Johan III's in the 1580s, one reprised by Axel Oxenstierna in the 1620s. In the last decades of the seventeenth century there was another renewal of interest in centralising the sale of iron, this time in conjunction with a restriction of exports. The mechanical genius and polymath Christopher Polhem had been a proponent of such a policy, identifying it as a way of promoting a vibrant metalware sector within Sweden. The restrictive policies of the Hats had, quite clearly, a lengthy lineage. But the polemicists of the 1740s did more than rehearse arguments that were familiar to earlier generations; Lars Salvius, Anders Nordencrantz and their peers demonstrated a more profound knowledge of international conditions and a more sophisticated conceptual understanding of the problems facing the Swedish iron industry. They appreciated that the international market was multi-dimensional and that it therefore merited

---

[37] Eric von Stockenström, *Tal om svenska Järn-Bruksnäringen, samt om Jern-Contoiret. Hållet för Kongl. Vetenskaps Academien, vid Præsidii Nedläggande Den 11 April 1767* (Stockholm, 1767), p. 29.

[38] Boëthius and Kromnow, *Jernkontorets Historia*, I, pp. 46–56; Ericsson, Andersson and Karlsson, 'Privilegiegivningen till järnbruk och järnmanufaktur', pp. 283ff.

[39] The effects of the capping policy were studied at micro-level by Eli Hecksher. His conclusion, surprisingly for one his liberal sympathies, was that the restriction on output had positive outcomes, both in terms of forest preservation and the income of *brukspatroner*. Eli F. Heckscher, *Sveriges ekonomiska historia. Från Gustav Vasa. Andra delen. Det moderna Sveriges grundläggning* (Stockholm, 1949), pp. 386–99.

careful study. For many earlier analysts, the world beyond Sweden had been an abstraction; the dimensions of the market, rather than its texture, commanded attention.

Nordencrantz exemplified the new approach of the mid-eighteenth century. As a young visitor to Britain in the 1720s he had shown no sign of the indefatigable collecting and classifying of data that was to mark his maturity. His assessments were brief and often bland. London, for example, was summarized in a single sentence: 'A Warehouse for all the Goods of the World used for necessity and pleasure, A Centre for human Rendezvous and trade'. By the 1740s, however, Nordencrantz was an inveterate data-gatherer, filling ream after ream with notes on foreign markets and overseas competitors.[40] Salvius, in the guise of *Mercurius*, also demonstrated the more searching critical approach of the mid-century. *Mercurius* evaluated iron making the world over, providing a varied political and institutional context for smelting and refining around the northern hemisphere. There was a distinction to be drawn, so Salvius claimed, between the superficial observer who saw only foreign threats and the seasoned expert who appreciated that American iron was produced in politically volatile circumstances and that Russian iron had defects of its own. These judgements reflected the sanguine outlook of a Hat partisan, but they were rooted in an investigative methodology that was rigorously empirical and comparative.

Samuel Schröder's tour of Britain in 1748–49 was undertaken in the same spirit. His studies of the British market and of British manufacturing techniques were lengthy, highly detailed and concrete—as his account of Birmingham demonstrates. His conceptual framework was innovative. Schröder spoke not of the iron trade, but of an 'iron system' that extended across Europe and the northern seas. The functioning of this 'iron system' was politically determined, as he well understood. For that reason, Schröder's trip to Britain was not only taken up with visiting manufactories and merchant exchanges; he scrutinized the journals of the House of Commons as well, making detailed extracts of the parliamentary proceedings that touched upon the iron trade. He was right to do so, for the fate of Britain's own iron industry had become a matter of public debate in the 1730s and 1740s.

---

[40] Anders Nordencrantz, Börstorpssamlingen, volym 82, signum E 3027, RA. See also Boëthius and Kromnow, *Jernkontorets Historia*, I, pp. 98f.

British anxieties

The dependence of Great Britain upon the 'Northern Powers' for the supply of iron and naval stores had long been an issue of official concern. Britain's negative balance of trade with the Baltic region—an apparently unalterable feature of British commerce—was another cause of disquiet. Britain's trade deficit with the Baltic was, in fact, a comparatively recent development. In the fifteenth and sixteenth centuries the flow of grain and naval stores westward through the Sound had been balanced by the shipment of English woollen textiles eastward. But during the seventeenth century there was a radical change in the composition of Britain's Baltic trade and the channels through which it flowed. English cloth was in decline on the once busy Danzig market, and the focal point of British mercantile activity shifted north to Stockholm and its *Järnvåg*. Yet there was little prospect of Stockholm, with its relatively impoverished hinterland, providing an alternative outlet for Yorkshire broadcloth. 'Sweden', it was lamented, 'takes... [only] a small Quantity of the Manufactures and Productions of *England*.'[41] While re-exported tobacco and a residual sale of textiles kept England's trade with the southern and eastern Baltic areas in rough equilibrium, the mounting tide of iron exports from Stockholm pushed England's balance of trade with Sweden into chronic deficit. The 'English Merchant or Factor's Busyness in Sweden', it was acknowledged, 'is only to buy up & Ship home Iron, Pitch and Tar, either for ready Money or (which is the same Thing) for Bills upon Hamb[urg] Amsterd[am] and London'.[42] By 1700 the value of English exports to Sweden amounted to just 29 per cent of the value of the iron and naval stores that moved westward.[43]

The trade deficit continued to deteriorate in the early eighteenth century. The value of British exports *vis-à-vis* the Swedish goods for which they were exchanged slipped to 27 per cent in the 1720s. The figure

---

[41] [Joshua Gee], *The trade and navigation of Great-Britain considered* (1729), p. 17.

[42] TNA: PRO, CO 390/12, fo. 90. See Jacob M. Price, 'Multilateralism and/or bilateralism: the settlement of British trade balances with "The North"', *Economic History Review*, XIV (1961), 254–74, for a discussion of credit flows, and H.C. Johansen, 'How to pay for Baltic products?', in Wolfram Fischer, R. Marvin McInnis and Jürgen Schneider (eds), *The emergence of a world economy 1500–1914. Part 1: 1500–1850* (Wiesbaden, 1986), pp. 123–42, for a comparison of British, Dutch and French trading patterns.

[43] Sven-Erik Åström, *From cloth to iron: the Anglo-Baltic trade in the late seventeenth century. Part 1: The growth, structure and organization of the trade* (Helsinki, 1963), p. 57.

plunged to just 12 percent in the 1730s.[44] This deterioration is reflected in Graffin Prankard's endless search for articles for which there might be a significant Swedish market. Luxuries and semi-luxuries from the Bristol area, such as glassware or 'Hotwells' spa water, were regularly sent to Stockholm.[45] Raw materials for Sweden's own industries were also shipped on the *Parham* or the *Baltick Merchant*. Calamine (zinc carbonate), for example, was a crucial input for Swedish brass smelters. It was mined in the Mendips, and Prankard came to an arrangement with the Bristol Brass Company, the major buyers locally, to obtain supplies at a preferential rate.[46] Salt found a ready market as well. The Baltic, into which the vast river systems of central and eastern Europe drained, lacked salinity. Salt, accordingly, was in short supply, and Prankard was in a position to make good the deficiency, being the proprietor of a large salt works at Droitwich in Worcestershire.[47]

Despite these efforts, Graffin Prankard could never assemble a cargo equal in value to the bar iron and timber for which it was to be traded. His account with Francis Jennings was in permanent deficit and could only be settled by bills of exchange drawn on Thomas Hyam, his banker in London. The surplus Prankard earned in his Atlantic trade had to be used to support his trade with the east. The situation of Prankard and other Baltic merchants was made worse by the determination of the Swedish authorities to exclude foreign—and especially British—commodities. The *Produktplakat* of 1724, which proved so disadvantageous to British shipping, was augmented by heavier duties on foreign manufactures in 1728, 1732, and 1734.[48]

---

[44] TNA: PRO, BT 6/185, [Sir Charles Whitworth] 'State of the Trade of Great Britain', 1776.

[45] GP to FJ, 17 July 1736. There was a sizeable trade in Hotwells water, but it was usually exported to the West Indies rather than Scandinavia. See Sylvia McIntyre, 'The mineral water trade in the eighteenth century', *Journal of Transport History*, VII, 1 (1973), 1–19, especially 5 and 13.

[46] GP to FJ, 7 July 1736.

[47] Prankard's father-in-law, William Alloway of Bridgwater, had been a considerable trader in salt, operating a variety of salt pans in Somerset and Cornwall. He was also an investor in the salt works that began to develop very rapidly in the 1690s around Droitwich. Prankard followed suit. See Hussey, *Coastal and river trade*, pp. 151–83, and A.P. Wakelin, 'Pre-industrial trade on the river Severn: a computer-aided study of the Gloucester port books c. 1640–c. 1770', (unpublished CNAA Ph.D. thesis, 1991), pp. 205–08.

[48] See the discussion in the Board of Trade correspondence for the 1720s and early 1730s, especially TNA: PRO, CO 388/25, which contains a memorial against the 'new Swedish ordinance on shipping', 13 April 1725; TNA: PRO, CO 388/27, Jackson to Lord Townshend, 10 July 1728, containing a translation of the 'Placart for laying the

All of this seemed to call for aggressive countermeasures. But Edward Finch, the British diplomat in Stockholm, warned against expecting any initiative from Britain's own Baltic merchants. 'The Iron Merchants in London & their Factors here, tho' each His Majty's Subjects, yet as to this Branch of Trade, should rather be lookt upon as Swedish than English Merch[an]ts since their whole Business lyes in buying up the Commoditys of Sweden, and revending them in England.' They were, he continued, 'in no ways concerned in purchasing English Manufactures, and selling them here'.[49] Having concluded that members of the Anglo-Scottish merchant community in Stockholm were essentially renegades (a conclusion given weight by the propensity of many foreign merchants to become naturalised Swedes), Finch looked for a solution to Britain's fraught relationship with Sweden elsewhere. He found it, so he thought, in Britain's Atlantic empire.

An Atlantic iron industry?

The answer to Swedish high-handedness, so Edward Finch believed, lay in encouraging iron smelting in the American colonies. If the plantations could supply large quantities of pig iron to the mother country then Britain's own blast furnace sector would be superfluous and the charcoal it consumed could be reassigned to a much enlarged forge sector. With this, the necessity for Baltic iron would fade away. Britain could become self-sufficient in bar iron if Parliament offered appropriate fiscal inducements. 'All these Grievances', wrote Finch, referring to the tariffs on British goods and the restrictions on non-Swedish shipping, 'will be redressed in one Shape or another the moment the Parliament thinks proper to discharge the five shillings per Tonn the Plantation Pig Iron now pays, & according to the Example here, Set a higher value on Swedish Iron with 5 per Ct Duty on it.' The fruits of this duty on Swedish iron, he went on, should be 'appropriated to pay such a Premium to Plantation Pigg Iron, which may then be furnished to the Forges in England, and if new ones are Set up, as they may be in the places

new Duty of 5 per Cent on several foreign Commoditys'; TNA: PRO, CO 388/31, Edward Finch to Lord Harrington, 5 January and 12 April 1732.

[49] TNA: PRO, CO 388/34, Edward Finch to Lord Harrington, 12 August 1735.

of melting Furnaces, We shall have both Forges and Wood enough in England to strike twice the quantity [of bar iron] We now do.'[50]

This was no abstract notion. Such an arrangement would have been a mercantilist commonplace. Had not the British state already put in place fiscal incentives for the production of ship's timbers and other naval stores in the north American colonies with the express intention of lessening national reliance on Baltic sources of supply?[51] Moreover, Edward Finch's suggestion built on tendencies already at work in the Atlantic economy. A colonial iron industry had developed strongly in the aftermath of the Peace of Utrecht. Progress was particularly marked in the mid-Atlantic region where ore beds, stands of timber, and good water-borne communication coincided. Furnaces sprang up along the Patapsco, the Potomac and other rivers that emptied into the Chesapeake; they studded the hinterland of Philadelphia; they encroached upon the pine barrens of New Jersey.[52] The indirect process of iron making was not a feature of the mid-Atlantic colonies at the end of Queen Anne's War, but by 1750 the region was home to 28 blast furnaces and 46 forges. Encouragement came from colonial assemblies that were keen to diversify one-dimensional plantation economies. Maryland's legislators, for example, passed acts in 1719, 1722, 1736 and 1750 granting lands to ironmasters and extending tax privileges to certain categories of iron worker; Virginia's legislators did likewise in 1727 and 1732. Capital was provided by colonial merchants and planters such as the Carrolls of Annapolis, the prime movers behind the Baltimore Company, established in 1731.[53] Finance was also to be had, as we shall see, directly from English entrepreneurs.

---

[50] TNA: PRO, CO 388/34, Edward Finch to Lord Harrington, 22 April 1735.

[51] R.G. Albion, *Forests and sea power: the timber problem of the Royal Navy* (1926); Joseph P. Malone, *Pine trees and politics: the naval stores and forest policy in colonial New England, 1691–1775* (London, 1964); Justin Williams, 'English mercantilism and Carolina naval stores, 1705–1776', *Journal of Southern History*, I, 2 (1935), 169–85.

[52] See Lester J. Cappon (ed.), *Atlas of early American history: the revolutionary era 1760–1790* (Princeton, 1976), pp. 105–06, for an overview of the industry's growth, and John Bezís-Selfa, *Forging America: ironworkers, adventurers, and the Industrious Revolution* (Ithaca NY, 2004), pp. 16–25, for a more wide-ranging interpretive narrative. Older regional accounts are still worth consulting: Arthur C. Bining, *Pennsylvania iron manufacture in the eighteenth century* (Harrisburg PA, 1938); Charles S. Boyer, *Early forges and furnaces in New Jersey* (Philadelphia PA, 1931); Kathleen Bruce, *Virginia iron manufacture in the slave era* (New York NY, 1931).

[53] Keach Johnson, 'The genesis of the Baltimore ironworks', *Journal of Southern History*, XIX, 2 (1953), 157–79.

Skilled English workers were recruited to implant the appropriate technology. This was the indirect process—the two-stage process involving a blast furnace at which ore was smelted and a forge at which the outcome was refined—that had come to maturity in the southern Netherlands in the late middle ages. Warfare and political instability in their native region drove Walloon furnacemen and forgemen who were schooled in these methods west and south in the course of the fifteenth century. By the 1450s they were at work in the forests of Bray in Normandy. From there, they crossed the English Channel to bring Walloon techniques to the Weald of Sussex and Kent, just as a later generation of francophone migrants would take their distinctive working habits to the *Vallonbruk* of Uppland. From their Wealden bridgehead, Walloon ironworkers spread across the British Isles in the sixteenth and seventeenth centuries, pushing aside the direct reduction techniques that had been employed by medieval smiths. Wherever the new style of forging was practised, workers bearing anglicised versions of Walloon surnames were to be found: Leonard, Vinton, Lambert, Tyler, Russell, Jarrett. At the start of the eighteenth century these old Walloon dynasties still loomed large in the workforce of the British iron industry.[54] It was fitting, therefore, that when Walloon techniques were introduced to the mid-Atlantic region in wake of the treaty of Utrecht, Walloon dynasts should be to the fore. James Jarrett (a corrupted echo of the ancestral name Gerarde), to name but one, left Gloucestershire in the early 1720s to become one of the first finers at the Principio ironworks in Maryland.

The finers and hammermen who set sail from Bristol in the 1720s and 1730s were keenly aware of their elite status and proved stiff-necked, fractious employees. James Jarrett and a fellow English finer refused point blank to train Africans in the Walloon ways: 'All ye Arguments yt Cou'd be used Cou'd not prevail to admit them of a clause to teach Negroes', the manager at Principio reported in 1725; 'they said they were murdering Rogues'.[55] But adepts of Walloon descent were a small minority of the workforce. Most of those who worked at

[54] See B.G. Awty, 'The continental origins of Wealden ironworkers, 1451–1544', *Economic History Review*, XXXIV (1981), 524–39, and Chris Evans, 'A skilled workforce during the transition to industrial society: forgemen in the British iron trade, 1500–1850', *Labour History Review*, LXIII, 2 (1998), 143–59.

[55] Quoted in Bezís-Selfa, *Forging America*, p. 76. Bezís-Selfa provides a very full analysis of the labour regimes in colonial iron making, on which we have drawn.

colonial ironworks, wrenching ore from the ground and coaling wood, suffered servitude in one form or another. Some were indentured servants, labouring at the ore pits to repay the costs of their passage from London or Rotterdam. James Sumners, a 'West Country Man [who] speaks thick', can stand for the thousands who passed through the Chesapeake. He was bound to the Bristol Company's furnace in King George County, Virginia. The life he found there was evidently not congenial, for he absconded in March 1737. A reward was offered for the capture of this 'short thick Fellow, with short black Hair, and a Ruddy Complexion'.[56] In an environment where labour was scarce it had to be fixed in place. For that very reason plantation ironworks also relied on English convicts who had been condemned to penal exile for periods of seven years or more. Transported felons were used on an extensive scale. The Bristol partnership of Stevenson, Randolph & Cheston, which shipped criminals across the Atlantic in the 1760s, sold 56 convicts to the proprietors of Northampton furnace, Maryland, in just one three-year period.[57] The position of such labourers was no enviable one. They were liable to be whipped or pilloried for acts of disobedience. The doctor who attended Northampton furnace spoke of 'acts of Cruilty... as would extort a Blush from a Turkish Bashaw'.[58] The work was arduous and transportees had no right to the 'freedom dues' that indentured servants could claim from their masters at the end of their term of service. Hence the regularity with which convicts took flight.[59] William Hatton, a stocking weaver by trade, could not endure the gruelling routine of the mine bank at the Principio Company's furnace at Kingsbury, Baltimore County. He slipped away in August 1768. Law-abiding inhabitants of the Chesapeake were duly warned against Hatton, a man with a 'down look, and a remarkable way of staring any person in the face that speaks to him'—or so his master claimed, to underline the incorrigible wickedness of his fugitive servant.[60]

When William Hatton fled Kingsbury furnace he did so in the company of Cyrus, an African slave. Indentured servants and convicts lacked freedom, but their lack of freedom was partial or of fixed duration.

---

[56] *Virginia Gazette*, 18 March 1737.

[57] Bezís-Selfa, *Forging America*, p. 80.

[58] Bezís-Selfa, *Forging America*, p. 85.

[59] Gwenda Morgan and Peter Rushton, *Eighteenth-century criminal transportation: the formation of the criminal Atlantic* (Houndsmill, 2004), pp. 106, 135–36.

[60] *Virginia Gazette*, 29 September 1768.

Africans, on the other hand, were chattels whose servile condition was absolute and likely to be life-long. Because of that, Africans figured heavily in the expansion of the mid-Atlantic iron industry. William Spotswood, the most substantial of Virginia's ironmasters in the 1720s and 1730s, thought that 120 slaves were needed for an ironworks to function well.[61] In those formative years Africans were largely restricted to the processing of ore and charcoal. By the mid-century, however, slaves featured in all departments. The reluctance of James Jarrett to reveal the mysteries of Walloon forging to Africans had been futile. By the 1750s one of his successors as a master finer, John Holloway, worked with 'Negro' assistants, Prince and Dick.[62] Indeed, in the era of the American Revolution the furnaces and forges of the mid-Atlantic were worked predominantly by African bondsmen. In America, as in Siberia, another iron making frontier, the development of the industry rested upon the exploitation of unfree labour.[63]

This burgeoning new sector catered in the first instance for local demand, but the export of pig iron to Great Britain began in the 1720s. The earliest exports scarcely attained statistical visibility (just 15 tons in 1723), but by the end of the decade exports had begun to move upwards as shippers became alert to the benefits of using iron to ballast cargoes of tobacco. The 886 tons that were shipped to Britain in 1728 became 1714 tons in 1730, then 2332 tons in 1732.[64] Much of this passed through Bristol, the commercial antechamber to the Midlands, the heartland of the English forge trade. Graffin Prankard could already speak of American pig as a regularly traded commodity on Bristol's quayside in 1730.[65] If smelting relocated to the colonies would it be doing anything more than following the example of

---

[61] Philip D. Morgan, *Slave counterpoint: Black culture in the eighteenth-century Chesapeake and Lowcountry* (Chapel Hill, NC, 1998), p. 230.

[62] Michael W. Robbins, *The Principio Company: iron-making in colonial Maryland* (New York, 1986), p. 109.

[63] This was a recurring feature of the industry in America. A pioneering attempt at iron making on Massachusetts Bay was only sustained by the arrival of Scottish prisoners of war in 1651. E.N. Hartley, *Ironworks on the Saugus: the Lynn and Braintree ventures of the Company of Undertakers of the Ironworks in New England* (Norman OK, 1957), pp. 146–47, 154–55. The range of unfree labour in colonial America is surveyed in Kenneth Morgan, *Slavery and servitude in north America, 1607–1800* (Edinburgh, 2000).

[64] Arthur C. Bining, *British regulation of the colonial iron industry* (Philadelphia, 1933), pp. 128–33.

[65] GP to 'Coz Paine', 17 June 1730. Prankard differentiated between the following pig irons: 'Virginia is [sold] for £5:10 English £6:10 to £6:15 but ye best Sorts yt is made at Colebrook in Shropshire is £7 per ton'.

shipbuilding, which was shifting wholesale to timber-rich New England at this very time?[66]

Graffin Prankard was scornful of American iron—'ye Virginia Pigg', he sniffed, 'is ye Worst of all for Castings or any other use'—but many forgemasters did not share his view. The Knight partnership, the most powerful group of ironmasters in the Midlands, made a heavy commitment to colonial iron in the 1730s. The Knights had five forges to supply along the river Stour: two at Mitton and others at Cookley, Wolverley, and Whittington. They had their own blast furnace at Charlcotte in Shropshire, but this was insufficient for their purposes, so they had to draw upon other suppliers in the Midlands and the Forest of Dean.[67] The arrival of colonial pig iron at Bristol offered a new, cut-price alternative, one that the Knights eagerly embraced. Thirty-four per cent of the pig iron refined at the Stour forges in 1735–36 was American; 45 per cent in 1736–37; and 43 per cent in 1737–38. Most of it came from the Virginia works of the 'Bristol Company', the creation of a consortium of Bristol tobacco merchants and ironmongers who had erected a blast furnace on the Rappahannock river with the express intention of shipping pig iron to Britain.[68]

In the North East the Crowleys were also interested in the possibilities raised by American iron. Colonial pig was being used in the forge at Swalwell by the mid-1720s, and by the mid-1730s the Crowley firm was buying large quantities from the two most important Maryland partnerships: the Principio Company and the Baltimore Company. The Crowleys took Baltimore Company pig iron at £6 5*s*. per ton, paying half in cash and half in ironwares 'at Ready Money Price' for the Baltimore partners to dispose of in the Chesapeake.[69] It was also at

[66] By the middle of the eighteenth century at least one quarter of Britain's merchant fleet was American-built. See Ralph Davis, *The rise of the English shipping industry in the seventeenth and eighteenth centuries* (Newton Abbot, 1962), pp. 66–68, and Joseph A. Goldenberg, *Shipbuilding in colonial America* (Charlottesville VA, 1976), especially chapter 3.

[67] The Stour works accounts show the forges taking pig iron from furnaces in Shropshire, Cheshire, the Forest of Dean, and Scotland.

[68] Worcestershire Record Office (St Helens Branch), MSS 810:399, Stour works general accounts; G. MacLaren Brydon, 'The Bristol iron works in King George County', *Virginia Magazine of History and Biography*, XLII, 2 (1934), 97–102. One of the 'Committee of the Bristol Company's Iron Works' in the late 1720s was William Donne, the Bristol ironmonger and proprietor of the slitting mill at Congresbury in Somerset who also bought Russian iron from Graffin Prankard.

[69] Suffolk Record Office (Ipswich), HAI/GD/5/13, 'An Inventory of the tools Implements and goods found in the Workmens Hands at Mill no. 1'; Keach Johnson, 'The Baltimore Company seeks English markets: a study of the Anglo-American iron

the initiative of Theodosia Crowley & Co that the Navy Board tested American bar iron in 1735–36. The smiths at the Royal Dockyards were ordered to 'make a Sufficient Tryal' of the iron 'Imported by Mrs Crowley from America, which she esteem to be equal in goodness to Sweeds Iron'.[70]

The comparability of American, Swedish, Russian, and British irons was indeed crucial. 'Orground' iron commanded a premium price because it was, quite literally, incomparable. Other Swedish irons had a durability that distinguished them from the standard English irons. Some Russian brands had 'tough' qualities, but others were coldshort after the English fashion. Determining where bar iron made from American pig lay on this spectrum was a critical question in the controversies that developed in the 1730s. The issue attracted parliamentary attention in March 1737 when a petition of the 'Merchants & Ironmongers of this kingdom' who traded to America was laid before the House of Commons. The petition complained that whereas there were 'premiums settled by Acts of Parl[iament] upon the import[atio]n of Tar, pitch, turpentine, Hemp, masts, yards & bowsprits' from the plantations, no incentive was offered for the importation of iron. Yet, it was claimed, the timber-rich colonies had it in their power to render Great Britain independent of Baltic iron. And they should be encouraged to do so. If they were not, there was every danger that the colonists would begin to manufacture ironwares of their own, precipitating a disastrous decay in the British metalware trades—trades that gave support to more of the 'poor laborious sort' than any other trade apart from woollens manufacturing. The petitioners called for a bounty to be paid on the importation of American pig and bar iron, and for a prohibition on colonial hardware manufacturing. It was a piece of rigid mercantilism: the colonists were to supply raw materials to metropolitan manufacturers whilst providing a captive market for British-made ironwares.

A similar appeal had been made in 1717 when a breakdown in diplomatic relations between Britain and Sweden, accompanied by a suspension of trade, had led to an acute shortage of bar iron on the British market. Swedish iron could only be obtained surreptitiously via the Dutch market, with predictable results: 'all commerce with Sweden

---

trade, 1731–1755', *The William and Mary Quarterly*, 3rd series XVI (1959), 37–60. See also Robbins, *The Principio Company*, pp. 211–18.

[70] National Maritime Museum, CHA/E/4, the Navy Board to the Chatham Dockyard officers, 11 July 1735.

[being] interrupted, has caused their iron to advance here from 16 to 24 pound a tone'.[71] Loud calls were made for the promotion of colonial iron making. Such calls had been no more than aspirational in 1717, but the development of ferrous metallurgy in British North America in the intervening twenty years lent the scheme more plausibility when it was revived in the 1730s. Joseph Farmer, the Birmingham gunmaker, had visited the plantations in 1718 and 1719 to assay ores. He had seen no furnaces or forges, just some primitive bloomery hearths that made malleable iron for local use. By the mid 1730s, however, the situation was much changed, or so Farmer told the House of Commons committee appointed to hear evidence in support of the ironmongers' petition. He had imported bar iron from both Maryland and Pennsylvania and found it good. Farmer was just one of a string of witnesses who testified to the quality of American iron. John Bannister, the chief agent of Theodosia Crowley, pronounced colonial iron equal to the Swedish product for every purpose save that of making steel, and even that could not be ruled out if careful use was made of 'Bar Iron made from Baltimore & Principio Pigs'.[72]

These assertions elicited a swift response from English ironmasters. They submitted counter-petitions claiming that the free admission of American iron threatened Britain's native iron industry with extinction. The suggestion that American iron could end British reliance on the Baltic was, they said, quite bogus. The ironmasters who were hostile to colonial imports suspected that it would be American bar iron, not pig iron, that was landed at British ports. The outcome would not be a selective closure of blast furnaces, as the advocates of a transatlantic iron industry maintained, but the extinction of the British forge sector. Colonial iron, its enemies asserted, was for the most part coldshort rather than tough. It resembled 'the second sort of English Iron', the sort used in nail making. It could not therefore act as a substitute for the tough Swedish brands. It would compete directly with the product of Midland forges instead. English coldshort iron was already menaced by cheaper Russian imports. If low-cost American producers were allowed to dispute the market for coldshort iron as well, it would be the makers

---

[71] J.D. Marshall (ed.), *The autobiography of William Stout of Lancaster 1665–1752* (Manchester, 1967), p. 177.

[72] SML, Weale MSS, 371/1, fo. 125.

of relatively expensive English bar iron who would be driven from the market, not the Russians.[73]

The repercussions would extend through the entire economy, with landowners, farmers, and other industrialists being adversely affected. As forges shut down, coppice woods would be grubbed up and the land they had occupied, being so often of poor quality, left idle, yielding no income to its owners.[74] All the charcoal burners, hauliers, and labourers who had been employed in supplying fuel to the forge trade would be reduced to poverty and become a charge on the parish. '[S]everal thousand *British* families', one propagandist expostulated, would be deprived of the means of subsistence, merely to benefit 'barbarous Herds of Criols and Negroes'.[75] Consideration should also be given to other trades that were interested in the preservation of woodlands. Leather tanners, for example, were dependent on the cut-price oak bark that was a by-product of charcoal making. If this accumulation of private interests was not enough to disabuse the legislature, then the ironmasters could point to the state's strategic requirement for iron. Was it not an essential public good that Britain enjoyed a secure domestic source of such an important instrument of war?

> ...how deplorable must be our Case *in Time of War*!...we divest ourselves at Home of *the necessary Means* of making Arms, and must trust to the *Curtesy of a Colony* to supply us; who may either furnish us or our Enemies, as best suits their Interest; or, perhaps, use them themselves, and, one Day, turn them against their Mother-country...[76]

The ironmasters and their allies revisited these sombre themes the following winter. They took the fight to the iron merchants and hardware manufacturers by petitioning Parliament for an *increased* duty on all varieties of imported iron. 'Our English Iron makers & Wood Gentlemen', Sampson Lloyd, the Birmingham ironmonger, reported

---

[73] SML, Weale MSS, 371/1, fo. 130. Graffin Prankard believed that the availability of colonial pig iron would depress the price of English and Welsh bar iron: 'as our Import of American Piggs increases...[it] will be a great Check on ye Iron masters in those neighbouring Cuntrys that they cannot advance the price of English Iron one shilling'. GP to Mackenzie & Grundy, 12 April 1732.

[74] See the petition from the borough of Monmouth: House of Lords Record Office, House of Lords main papers, 1750, fo. 1131.

[75] *The state of the trade and manufactory of iron in Great-Britain considered* (1750), p. 15.

[76] *National considerations on importing iron in bars from America, &c* (nd). Compare the views of Salvius in *Tanckar öfwer den Swenska Oeconomien*.

in February 1738, 'are very busy to form a Strong Interest this Sessions to get a further Duty laid on Foreign Iron in order to advance their own'.[77] Once again, a committee of the House of Commons was appointed to investigate the claims made. Many of the witnesses and counter-witnesses of the previous year reprised their testimony. Once again, much was made of the supposed qualities of colonial iron, which was represented both as the salvation and the damnation of the English metalware trades. The outcome was confused. All sides were agreed that the colonists should be prevented from erecting slitting mills, steel furnaces, and other facilities that would allow them to develop a hardware manufacturing sector of their own, but there was agreement on little else. Leave was given for a bill to be brought in that embodied the resolutions of the committee, but no firm, uncontested resolution had emerged. Not surprisingly, no bill was ever brought before the House of Commons.

There was a lack of clarity in the parliamentary investigations of 1737–38, one that faithfully reproduced the lack of clarity among the protagonists as to what was meant by 'American iron'. Should colonial *bar iron* be admitted duty free to the British market, or even receive a bounty upon its importation? This was what the merchants and manufacturing ironmongers had requested in their petition of March 1737. This was contrary to the interests of British ironmasters, plainly enough. Yet the ironmasters were by no means automatically opposed to the introduction of American *pig iron*. *Pace* T.S. Ashton, who, in his classic *Iron and steel in the Industrial Revolution*, wrote that 'ironmasters in both furnace and forge branches of the industry were anxious to shield their product from the rude breath of American competition', the reaction of the iron trade to the spectre of American iron was, as we have seen, ambivalent.[78] Many forgemasters were quite content for American pig iron to enter duty free. Many ironmasters, like the Knights, already made extensive use of pig iron from the Chesapeake. Many more were ready to follow their example. 'All the Iron-masters

[77] Religious Society of Friends Library, TEMP MSS 210/2/42, Sampson Lloyd to Thomas Kirton, 11 February 1738. Lloyd was not impressed: 'it is not for every Man that hath a bad Trade to apply to Parliament to make good the deficiency'. Graffin Prankard urged his London banker to take counter steps: 'would it not be of Service', he asked Thomas Hyam, 'to Represent the Case to the Spanish Minister residing in London also those from Muscovy & Sweden'? GP to Thomas Hyam, 18 February 1738.

[78] T.S. Ashton, *Iron and steel in the Industrial Revolution* (Manchester, 1963), p. 117.

in these parts have met and consulted about the affair', it was reported from Yorkshire in March 1737, 'and desire the Metal [pig iron] may come in duty free.'[79] This was to recommend the fostering of a truly imperial iron industry, as Edward Finch in Stockholm had envisaged, bridging the Atlantic. American smelters would sustain a rejuvenated forge sector in Britain. If ironmasters were to 'Distroy the furnaces in England', two representatives of the Yorkshire iron trade reckoned, echoing Finch, '& consume the wood now used by them at Forges in making american pigg Iron into barr Iron...this Nation would be able to make near double the quantity of Bar Iron it dose at present'.[80] By doing so, English ironmasters would be doing no more than extending a spatial logic that was already at work, pushing smelting to ever more westerly, seaboard locations. The last generation of charcoal blast furnaces in the British Isles, built between the 1720s and the 1750s, were almost all located on the northern or western fringes of the British Isles, at remote coastal sites where wood was abundant and from which pig iron could be shipped to forges in central England.[81] The furnace at Invergarry in the Scottish Highlands, for example, first blown-in in 1729, supplied pig iron to the Knights' Stour forges in the 1730s.[82] If pig iron could be freighted from Highland Scotland, why not from the Chesapeake?

The parliamentary stalemate of 1737–38 was repeated in 1749–50 when the arguments over Baltic and colonial iron were rehearsed once more. This was unavoidably so, given that the contending parties in the debate were not clearly demarcated. Because of the fluid organisational patterns of the iron trade, furnace owners and forgemasters could not be neatly separated, nor could ironmasters be easily distinguished from ironmongers and merchants. All found a refuge in the capacious partnerships that characterised the iron trade. The seamlessness of the iron trade is well illustrated from amongst Graffin Prankard's customers. The Homfrays of Stourbridge were simultaneously steel manufacturers,

[79] West Yorkshire Archive Service, Bradford, SpSt/5/5/1/4, John Watts to William Spencer, 9 March 1737.

[80] Sheffield Archives, Wharncliffe Muniments, 118/15, John Cockshutt and Joseph Broadbent to unidentified correspondent, 17 February 1750. Compare [Gee], *Trade*, p. 71.

[81] Philip Riden, 'The final phase of charcoal iron-smelting in Britain, 1660–1800', *Historical Metallurgy*, XXVIII (1994), 14–26.

[82] Philip Riden, *A gazetteer of charcoal-fired blast furnaces in Great Britain in use since 1660* (Cardiff, 1993), p. 151. This was the 'Scotch' pig listed in the Stour accounts.

forgemasters, iron slitters, putters-out in the nail trade, and export ironmongers.[83] Their attitude to foreign iron was correspondingly mixed. As steel manufacturers they depended completely upon 'Orground' imports, and were therefore opposed to higher duties on Swedish iron. As the owners of a forge at Swindon in Staffordshire they were opposed to bar iron imports from both the Baltic and America, yet welcoming of pig iron from the plantations. But as proprietors of a slitting mill, the Homfrays required regular supplies of bar iron, and if they could not produce it themselves they would buy Russian iron from Prankard.

Such was the skein-like organisational complexity of the British iron trade, and such the difficulty of establishing a single iron 'interest', that there was little likelihood of a clear policy emerging from the renewed parliamentary deliberations of 1749–50. An Act of Parliament was passed in 1750 (23 Geo.II c. 29—the Iron Act) but its effects were limited. The duty on colonial bar iron was removed, but only on iron shipped to London, and this iron could not be shipped coastwise or carried by land outside a ten-mile radius of the capital. It was another seven years before further legislation (30 Geo.II c. 16), prompted by a petition of the Merchant Adventurers of Bristol, allowed colonial bar iron to enter outports duty-free.

Potentially, this was the stuff of Swedish nightmares. Had colonial iron been underestimated in Stockholm? Pehr Kalm, professor of natural history and economy at the university of Åbo, who toured British North America in 1750, reflected gloomily on the mineral abundance of the New World: '*Iron* is dug in such great quantities in Pennsylvania and in the other American provinces of the English, that they could provide with that commodity not only England but almost all Europe, and perhaps the greater part of the globe'.[84] The way was now clear for the colonial iron industry to assume a major role in the international 'iron system', either by supplying pig iron to British forges or by furnishing bar iron in sufficient quantities to expel Baltic iron from the British market. Neither eventuality came to pass. A few American ironworks were dedicated to supplying the English market, it is true, and some new-built processing plants of the 1760s appear to

[83] Laurence Ince, 'Homfray family (per. 1702–1833)', *Oxford Dictionary of National Biography*, Oxford University Press, 2004 [http://www.oxforddnb.com/view/article/47499, accessed 9 April 2006].

[84] Adolph B. Benson (ed.), *The America of 1750: Peter Kalm's travels in North America. The English version of 1770* (2 vols, New York, 1937), I, p. 159.

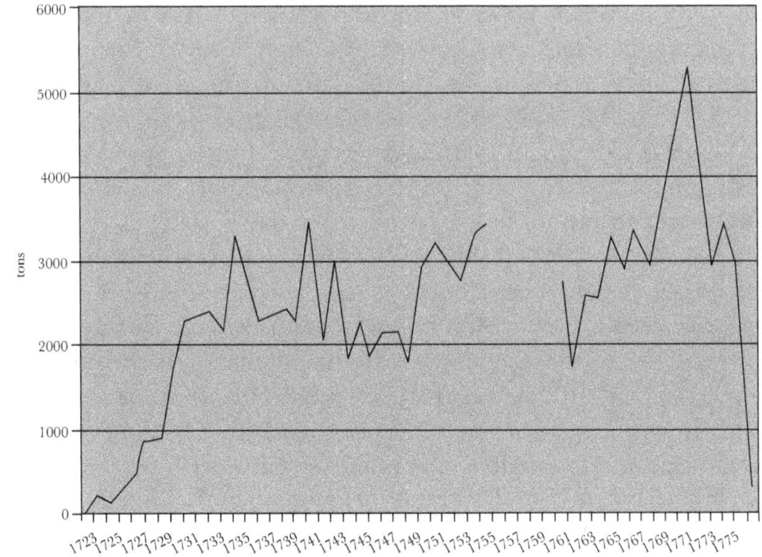

Source: A.C. Bining, *British regulation of the colonial iron industry* (Philadelphia, 1933), pp.128–33.

Figure 3.1. North American pig iron exports to the British Isles.

have been premised upon the exploitation of American pig iron. The Melingriffith tinplate works in south Wales was one such. In 1771–72 its forge took over 80 per cent of the pig iron it consumed from colonial furnaces: from the Forest of Dean furnace in New York, from Batsto and Aetna in New Jersey, from Occoquon in Virginia, from Pine Grove in Pennsylvannia, and from Elk Ridge in Maryland.[85] But Melingriffith was an exception. The export of pig iron from the colonies, by which forgemasters like the Knights set such store, did not seriously exceed the volume achieved in the mid-1730s until the eve of the American Revolution. (See Figure 3.1)

The record of bar iron was even less impressive. Imports into London were negligible in the 1750s, whilst the opening of the outports to colonial bar iron, by coinciding with the onset of the Seven Years' War, had little immediate impact. Unhindered transatlantic commerce did not resume until after the Peace of Paris in 1763. From the mid 1760s

[85] Chris Evans, 'Global commerce and industrial organization in an eighteenth-century Welsh enterprise: the Melingriffith Company', *Welsh History Review*, XX (2001), 413–34, especially 418.

there was a noticeable upturn in the importation of American bar iron, but the quantities remained rather small. Only in 1772, the peak year for imports, did the volume of colonial bar iron entering British ports top 2,000 tons. This was not the counterweight to Baltic iron that the proponents of free importation had promised. Indeed, it paled besides the contemporaneous growth of Russian iron imports, which averaged 24,000 tons per annum in the last years of the 1760s.[86]

The course of events in the 1750s and 1760s revealed that it was not the much disputed tariff barriers that had kept American iron from British markets; it had been the lure of local demand in the colonies themselves. Given the steepling rise in the population of British North America, colonial ironmasters had customers enough in their immediate vicinity, without resorting to overseas markets. The American iron industry grew at a prodigious rate in the generation before the Revolution, with 47 furnaces and 59 forges starting up in the middle colonies alone between 1750 and 1776, but the output was retained within the colonies, not exported.[87] In 1752, when just 156 tons of pig iron was exported from Pennsylvania to Britain, 4,600 tons of bar and rod iron was shipped from Philadelphia, the principal port of that colony, to other harbours in British North America and the West Indies.[88] Imperial authority and commercial logic were at odds. The mercantilist regulations that aimed to restrain ironware manufacturing in the plantations were simply disregarded by the colonists.

Transatlantic failure

When the *Griffin*, an English East Indiaman of 600 tons, struck a reef in the Sulu Sea early in 1761, three weeks after quitting Whampoa in China, she was laden with 4,087 chests of tea, 800 pieces of Nanking cloth and nearly 200 chests of chinaware. She was also carrying pigs of iron from the Elk Ridge furnace in Maryland, presumably as ballast.[89] When marine archaeologists recovered them from the sea bed

---

[86] Kaplan, *Russian commerce*, p. 58.

[87] Bezís-Selfa, *Forging America*, p. 19.

[88] Bining, *British regulation*, p. 132; S.D. Smith, 'The market for manufactures in the thirteen continental colonies, 1698–1776', *Economic History Review*, LI, 4 (1998), 687.

[89] Charles Daggett, Evelyne Jay and Frederic Osada, 'The *Griffin*, an English East Indiaman lost in the Philippines in 1761', *International Journal of Nautical Archaeology and Underwater Exploration*, XIX, 1 (1990), 35–41.

off Mindanao, 225 years after the *Griffin* had foundered, the pigs were easily identified. They bore the legend 'Elk Ridge' in conformity with the Iron Act of 1750, which stipulated that colonial iron, whether in pig or bar form, should bear a mark that would allow it to pass duty-free into Britain. To those who had watched the Elk Ridge pigs being stowed aboard the *Griffin* at the East India Company's great Thames-side yard at Blackwall, their presence might have portended the global prospects of American iron. American furnaces, so it seemed, were on the brink of being integrated into a vastly expanded 'iron system' under British tutelage. That prospect was, as we now know, an illusion. Elk Ridge pigs were shipped from Annapolis to London, Bristol and Glasgow in the first years of the furnace's existence, but exports did not endure.[90] The immense transoceanic division of labour anticipated by many members of the British iron trade never came to pass. The future of American iron was continental, not transoceanic.

The American Revolution settled the question definitively, putting paid to whatever chance there had been of colonial ironmasters being subordinated to their British counterparts. That subordination had been widely denounced by Patriots in the years prior to the Revolution. The Iron Act of 1750, by prohibiting the construction of steel furnaces, plating forges and slitting mills in the colonies, was a focus for resentment. 'This country abounds in iron', the Rhode Island lawyer Silas Downer complained, 'yet there is an act of parliament, passed in the late King's reign to restrain us from manufacturing it into plates and rods by mill work.' He continued:

> Be astonished all the world, that the people of a country who call themselves Christians and a civilized nation, should imagine that any principles of policy will be a sufficient excuse, for their permitting their fellow subjects on a distant part of the earth from making use of the blessings of the GOD of nature.[91]

[90] Ronald W. Fuchs II, '"At Elk Ridge Furneis as you See, William Williams He Mad Me": the story of an eighteenth-century Maryland iron furnace', *Journal of Early Southern Decorative Arts*, XXII, 2 (1996), 40–59. See National Archives of Scotland, GD58/6/1/4, Carron Company to John Glassford & Co., 7 September 1764, for sales to Glasgow.

[91] 'A Son of Liberty' [Silas Downer], *A discourse at the dedication of the Tree of Liberty* [1768], in Bruce Frohnen (ed.), *The American Republic: primary sources* (Indianapolis, 2002), pp. 143–44.

John Dickinson took up the refrain in his *Letters from a farmer in Pennsylvania* (1767–68): 'Great Britain has prohibited the manufacturing *iron* and *steel* in these colonies, without any objection to her *right* of doing it'.

> If *Great Britain* can order us to come to her for necessaries we want, and can order us to pay what taxes she pleases before we take them away, or when we land them here, we are as abject slaves as *France* and *Poland* can shew in wooden shoes and with uncombed hair.[92]

The provisions of the Iron Act were, in fact, routinely flouted, yet a perception of the Act's iniquity contributed to the collapse of imperial authority in the 1770s.

The outbreak of hostilities in America in 1775 severed the ties between American furnaces and British forges. Colonial production slumped as troops marched and counter-marched across the iron industry's mid-Atlantic heartland. New Jersey and south-eastern Pennsylvania was a major theatre of war in the early years of the conflict, causing massive disruption in iron making communities. The British demolished more than one ironworks, and those left standing were subject to the depredations of foragers who stripped the countryside of food and movable goods. Many workers—like those at Hibernia, New Jersey—exited the industry 'because they were naked and could work no longer'.[93] Iron making in the Chesapeake also stuttered. Thousands of slaves—iron workers amongst them—responded to the November 1775 proclamation of Lord Dunmore, Virginia's royal governor, who promised freedom to those who left rebel masters to bear arms for the King.

That said, the war also stimulated fresh production, as the revolutionary authorities actively encouraged the setting up of works to make munitions for the Continental Army. Premiums were offered to prospective ironmasters by the provincial congresses of Pennsylvania, Maryland, Virginia, North Carolina and South Carolina. Scots-Irish entrepreneur William Hill was one of those who took up the offer of public subvention to build an ironworks in the Carolina backcountry. The furnace at Aera supplied 'Cannon Balls, Shells, Camp Kettles and Other utensils for the Army' before its destruction by British forces in June 1780. Hill also cast firebacks. A surviving example bears the legend 'LIBERTY OR DEATH' above the initials of Hill and his partner

---

[92] Quoted in Bining, *British regulation*, pp. 97–98.
[93] Quoted in Bezís-Selfa, *Forging America*, p. 180.

Isaac Hayne.[94] For Hayne, the rhetorical courting of death proved fatal; he was hanged by the British at Charleston in August 1781. For William Hill, however, the Revolution was a time of opportunity. He fought with distinction in the partisan war that wracked the Carolinas, held public office as a state assemblyman and senator, and rebuilt his industrial concerns when the war ended.[95] A new furnace was erected at Aera in 1787; a second followed at Aetna in 1788. They furnished cast wares and farm tools for a local clientele. No mention was made of transatlantic markets.

These developments bore out the misgivings that *Herr Mentor* and *Fru Swea* had expressed about colonial revolt. Swedish complacency, it seemed, had been justified. Baltic iron remained dominant on the British market in the wake of the American War. Indeed, Baltic imports exceeded domestic bar iron production by a ratio of roughly two to one in the late 1780s. *Herr Mentor* had been less prescient with respect to Russia, however. Siberian iron was not the phantom threat that Lars Salvius had imagined. On the contrary, exports from St Petersburg to Britain grew with such speed in the 1760s and 1770s that they soon came to exceed Swedish shipments. The balance of power in the Baltic had shifted, but Swedish officials and *brukspatroner* had no immediate cause for alarm. Their share of the British market may have slipped, but the absolute quantity of Swedish iron sold in Britain was undiminished. The growth in demand for bar iron was such that Swedes and Russians could co-exist peacefully. Moreover, Swedish policy makers could rest assured that the Russians posed no threat on the market for steel-making irons. There, 'Orground' iron remained pre-eminent.

Baltic hegemony was not secure, however, in the wake of the American War. It was about to be capsized by the wholesale technological transformation of Britain's own iron industry. The advent of new coal-based working methods, allied to new political conditions in the Revolutionary Age, reduced Baltic iron to a subaltern role on the British market in little more than a generation. The rival mercantilist systems in which the Swedes and the British alike had invested so many hopes were swept brusquely aside.

---

[94] Thomas Cowan, 'William Hill and the Aera ironworks', *Journal of Early Southern Decorative Arts*, XIII (1987), 1–32.

[95] Keith Krawczynski, "Hill, William"; http://www.anb.org/articles/01/01-01286.html; *American National Biography Online* Feb. 2000. Access Date: Thu Jun 29 09:49:42 UTC+0100 2006.

CHAPTER FOUR

AN INDUSTRIAL REVOLUTION IN IRON—TECHNOLOGY, ORGANISATION AND MARKETS, 1760–1870

Baltic iron had come to dominate the British market because of the incapacity of Britain's own forge sector. British ironmasters lacked the energy resources to keep pace with the heightening demand for malleable iron on their domestic market. Some ironmasters sought to overcome this deficiency by organisational means. They hoped to raid the abundant energy reserves of British North America by transferring the preliminary stages of the production chain to the colonies. An Atlantic iron trade, with smelting out-sourced to the charcoal-rich plantations, would be a reproof to the 'Ignorance & wrong reasonings' of those Swedish ministers who maintained 'that England can not be without their Iron'.[1] That hope was, as we have seen, thwarted.

The alternative to organisational re-jigging was technological transformation. Technological revolution there was, as every textbook on British economic history makes clear. Smelting with coke and the development of coal-fired refining methods, most notably Henry Cort's puddling technique, freed the British iron industry from its dependence upon vegetable fuel in spectacular fashion. Yet technological change could not be conjured up at will. The development of effective coal-based technologies was a drawn-out, tortuous business. Some elements of the 'coal technology package' were present by the first decade of the eighteenth century, but it was not until the 1790s that the iron industry turned fully to a mineral fuel platform. Indeed, it was not until the Napoleonic era that the combination of coke smelting, puddling furnaces, rolling mills, and steam power became the industry standard. Because of this, Baltic iron remained fundamental to the British economy in the early stages of industrialisation. Baltic iron, it should be recalled, did not reach its peak on the British market until 1793. It was only in the years after 1800, when tariff barriers against foreign iron were ratcheted up, that the fate of Russian and Swedish iron was sealed.

[1] TNA: PRO CO 388/34, Edward Finch to Lord Harrington, 22 April 1735.

The introduction of coal technology in Britain

The revolutionizing of the British iron industry is conventionally dated to 1709, when Abraham Darby mastered smelting with coke. It was indeed a signal event. But that is not to say that mineral fuel had no earlier role. Far from it. Pit coal had infiltrated some parts of the production process almost as soon as the indirect process (blast furnace plus forge) was introduced to the British Isles. Coal could not be used in a finery hearth because sulphurous impurities in the fuel would be imparted to the iron as it liquefied, but pit coal could be used in the chafery without too adverse an effect. Here, the blooms of refined metal were merely being raised to a red heat before being drawn out under the forge hammer into bars. No chemical transformation was involved and so the risk of contamination was far lower. There is evidence of coal being used in this way from the second half of the sixteenth century.[2] By the eighteenth century 'Mill-bar', the type of brittle bar destined to be slit into nail rods, was routinely drawn out in a coal-fired chafery; it was only the tough 'Merchant-Bar' that had to be 'drawn out with a Charcoal Fire only'.[3] Furthermore, mineral coal could be used quite safely to reheat iron in rolling and slitting mills, and was so from the very introduction of that technology into the British Isles.

In the course of the seventeenth century mineral coal insinuated its way into every part of the production chain for which it was eligible. But the reduction of iron from its ore remained proof against it. All attempts to use mineral coal in a blast furnace ended in failure. It was not until Darby tried subjecting coal to a prior air-free combustion, one that yielded coke, that a satisfactory pig iron could be coaxed from the furnace.

The breakthrough at Coalbrookdale in 1709 was a landmark in the history of humankind. Even so, coke-smelted pig was not an exact substitute for its charcoal-smelted equivalent. Coke pig had a high silcon content. This had its advantages: silicon-heavy pig iron was especially fluid in its molten condition, and therefore well suited to the production of castings. Yet there were countervailing disadvantages. Coalbrookdale

[2] Richard S. Smith, 'Sir Francis Willoughby's ironworks, 1570–1610', *Renaissance and Modern Studies*, XI (1967), 90–140; A.C. Jones and C.J. Harrison, 'The Cannock Chase ironworks, 1590', *English History Review*, XCIII (1978), 795–810.

[3] *The state of the trade and manufactory of iron in Great-Britain considered* (1750), p. 4.

pig iron was difficult to refine. Finers had to work longer and harder, using more charcoal. Abraham Darby's pig iron had, then, contradictory properties. It was cheap to smelt, but expensive to refine. For Darby, who was a specialist in the foundry trade, making cast iron pots for Graffin Prankard to ship to the New World, this was of little importance.[4] He was not greatly interested in making bar iron. But for other ironmasters there was no clear benefit to be had from using coke-smelted pig iron. Hence the paradox of one of the most celebrated technological innovations of the early industrial age: Darby's coke smelting process was restricted to the Coalbrookdale works and a handful of sister works for its first forty years. Forgemasters like the Knights continued to buy charcoal-smelted pig iron.

Coke smelting remained marginal until the early 1750s. Then there was a burst of investment in new coke-smelting plant in the Shropshire coalfield, followed by the rapid spread of the technique to other parts of the Midlands and to south Wales. The reasons for this sudden flowering are disputed. It was long thought that the delayed take-up of coke smelting was due to deficiencies in the pig iron that were rectified in the course of experiments at Coalbrookdale in the 1740s. The desirability of smelting with mineral fuel was so obvious that only some difficulty with the end product could explain the hesitation of ironmasters outside Darby's immediate circle. More recently, however, attention has focused upon the price differential between pig iron made with coke and iron made with charcoal: if coke-smelted iron was sufficiently cheap forgemasters could compensate themselves for the heavier costs of refining it (a greater consumption of fuel, and a greater wastage of pig iron in the conversion process) and would happily buy it. The difficulty was that there was no such price advantage in the first half of the eighteenth century, so ironmasters, behaving in accordance with strict economic rationality, remained loyal to iron made using vegetable fuel. The eventual spread of coke smelting was not the result of a supplementary technological advance in the 1740s; it arose

[4] Hussey, *Coastal and river trade*, p. 106; N.C. Cox, 'Imagination and innovation of an industrial pioneer: the first Abraham Darby', *Industrial Archaeology Review*, XII (1990), 127–44. The Coalbrookdale Company continued to specialise in castings and the sale of pig iron to founders during the 1730s, as is made clear by the correspondence of Richard Ford, the works manager: Ironbridge Gorge Museum, 1992.11941, letterbook of Richard Ford, 1733–1736.

from an upswing in charcoal prices in the 1750s, overturning the cost advantages that charcoal furnaces had previously enjoyed.[5]

However the timing of the process is to be explained, there can be no doubt that the iron industry underwent a tectonic shift in the 1750s and 1760s. Coke blast furnaces began to stud the landscape wherever coal and iron ore outcropped. There were just three such furnaces in 1750, but by 1760 their number had grown to 14, and by 1770 there were 28 in blast.[6] There were initial problems with supplying a sufficiently strong air blast—the wooden bellows of old had to be replaced by more robust cast-iron blowing cylinders[7]—but once these difficulties were overcome the share of national output coming from coke furnaces bounded upwards, not least because coke furnaces could be built higher, to a greater capacity than their charcoal-burning rivals.[8] The output of coke furnaces exceeded that from charcoal furnaces for the first time in 1776.[9]

There was a rich irony to the timing of this epochal breakthrough. The first great surge of investment in coke smelting followed hard on the passage of the 'Act to encourage the importation of Pig and Bar Iron from His Majesty's Colonies in America' in 1750. It served to undermine the purpose of that Act. By releasing a flood of coke-smelted pig onto the market, the proprietors of the new ironworks at Horsehay (1753), Dowlais (1759), or Carron (1760) rendered null and void the new legislative framework governing the Atlantic iron industry. Colonial pig

---

[5] The rival interpretations are to be found in T.S. Ashton, *Iron and steel in the Industrial Revolution* (Manchester 1963) and C.K. Hyde, *Technological change and the British iron industry 1700–1870* (Princeton, 1977). For a critical discussion of the debate see Chris Evans, 'The Industrial Revolution in iron in the British Isles', in Chris Evans and Göran Rydén (eds), *The Industrial Revolution in iron: the impact of British coal technology in nineteenth-century Europe* (2005), pp. 15–28.

[6] S. Pollard and R.S.W. Davies, 'The iron industry 1750–1850', in C.H. Feinstein and S. Pollard (eds), *Studies in capital investment in the United Kingdom 1750–1920* (Oxford, 1988), p. 75.

[7] Coke is a less reactive material than charcoal, and requires a stronger air blast to support combustion: J.E. Rehder, 'The change from charcoal to coke in iron smelting', *Historical Metallurgy*, XXI (1987), 37–43. See also Trevor Daff, 'The introduction of furnace blowing cylinders', *Steel Times*, (May 1973), 401–02.

[8] Coke is a structurally stronger substance than charcoal. It was therefore able to support a greater mass of superincumbent material within the blast furnace. Friable charcoal was apt to crumble, so that charcoal-fired furnaces could not be built to a height of more than 30 feet.

[9] Pollard and Davies, 'The iron industry', p. 80.

iron would be little needed when the productive capacity of Britain's own smelting sector had been boosted so significantly.[10]

The sudden profusion of pig iron posed questions for forgemasters. Coke-smelted pig iron was both plentiful and economical, but was there fuel enough to convert it to bar iron? The answer, it seemed, was that there was not. Not, at least, while Walloon forging, with its charcoal-fuelled finery hearth, remained the only technique available at the 120 to 130 forges operating in England and Wales in the second quarter of the eighteenth century.[11] The Walloon method as practised in Britain shared a common provenance with that used in Uppland, having been transplanted to southern England by francophone forgemen at the very end of the fifteenth century. Many of the finers and hammermen at work in the eighteenth-century forge trade—a corps of little more than one thousand individuals—were of Walloon ancestry.[12]

Yet for all the shared provenance, the division of labour in British forges differed in important respects from that in Uppland. In Swedish Walloon forges it was conventional for a single finery hearth to be worked in concert with a forge hammer and a chafery, whereas in Britain two fineries, or sometimes three, would be teamed with just one chafery hearth. The standard layout was that sketched by Angerstein at Bromford in the 1750s.

The quotidian work regime at such a forge can be reconstructed from the abundantly documented site at Melingriffith in Glamorgan. Melingriffith forge in the 1770s was laid out in the classic British fashion, with two fineries serving a single chafery. A three-man team worked at each finery: a master finer and two 'bloom makers'. The chafery was

---

[10] American pig iron, which had played an increasingly prominent role in the Stour forges in the 1730s, had been sidelined twenty years later. In 1754 just 275 tons was consumed, less than 12 per cent of the total. On the other hand, the profile of coke-smelted pig grew steadily from the late 1750s. By 1765 nearly 40 per cent of the pig iron used at Wolverley, Cookley, Whittington and Mitton came from the coke-fired furnaces at Ketley, Horsehay and Lightmoor on the Shropshire coalfield.

[11] E.W. Hulme, 'Statistical history of the iron trade of England and Wales, 1717–1750', *Transactions of the Newcomen Society*, IX (1928–29), pp. 25, 27; P.W. King, 'Early statistics for the iron industry: a vindication', *Historical Metallurgy*, XXX, 1 (1996), 23–46.

[12] See B.G. Awty, 'The continental origins of Wealden ironworkers, 1451–1544', *Economic History Review*, XXXIV (1981), 524–39, and Chris Evans, 'A skilled workforce during the transition to industrial society: forgemen in the British iron trade, 1500–1850', *Labour History Review*, LXIII, 2 (1998), 143–59.

Illustration 4.1. Bromford forge.

Courtesy of Jernkontoret.
Caption: Angerstein shows two finery hearths on the left, each with wooden 'box' bellows. The rather larger chafery hearth lies to the right, equipped with bellows that were 'breeched' with leather. The hammer, with its huge cam drum, occupies the far corner.

attended by a master hammerman and a 'hammerman's man'.[13] The composition of these forge crews was, it should be stressed, a matter for the master forgemen, not the ironmaster. The finer selected his bloom maker, and the hammerman his man. Employers might cajole, but not with any certainty of success. Indeed, ironmasters' letters of this period resound with complaints about the capriciousness of forgemen in this respect. Authority rested with the master workmen, and with good reason. The fining of pig iron was no simple matter. The finer and his bloom maker had to judge the progress of the operation by the texture of the cinder exuded by the metal: during the early stages of fining the cinder 'fumed and sputtered like liquors in fermentation', but towards the end of the process it was 'scarce so fluid as bricklayers mortar'. Workers had to interpret signs and portents such as the 'snuffling of the blast' that began as accumulating cinder impeded the flow of air into the hearth.[14] Long years of crouching at the hearth were needed to school a neophyte in the ways of iron working. To be 'bred up' in the trade meant not just the acquisition of bodily strength and dexterity but the steady accretion of craft lore. The Swedish traveller Bengt Andersson Qvist told of how finers had their own argot to describe how iron reacted to the battering of the hammer. When too little heat had been applied it was said that 'the iron don't batter', or 'it don't feel the weight of the hammer'; when iron came too hot from the finery it was liable to 'read sear'.[15]

Work began in the early hours of Monday morning when the charcoal was lit in the finery hearths and the bellows set on. It would continue through the week until the hearths were extinguished on the following Saturday afternoon. Forgemen, it was well understood, laboured for six days. Nevertheless, control over the pace of work always eluded the ironmasters. Their influence came at one remove, through a strict audit of the materials that were issued to forgemen. In effect, pigs or blooms were 'put out' to forge crews. The thoroughness

---

[13] At Cookley and Wolverley forges, owned by the Knight Partnership, the hammer crew was three-strong. See Göran Rydén, *Production and work in the British iron trade in the eighteenth century: a Swedish perspective* (Uppsala, 1998), pp. 34ff.

[14] CCL, MS 3.250, p. 172.

[15] Anmärkningar uti Hvarjehanda förefallande Ämnen samlade på resan i England åren 1766 och 1767. Benct Qvist Andersson, Jernkontorets Arkiv, Fullmäktiges Arkiv, Handlingar ordnade efter Ämne Vol. F II a:20, RA. 'Read sear' was Andersson's rendering of 'red short', a condition in which iron crumbled under the blows of the hammer.

with which they worked up these materials was measured, not the speed at which they did so.[16] Each set of finers acted independently. When the stocktaker at Melingriffith weighed out pigs to the master finers in August 1772 the type and quantity of iron advanced to each of them was quite distinct: William Dawes was issued with iron from the charcoal furnace at nearby Caerphilly, as well as pigs from Batsto (New Jersey), Pine Grove (Pennsylvania), and Occoquon and Neabsco (both Virginia); Richard Coley was supplied with a rather different mix—a preponderance of coke-smelted iron from Calcott in Shropshire, together with two varieties of colonial pig. Moreover, each set of finers worked to its own rhythm. Dawes melted down 34 tons of his assortment in the month that followed, Coley just 25 tons of his. In the following accounting period it was Coley who was the more industrious, sinking 15 tons of mostly coke-smelted pigs against the 11 tons of American pig iron melted by Dawes.[17]

Yet despite the automony shown by individual forge crews on a week-by-week basis, there was an underlying consistency in their performance. When averaged out over the year, the weekly output of blooms per hearth was remarkably stable.The output achieved by Melingriffith forge crews remained steady between the 1770s and 1790s, almost always falling within the range 2.5 to 3 tons per week.[18] These figures are congruent with what is known of the charcoal forge trade as a whole on the eve of the coal technology revolution.[19]

Such consistency was all the more noteworthy in that forge crews did not always comprise three members. The presence of a three-strong team at each hearth was intended to allow round-the-clock

[16] The finers were allowed a wastage of 25 per cent in 'sinking' pigs into blooms. For hammermen—who were reshaping the blooms rather than effecting a major chemical change—the margin was narrower. Blooms were delivered to a hammer crew in 'longweight' tons of 2,400 pounds. For every ton so received, they had to return a 'shortweight' ton of 2,240 pounds.

[17] NMW, 89.76I/45, 'Day Book D', 1771–72.

[18] NMW, 89.76I/10, 16–20, forge accounts 1772–74, 1778–79, 1780–81, 1783–84, 1785–86; NMW, 1994.120/642, quarterly stock accounts 1777–78; GA, D/D X 809, forge accounts 1779–80; NMW, 89.76I/7–8, 'reckoning' ledgers 1782–84; NLW, EL Chappell MSS, box 5, forge accounts 1786–87; NMW, 1991.25/3–5, quarterly stock accounts 1790–91, 1792–93; NMW, 89.76I/66, yield book 1791–92; Carmarthenshire Record Office, Trostre 34, forge accounts 1795–96.

[19] See Chris Evans, 'Work and workloads during industrialization: the experience of forgemen in the British iron industry, 1750–1850', *International Review of Social History*, XLIV (1999), 197–215, especially 204–06, and Rydén, *Production and work*, pp. 44–53.

production—working 'double hand', as it was known—wherein a rota of overlapping shifts ensured that two of the three finers were always present at the hearth. Nevertheless, 'single-hand' working, involving a master forgeman and just one assistant, was a recognised option. Naturally enough, 'single-hand' working necessitated a break in production. Even so, it was possible for a two-man team to match the performance of a 'double-hand' forge crew by extending the working day to its very limits.[20] The implication is plain enough: except for bursts of extraordinary effort on the part of single-hand forge crews, the workforce of the British forge sector was not utilised to anywhere near its physical limits.[21]

The availability of charcoal set an absolute limit on the quantity of bar iron that might be drawn out, but there was also a socially determined ceiling to production. Walloon forgemen in Britain, like their counterparts in Uppland, observed an unspoken protocol governing the effort they would devote to the making of blooms and the drawing of bars. *Directeur* Touscher at Leufsta had written that 'when it functions well [weekly production] will always be 40 skeppund and a little more' whilst knowing that his finers and hammermen could turn out considerably more. Indeed, *Leufstawerken*'s forgemen were capable of spells of Stakhanovite exertion, but they were loath to do so for long. Like their distant cousins in Britain, when measured over the long term, they displayed an unnatural regularity, suggesting a rather more stately rhythm of production.

The rigidity of working habits among forgemen thoroughly exasperated British ironmasters. John Bedford, the ironmaster of Cefn Cribwr in south Wales, dwelt obsessively on this issue. His memoranda from the 1780s return again and again to the question of how to downgrade the labour of 'that proud Rogueish & Ignorant Sett of men Calld forgemen'.[22] Bedford's solution was impose an entirely new division of labour on finers: 'in order That These Refiners Should See Themselves of The Less Consequence, & by which They will be proportionately Less Insolent I have settled to Keep The Refining in 2 parts & Teach one

[20] See Evans, 'Work and workloads', p. 204, for further documentation of this phenomenon.

[21] A detailed examination of output at the Cookley and Wolverley forges of the Knight family reveals that they ran at 75 per cent of capacity in the 1750s. See Rydén, *Production and work*, pp. 64–67.

[22] NLW, Bedford MSS, 'Anchors and Pillors', 17 January 1788.

man to one part & a different man to the other'. By doing so Bedford hoped to exert an authority over his forgemen that was not possible as long as 'one man held The whole Branch of Refining as anciently'. And with that, he would no longer be in thrall to forgemen who would 'admit no other man to work at the Refinery but what have been Bred up to it from Their Cradles'.[23]

The search for new forge techniques in the eighteenth century was not then just a quest for energy-saving measures. That there was a need for a coal-fired refining technique to redress the growing imbalance between a rampant smelting sector and a sluggish forge sector was apparent to every ironmaster, but there was also an enthusiasm among many of them for a total reconfiguration of working habits and practices in the forge. It is in this context that Henry Cort's puddling and rolling process, patented in 1783–84, must be understood. There was more at stake than the substitution of coal for charcoal. After all, usable coal-fired refining techniques predated Henry Cort by some margin and were widely taken up in the 1770s and 1780s.[24] Indeed, the 'stamping and potting' method, patented by the brothers John and Charles Wood in 1753, had seemed to offer a definitive solution. By 1788 there were an estimated 60 'melting fineries' on the Wood brothers' model at work in Britain, making close on 16,000 tons of bar iron, nearly as much as was made at the country's 105 Walloon forges. Given this, why did Henry Cort persist with his puddling process? And why did several other technicians continue to experiment with coal-based systems of bar iron production?[25] There was, of course, reason to suppose that the Wood brothers' method could be bettered. Although they had succeeded in finding a way of protecting the iron from contamination as it underwent decarburisation, their solution—to enclose the iron in clay urns—was rather ponderous. Yet there were other compelling reasons that spurred on Cort and his sponsors. One of these was disciplinary—the hope of devising a production system that would allow ironmasters to circumvent the 'Ignorance and vile wickedness of forgemen'. Another concerned

[23] NLW, Bedford MSS, 'Forge Rule Settled to Employ Carefull Labourers for forgemen', 26 March 1787.

[24] G.R. Morton and N. Mutton, 'The transition to Cort's puddling process', *Journal of the Iron and Steel Institute*, CCV (1967), 722–28.

[25] R.A. Mott (ed. Peter Singer), *Henry Cort, the great finer, creator of puddled iron* (1983), chap. 1; Richard Hayman, 'The Cranage brothers and eighteenth-century forge technology', *Historical Metallurgy*, XXXVIII, 2 (2004), 113–120.

Britain's relationship with Sweden—the dream of finding a home-grown alternative to 'Orground' iron.

The south Walian ironmaster Richard Crawshay, the man who did most to make Cort's process a commercially viable proposition, made no secret of his disciplinary ambitions. Crawshay believed that puddling could crack open the intractable workplace culture of his forgemen. Cort's method, it was hoped, would dispel the 'mystery' of iron making to which forgemen owed their strength. The use of rolls, for example, to reshape the refined metal would allow the production of bars in an unvarying, standardised format. Drawing out bars under the forge hammer, a process entirely at the discretion of the workman, would give way to a procedure that combined celerity with exactitude. Instead of the workman using the hammer, the workman would be used to feed the rolls. Small wonder that Crawshay was quick to recommend the puddling system to his friend William Reynolds, the leading Shropshire ironmaster of the day. He did so knowing Reynolds to be 'in a very painful situation being quite at the mercy of his workmen'.[26] Crawshay intended that the introduction of Cort's methods would be accompanied by a complete turnover of personnel. He would bypass the existing workforce in the forge trade by having the new technique taught to complete novices. When Henry Cort was dispatched to Crawshay's Cyfarthfa works he was exhorted to 'teach the Welch your mode of making Iron'. Members of the old forge dynasties, accustomed to tramping from forge to forge, would be excluded. The 'created servants of the vicinity', Crawshay averred, 'will be the best security for peace & performance'.[27] Local recruits would not be sullied by the self-assured pride of regular forgemen and would therefore prove obedient, pliable and productive.

The regime to which this first generation of puddlers was subjected was punishing. Cort's method, as brought to perfection at Cyfarthfa at the start of the 1790s, began with the melting down of pig iron to burn off excess silicon. The iron was let out into moulds and allowed to cool into plates of 'finers metal'. It was this 'finers metal' that was the puddler's raw material. Cort, like others in the forge trade, had wrestled with the problem of how to use mineral coal as

[26] SML, Weale MSS, 371/3 fo. 89, A. Jellicoe to S. Jellicoe, 22 May 1787.

[27] SML, Weale MSS, 371/3 fo. 192, Richard Crawshay to Henry Cort, 3 November 1787.

Illustration 4.2. Richard Crawshay of Cyfarthfa, *c.* 1796, by Wilson of Birmingham.

Courtesy of Cyfarthfa Castle Museum and Art Gallery, Merthyr Tydfil. Caption: Crawshay was a man of unquenchable ambition. 'He is in high feather', a neighbouring ironmaster reported in 1792, just as the teething troubles of the puddling process were being overcome, 'and is determined not to be easy, till he sends down 10,000 tons of bar iron a year.'[28]

[28] Gloucestershire Record Office, D1086/F122, William Lewis to John Blagden Hales, 26 April 1792.

an energy source without exposing iron to its damaging impurities. His solution was to use a reverberatory furnace in which a brick firebridge, separating the grate from the hearth, prevented the fuel coming into contact with the metal. The strong draught provided by the chimney ensured that the combustion of the coals was fierce, whilst the sloping roof of the furnace deflected (or reverberated) the heat issuing from the grate onto the materials in the hearth.

The heat generated was so intense that the charge of 'finers metal' was soon reduced to molten form. The task of the puddler was to stir about the liquid metal, ensuring that the iron was evenly exposed to the current of air that oxidised out the carbon. This was a feat of herculean exertion, and one that got progressively more arduous as the process reached its climax. As the carbon level in the iron diminished, the metal grew viscous and coagulated—or 'came to nature', to use the parlance of the trade. The decarburised iron was then divided into 'loops' that were hauled, radiant with heat, from the furnace to the block of a shingling hammer, where liquid slag was expelled and the metal consolidated into slabs.[29]

The sinew-snapping struggle of the puddler with his materials remained the central drama in the making of malleable iron for over eighty years. Indeed, it went unchallenged until the advent of mild steel, made by bulk production methods, rendered malleable iron obsolete. As all who witnessed it were to agree, puddling was a brutal species of labour. The puddling furnace housed a chemical transformation of elemental fury, yet a transformation that required careful and uninterrupted management by the puddler. To stand at the furnace door and rake the seething metal back and forth made the most grievous demands on the human frame, hastening puddlers towards a hunched and broken decline.[30]

That puddling represented an appalling intensification of human labour is readily demonstrated. It requires nothing more than a comparison between the throughput required of puddlers and the average production of finers at Walloon forges. The customary three-man forge crew at a finery hearth usually made about three tons in a week, but once Cort's system was introduced an entirely new production landscape

---

[29] W.K.V. Gale, 'Wrought iron: a valediction', *Transactions of the Newcomen Society*, XXXVI (1963), 1–11.

[30] See some of the assessments gathered in Chris Evans, 'Iron puddling: the quest for a new technology in eighteenth-century industry', *Llafur*, VI, 3 (1994), 44–57.

Illustration 4.3. A loop of puddled iron being turned under the shingling hammer at Cyfarthfa ironworks, 1792. By J.C. Ibbetson.

Courtesy of Cyfarthfa Castle Museum and Art Gallery, Merthyr Tydfil.

Table 4.1. Average weekly output of puddled bars per puddling team in selected forges (in tons).

1798–1801	Kirkstall, Yorkshire	5.01
1812	Cradley, Worcestershire	5.10
1819	Hampton Loade, Shropshire	5.49
1832–33	Old Park, Shropshire	4.96

Sources. Kirkstall: West Yorkshire Archive Service, Leeds, K/F 5/1, Kirkstall forge wages book 1794–1802. Cradley: Dudley Archives, Z121, Cradley forge stock and yield accounts 1805–12. Hampton Loade: Shropshire Records and Research Unit, 5686/1, Hampton Loade stock accounts 1803–36. Old Park: John Rylands Library, Manchester, BOT2/8/4, Old Park furnaces and forge wage accounts 1832–39.

was opened up. As early as May 1788 Richard Crawshay was able to boast of dramatic advances: 'we have got six setts of Finers [i.e. puddlers] to produce near thirty Tons a week'; that is, almost five tons per forge crew. Within a few years this sort of performance, once thought of as freakishly high, had become standard, as Table 4.1 reveals.

The rupture in working practice that puddling brought about becomes all the more stark once the composition of individual forge crews is taken into account. A crude measure of workload can be arrived at by dividing the tonnage handled on a weekly basis by the number of those who worked together at a hearth. For sites that practised Walloon forging in the orthodox fashion (that is, three men working 'double-hand') workload can be gauged by dividing the weekly output by three. This gives an average workload per finer at Melingriffith of 0.92 tons for the years 1772–1796.[31] At Pentyrch forge the equivalent figure for the period 1790–1792 was 1.06 tons per week.[32] But because the puddling furnace was usually worked by a two-man team, workload per forgeman was arrestingly higher than under the old Walloon regime: 2.50 tons at Kirkstall, 2.55 tons at Cradley, 2.74 tons at Hampton Loade, 2.49 tons at Old Park.

The contrast between the traditional Walloon method and the new coal-based procedure was even more pronounced than the difference in weekly output per forgeman might suggest. The working day in a Walloon forge had an elastic quality to it. Each hearth was the province of a single forge crew, usually working double-hand. There was no fixed shift system; the three members of the forge crew swapped

[31] This excludes the aberrant year 1778–1779 when the forge was being extended.
[32] GA, D/D Xn 3, Pentyrch cashbook 1790–93.

roles at their own discretion. With puddling, the situation was different. It soon became standard for puddlers to work shifts of twelve hours, so that two teams could alternate at the same furnace. As one set of puddlers left off, another took its place, allowing the furnace to be used with unrelenting intensity. An old-style finery hearth yielded about three tons of refined iron per week; under the new system as much as ten tons could be puddled at a single furnace.[33]

The intensification of work was just as evident in the next phase of Cort's process: the shaping of the refined metal by means of powered rolls. The rolling of iron enhanced throughput enormously. 'One pair of rollers', a supporter of Cort asserted in 1784, '...will roll shingled slabs into bars, at the rate of a ton in 30 minutes'.[34] As the hammer crew in a traditional Walloon forge would seldom produce more than ten tons of bars in a week, and usually far less, this represented a staggering advance. The chafery hearth in a Walloon forge served two or occasionally three fineries. A set of rolls, however, had the potential to process the output from half a dozen or more puddling furnaces. The mill at Penydarren, just a mile to the east of Cyfarthfa, serviced 'near 20' puddling furnaces in the first decade of the nineteenth century.

By the end of the Napoleonic wars the largest rolling mills on Cort's model—those at Cyfarthfa and Dowlais in south Wales—might reach an output in excess of 200 tons a week.[35] This was achieved through a far more elaborate division of labour. For example: merchant bar had been made at Pentyrch forge in the early 1790s by two hammerman and their respective assistants; a generation later bar fabrication at Pentyrch

[33] Eric Thomas Svedenstierna recorded an average daily output of 1.2 tons per puddling furnace at the works he visited in South Wales and Staffordshire in 1802. This implies a weekly output of above seven tons. Svedenstierna did, however, report higher daily outputs at individual forges, and later Swedish reports refer to weekly outputs of 10 tons and above. E.T. Svedenstierna, 'Om Puddlingsprocessen', *Jernkontorets Annaler* (1817), pp. 130, 138. See also A.G. Tamm, *Anteckningar öfver Främmande Länders Jernhandtering, gjorde under en Resa i Tyskland, Frankrike och England, åren 1830 och 1831* (Stockholm, 1832), pp. 75ff.

[34] *Address & proposals from Sir John Dalrymple, Bart. on the subject of the coal, tar, and iron, branches of trade* (1784), p. 5. A French observer thought that a bloom of puddled iron could be transformed into a bar eleven or twelve feet long in just 40 seconds: A.H. de Bonnard, 'Memoire sur les procédés employés en Angleterre pour le traitement du fer par le moyen de la houille', *Annales des arts et manufactures*, XXIV (1806), 47. Compare Svedenstierna 'Puddlingsprocessen', pp. 127f., where it is made plain that it is the processing of rough bars at the forge train that is under discussion.

[35] See the figures given for rolling mill performance in South Wales in 1812 and 1817 in GA, D/D G 1817 (3) G, Gilbert Gilpin to William Wood, 15 September 1817.

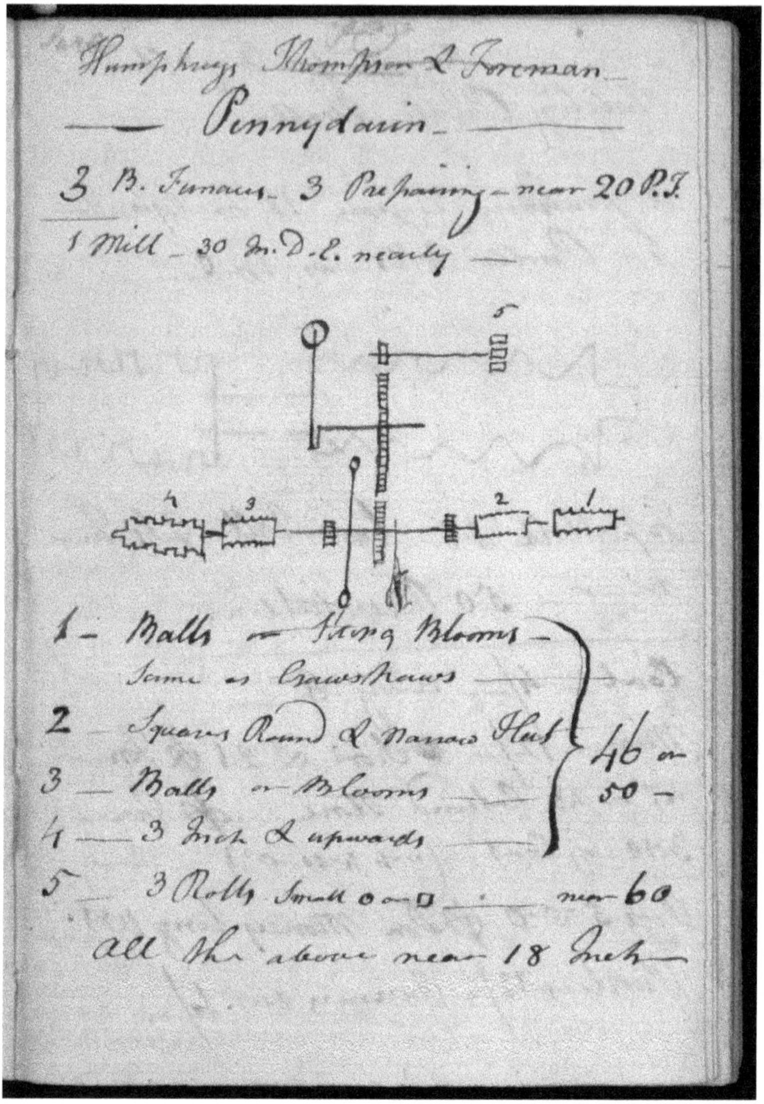

Illustration 4.4. The layout of the rolling mill at Penydarren ironworks, Merthyr Tydfil, *c.* 1805.

Courtesy of Birmingham City Archives (MS 1513/3).
Caption: The rolls labelled 1 and 3 were used for rolling blooms into rough bar. Roll number 2 was used to make the rough bars into square bars, round bars and narrow flats. Roll 4 was employed in making flat bars that were '3 Inch & upwards'. All these rolls were turned by the main drive shaft at between 46 and 50 revolutions per minute. Three smaller rolls (labelled 5), driven by a separate shaft at 60 revolutions per minute, turned out small round or square bars.

was carried out at a rolling mill where a master roller presided over eighteen workers, distributed between seven occupational subgroups.[36] Gustaf Ekman, who visited Merthyr Tydfil in 1830, said that it would take 'at least 20 workers' to run a large rolling mill.[37]

The French engineer Auguste-Henri de Bonnard, who witnessed puddling and rolling at Cyfarthfa in the first years of the nineteenth century, hailed Cort's system as a much needed rationalisation of forge labour. It was axiomatic, he declared, that 'to divide labour is to abridge it, and that to multiply the number of operations is to perfect them'. This principle, he asserted, had been amply demonstrated in the 'mechanical arts'. Now it could be extended to the 'chemical and metallurgical arts'. When 'each operation has but a single intended outcome' different operatives could be deployed in 'the best, most rapid and most economical manner', but if the division of labour lacked clarity the productive sequence could become disjointed and the forgemen might 'even become antagonistic to one another at times, hindering the progress of the operation and lessening or spoiling the outcome'.[38] (De Bonnard, had only he known it, might have been describing the rift that opened up between the finers and hammermen under the Walloon system at Leufsta in the 1730s.)

It was the rationalisation of production, quite as much as the use of mineral energy, that distinguished the new technological package. Samuel Smith, the Yorkshire ironmaster who toured South Wales, Shropshire and Staffordshire in 1794, focused upon the organisational novelties that he encountered, not the chemistry of puddling. His diary dwells upon the sequential celerity of what he found. His shorthand descriptions and rough drawings emphasised the interconnected character of the different processes and mechanical devices. Particular stress was laid on the correct ratio between different items of plant. The Dowlais works, Smith reported, was equipped with '3 [blast] Furnaces—4 Preparing Fires—26 P[uddling] F[urnaces and] 6 Balling [furnaces]', sufficient for '8 or 9 pair of Rolls' to function.[39]

One reason, then, for the take-up of Henry Cort's patent methods was the potential they held for a streamlined production chain, one

[36] GA, CL/MS 1.170, 'B. Haddock's Memorandum Book Jany 1st 1818'.
[37] Gustaf Ekman, 'Bihang om Valsverk', included in Tamm, *Anteckningar*, p. 135.
[38] Bonnard, 'Memoire sur les procédés employés', 53–54.
[39] See Smith's notes from 1794 in BCA, MS 1513, vol. 3, and vol. 5 for a set of observations from *c.* 1800 that make the same point.

Illustration 4.5. A rolling mill at Cyfarthfa ironworks, *c.* 1830. By Penry Williams.

Courtesy of Cyfarthfa Castle Museum and Art Gallery, Merthyr Tydfil.
Caption: Rolling hastened the production of bar iron so much that a set of reheating furnaces was required to keep the rolls fully occupied. Because the reheated iron could not be dragged over long distances without losing its optimal heat, the furnaces had to be kept close to the rolls. This made for a considerable concentration of plant, raising the ambient temperature in the mill to stifling levels. It was to alleviate this that rolling mills, as seen here, were often high, open-sided structures.

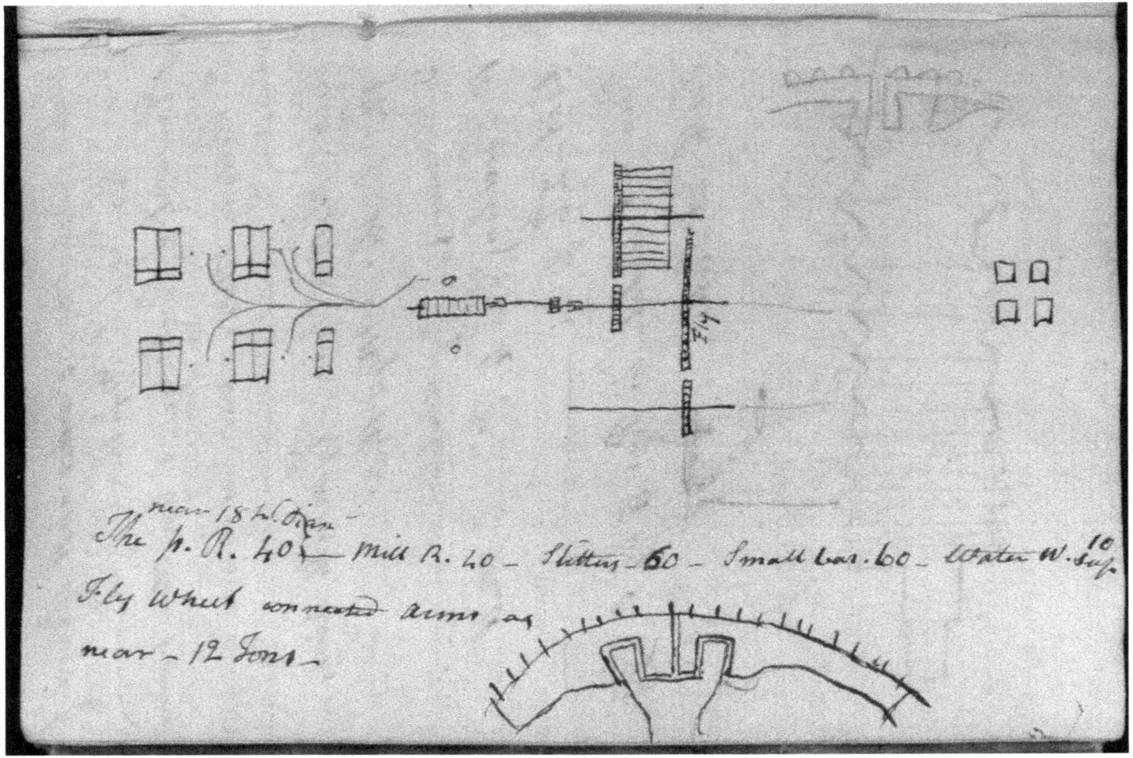

Illustration 4.6. The rolling mill at Cyfarthfa.

Courtesy of Birmingham City Archives, (MS 1513/3).
Caption: Samuel Smith made this sketch plan of the puddle rolls at Cyfarthfa in 1794. To the left are the ten balling furnaces in which the shingled blooms of iron were reheated. Smith has indicated the route along which the blooms were dragged before being handed over to the roller. The water wheel, the drive shaft and the gearing that powered the rolls are seen to the right.

purged of the craft idiosyncrasies that beset Walloon forging. Puddling had one further attraction: that it might make coal-refined bar iron to stand comparison with the best Swedish brands. Cort's professional background is revealing in this respect. He was not an ironmaster; he made his living as a naval agent, disbursing pay and allowances to crews in the Royal Navy. However, the family connections of his second wife brought him the management of the Fareham ironworks (a forge at Fontley and smiths' shops at Gosport) in the mid-1770s. These works were hard by the Royal Dockyard at Portsmouth and were concerned with supplying ironwares to the Yard. Cort could hardly have been unaware of how lucrative the market for naval ironwares was, nor that the Navy Board insisted on its contractors using high-grade Swedish iron—'Orground' iron for the most part. It would have required little imagination in a man of Cort's energy and resource to conceive of the rewards that would accrue to the British ironmaster who could provide a coal-made iron of sufficient quality to be made into anchors, bolts and hoops for His Majesty's ships. Accordingly, samples of puddled iron were put to the test at the Royal Dockyards soon after Cort's method reached technical stability. There it was 'wrought into anchors, and other naval implements, [and] underwent a series of experiments, against the like instruments wrought from the best Orground's Iron, the marks of which were P.L. & Crown [Åkerby], Double Bullet [Österby], Hoop & L [Leufsta].'[40] The results were favourable. Intensive lobbying by Cort and his supporters paid off when, in 1791, the Navy Board announced that it would only accept tenders from those who made bar iron according to Cort's patent methods. From this point of view, Cort was not trying to supplant bar iron made by stamping and potting, he was trying to supplement it with a high-quality, 'tough' iron intended for a niche market. Significantly, Cort's key associate in this, Richard Crawshay, was himself a former munitions supplier to the British state who shared Cort's vision of driving Swedish iron from one of its most entrenched and prestigious markets in Britain. Hence Crawshay's triumphant exclamation to Cort: 'this will do—we shall now make Swedish Iron for Free in England which heretofore we were obliged to have from Sweden.'[41]

[40] *A brief state of facts relative to the new method of making bar iron, discovered and brought to prefection by Mr Henry Cort* (nd).

[41] SML, Weale MSS, 371/3 fo. 191.

Some of the more extravagant hopes invested in puddled iron remained unfulfilled. It never supplanted 'Orground' iron as the essential input for steel making.[42] Nevertheless, puddling proved to be a 'method of converting pig into bar iron much preferable, in respect both of yield and quality, to their former methods'.[43] Together, Cort and Crawshay established a system of bar iron manufacture that was economical and expeditious. So expeditious, in fact, that it was soon preferred to the rather less wieldy stamping and potting. This was to bring to a conclusion the search for a streamlined, entirely coal-fuelled form of iron making. Richard Crawshay bristled with pride: 'we work all with Fossel Coal', he announced in 1793. He elaborated: 'my Blast Furnaces are 60ft high, each Furnace produces about 1400 Tons per Annum—we make use of Air Furnaces [i.e. puddling furnaces] instead of Finerys, when the Metal is brought to nature, instead of Hammers, we put it between a pair of Rolls, & crush it like a paste'.[44]

Baltic iron in crisis

Crawshay's remarks were addressed to a Russian correspondent, Grigorii Aleksandrovich Demidov, who had visited Britain the previous year. This Demidov was, of course, a member of the well-known family of Urals ironmasters.[45] For him, Crawshay's announcement augured ill. The technological breakthrough made at Crawshay's Cyfarthfa ironworks spelt danger for the Urals iron industry. The British iron industry had at last acquired the capacity to drive back Baltic imports. South Wales, home to the Cyfarthfa ironworks, was in the van. In 1788 a mere 12,500 tons of pig iron were cast in the region, but by 1796 production had

[42] 'Mr [William] Reynolds has at Ketley made as good Steel from English iron as ever was made from Swedish iron', it was said in 1791, 'but he says the process sometimes misgives with him without his yet being able to account it': Royal Society of Arts, C3/102, John Grieve to Samuel More, 27 July 1791. The Shropshire ironmaster was, De Bonnard wrote some years later, 'obliged to abandon his enterprise. The iron intended [for steelmaking]... is brought from Sweden and Russia': Bonnard, 'Memoire sur les procédés employés', 58–59.

[43] Royal Society of Arts, C3/102, John Grieve to Samuel More, 27 July 1791.

[44] Gwent Record Office, D2.162, fo. 93, Richard Crawshay to Baron Demidov, 3 March 1793. Charcoal furnaces seldom exceeded 30 feet in height and rarely produced more than 700 tons a year.

[45] A.G. Cross, *"By the banks of the Thames": Russians in eighteenth-century Britain* (Newtonville MA, 1980), p. 322.

topped 34,000 tons, and in 1805 exceeded 78,000 tons.[46] Most of this was refined into bar iron by Cort's method. The conclusion drawn by one seasoned observer of the iron trade was just: 'In short it appears to me that South Wales must in a very few years be the Siberia of this Kingdom'.[47]

Siberian iron had entered Britain in ever greater quantities in the late 1780s, buoyed by the economic upswing that followed the end of the American War. Despite the spread of stamping and potting, the output of the British forge sector was nowhere near equal to domestic demand. Indeed, it accounted for not much more than a third of the quantity consumed. With Cort's puddling method beset by teething troubles, foreign makers continued to hold the upper hand. Thus, Baltic iron reached its apogee on the British market as late as 1793 when 59,000 tons was imported. Thereafter a rapid decline set in, as puddling achieved technological stability and commercial viability. British bar iron production had been pegged at around 20,000 tons per annum for the first half of the eighteenth century. The spread of stamping and potting had allowed for an upward shift to 32,000 tons in 1788, but by 1810 national output stood at 130,000 tons.[48] The post-1790 growth was attributable almost entirely to Cort's method. Yet the expulsion of Baltic iron, which occurred with such rapidity in the years around 1800, was not solely due to the cost superiority of Cort's iron. It was also a matter of policy. The tariff on foreign bar iron was more than doubled in the course of the wars with revolutionary France, from £2.81 per ton in 1790 to £6.49 in 1813.[49] Now that the domestic forge sector had escaped the fetters imposed by vegetable fuel, Baltic iron, for so long the mainstay of metalware manufacturing in Britain, could be ruthlessly suppressed.

The effect on the Russian iron trade was devastating. Total bar iron exports fell from over 56,000 tons in 1793 to little more than 14,000 tons in 1817. The collapse reflected developments on the British market,

---

[46] M. Atkinson and C. Baber, *The growth and decline of the south Wales iron industry, 1760–1880: an industrial history* (Cardiff, 1987), p. 5.

[47] Shropshire Records Unit, 1781/6/22, Gilbert Gilpin to William Wilkinson, 24 October 1796.

[48] Pollard and Davies, 'The iron industry', pp. 86ff.

[49] Ten years later, when British ascendancy on the world market seemed assured, tariffs were lowered to below £2. See Eli Heckscher, *Sveriges ekonomiska historia. Från Gustav Vasa. Andra delen. Det moderna Sveriges grundläggning* (Stockholm, 1949), figure XV.

once the most important of all recipients, but now effectively closed to Russian iron.

Some respite came from the opening of new markets, most notably in the United States and in southern Europe. In the 1820s and 1830s Russian iron became an integral part of the 'sugar triangle' that connected the Baltic to north America and the Caribbean. American ships would carry sugar or coffee from Cuba to St Petersburg, returning with bar iron and hemp to New England. The US tariffs introduced in 1816 discriminated between rolled iron from Britain and 'hammered' iron from the Baltic, very much to the advantage of Baltic producers. As a result, the United States was taking over half of Russia's iron export in the early 1830s. Yet that export was but a shadow of its earlier self, running at less than 10,000 tons annually in the early 1840s. By the end of that decade the Russian presence on the American market had ended. Changes in the tariff system allowed puddled iron from Britain to sweep across the Atlantic, squeezing out imports from St Petersburg.[50]

The outcome was the withdrawal of Russian iron from the international stage. In 1782 over 60 percent of Russian bar iron output was exported. By 1822 the proportion had fallen to 16 per cent, and in 1851 reached a nadir of just 6 per cent. To compensate, Russian ironmasters turned to domestic customers. There was no slackening of output—Urals production climbed steadily in the first half of the nineteenth century—but it was no longer directed to London, Hull or Bristol, nor to Boston or New York. It was sent instead to the great Nizhnii Novgorod fair and other domestic outlets for distribution among the artisanal metal workers of northern and central Russia.[51]

Swedish iron makers were also thrown into crisis by the new conditions on the British market. Yet their predicament was not so acute as the Russians'. Although Swedish 'common sorts' seemed fated to follow Russian iron into desuetude, 'Orground' iron remained indispensable for steel making and retained its position on the specialised Sheffield market. The crisis afflicting Swedish ironmakers should therefore be seen as two crises, differing in severity and following different chronologies.

[50] Kalevi Ahonen, *From sugar triangle to cotton triangle: trade and shipping between America and Baltic Russia, 1783–1860* (Jyväskylä, 2005), pp. 362–86.

[51] Ian Blanchard, 'Russia and international iron markets, ca. 1740–1850', at http://www.esh.ed.ac.uk/Research%20IB/Rus_Ind/R_market.pdf

The makers of 'Orground' iron were fortunate in that there was an unfailing demand for their product. Prices for the top 'Orground' brands rose steadily in the second half of the eighteenth century, for the growth in demand among British steel makers was not matched by an increase in supply from the *Vallonbruk* of Uppland.[52] The misbegotten efforts to force up production at Leufsta in the late 1730s and early 1740s had proved chastening. From the 1750s the *brukspatron* of Leufsta preferred to work in concert with his brothers Antoine and Louis to monopolise the elite 'Orground' marks.[53] The upward flotation of prices allowed the De Geers to reap profits enough. There were some attempts to diversify the product range in the 1760s by building cementation furnaces at Österby and Åkerby and a slitting mill at Johannisfors (an outlier of Forsmark), but these did little to deflect the now well-established orientation on the British market.[54]

The last years of the eighteenth century saw continued prosperity at the *Vallonbruk*, despite the onset of the revolutionary wars in 1792. The Napoleonic period, however, was one of increased difficulty. The wartime blockades and counter-blockades that impeded maritime trade had a serious impact on the Uppland works. Export figures for 'Orground' iron are not available, but the decline in output at Leufsta tells its own story. The four forges made 1,161 tons in 1800; just 782 tons was made five years later. Production capacity was cut permanently in 1808 when the Upper Forge was closed. The three remaining forges made a mere 612 tons in 1815.[55] The coming of peace brought respite. The demand for 'Orground' spiralled upwards on the Sheffield market, with the result that the high wartime prices—£35 a ton or higher—were maintained. Indeed, the trend was upwards until the mid 1830s.[56] It was not until the second half of the nineteenth century that

---

[52] Heckscher, *Sveriges ekonomiska historia*, pp. 400ff., and figures XII and XIV.

[53] K.-G. Hildebrand, *Fagerstabrukens historia. Sexton- och sjuttonhundratalen* (Uppsala, 1957), pp. 205f.

[54] Marie Nisser, 'Forsmark—ett av vallonbruken kring Dannemora gruvor', in *Forsmark och vallonjärnet* (Forsmark, 1986), pp. 54ff., and B. Molander, 'Forsmarks stångjärnsstämpel. Vallonbrukens stångjärnsstämplar i över 250 år', in *Forsmark och vallonjärnet*, pp. 78ff. Hildebrand, *Fagerstabrukens historia*, pp. 430ff. See also Sven Rinman, *Tjänsteberättelser rörande den grövre järnförädlingen 1761–1770*, ed. E. Malmborg (Stockholm, 1935).

[55] Heckscher, *Sveriges ekonomiska historia*, pp. 443ff., and figure XIV. For the output figures for Leufsta see different volumes at Leufstabruks arkiv, Leufsta Bruk.

[56] This was the conclusion of Artur Attman, based upon the experience of Österby and Åkerby. See his 'Vallonjärnets avsättning på världsmarknaden 1800–1914', in *Forsmark och vallonjärnet*, pp. 188ff.

new competitive pressures forced changes in technology and organization upon the Walloon ironworks.

The prospects for the 'common sorts' of Swedish iron at the start of the nineteenth century were entirely different. The Napoleonic age was one of unambiguous crisis. Total iron exports were running at an annual average of 55,000 tons in the years 1795–1800; in the five years after 1806, as the Continental System took hold, exports slumped to an annual 37,000 tons.[57] Eric Thomas Svedenstierna, a leading official of *Jernkontoret*, wrote in 1810 of his longing for a time 'when the sea regains its freedom'. That way lay recovery; otherwise the Swedish iron trade would 'plunge violently' into the abyss.[58] Yet the resumption of international peace brought no relief. The price of German-forged Swedish iron on the British market, which had peaked at over £22 per ton in 1801, had dipped to £16 in 1815. A postwar boom did something to raise prices, but the boom was shortlived. Ordinary grades of Swedish iron fetched just £15 on the British market in 1822. Faced with such low prices, Swedish exports to Britain halved in the course of the Napoleonic Wars, dwindling to less than 10,000 tons. There they remained; Swedish exports to Britain rarely exceeded 10,000 tons in the 1820s.[59]

Salvation for the Swedish iron industry in the post-Waterloo era came from across the Atlantic, just as it had for Russian iron makers. Exports to the United States compensated for the collapse of the British market. Gothenberg had become an important port of call for American shipping after 1809 when the blockades and counter-blockades of the war years inhibited trade to Hamburg and other northern harbours. Because of this, iron from the western county of Värmland became an important article of trade with America, where domestic production could not keep pace with demand. 'As rapidly as shipments to Britain dwindled', one Gothenberger would later recall, 'so did those to America increase, and exports to the latter country were so brisk, that so long as the bar was even it was accepted and paid for, irrespective of whether it was thin or thick, well or badly forged.'[60] Nearly 18,000 tons of bar iron

[57] Heckscher, *Sveriges ekonomiska historia*, p. 398 and graph XIII.

[58] E.T. Svedenstierna, 'Tal om den svenska jernhandteringen i äldre och nyare tider', *Kungliga Vetenskap Akademins handlingar* (Strängnäs, 1810), p. 52.

[59] Artur Attman *Fagerstabrukens historia. Adertonhundratalet* (Uppsala, 1958), pp. 9ff. For bar iron prices see Heckscher, *Sveriges ekonomiska historia*, diagram XV.

[60] Quoted in Rolf Adamson, 'Swedish exports to the United States, 1783–1860', *Scandinavian Economic History Review*, XVII (1969), 70, n. 30.

was shipped to American ports in 1815. With this, America replaced Britain as the most important recipient of Swedish bar iron. Indeed, by the 1830s the United States was regularly absorbing 50 per cent of Swedish iron exports.[61] The US tariffs that discriminated against rolled iron from Britain were a boon to the Swedes, just as they were to the Russians. The duty of $30 per ton that was charged against rolled iron in 1816 allowed the more expensive charcoal-made iron of the Baltic, which was subject to just $9 per ton, to compete. The nail industry of New England, which was expanding rapidly in the early nineteenth century, offered a ready market.

The respite was temporary, however. The tariff differential between rolled (British) iron and hammered iron narrowed between the late 1820s and early 1840s, eventually disappearing. Imports of rolled iron to the United States sprang up in response. They amounted to just 3,300 tons in 1829; ten years later they were 60,300 tons. More worryingly still, American iron producers were adopting puddling and rolling technology for their own use. Puddling had proved difficult to implant on the eastern seaboard because the long-flame bituminous coal that the British used was only to be found in the small coalfield around Richmond, Virginia. All other deposits of coal east of the Appalachian divide were of anthracite, whose short flame militated against its use in a reverberatory furnace. It was not until the 1830s that engineers in Pennsylvania developed blowers that could successfully spread the heat of anthracite across the hearth of a puddling furnace. But once they had done so, market conditions fast became unfavourable for iron imported from Sweden.[62] The Fall River Iron Works Company, set up in south-eastern Massachusetts in 1821, had used Swedish iron in its early years. In the early 1840s, however, its chief supplier, Olof Wijk of Gothenberg, was told that rival nail manufacturers were underselling Fall River products on the basis of American-made puddled iron. The Fall River managers drew their own conclusions: they installed puddling hearths of their own in 1842.[63]

---

[61] Attman, *Fagerstabrukens historia*, pp. 17–21; Artur Attman, *Svenskt järn och stål 1800–1914* (Stockholm, 1984), pp. 12–14; and Heckscher, *Sveriges ekonomiska historia*, table 15.

[62] Robert B. Gordon, *American iron 1607–1900* (Baltimore MD, 1996), pp. 135–38.

[63] This paragraph is based upon Adamson, 'Swedish exports to the United States'.

The Swedish response to British coal technology

The prospects for Swedish ironmasters seemed bleak. Expulsion from the American market appeared imminent, and unlike their Russian counterparts they had no substantial domestic market on which to fall back. By the 1830s it was quite evident that counter-measures were required if Swedish iron was not to be chased from one overseas market after another by puddled iron. Swedish engineers realised that they would have to mimick British technology or be defeated by it.

Swedish experts had been aware of coal technology in Britain since its inception. Indeed, one of the earliest references to the puddling process by a non-British author was made by the metallurgist Sven Rinman, whose *Bergwerks lexicon* of 1788–1789 reported that reverberatory furnaces were being used in England to refine pig iron to a 'forgeable bloom' using only the 'flames from mineral coal'.[64] Rinman's account was based on second-hand information, not direct observation, but a detailed, first-hand account of the new production system came early in the new century when Eric Thomas Svedenstierna, together with Auguste-Henri de Bonnard, the French engineer, embarked on a tour of British iron making districts. Svedenstierna's journal, published as *Resa, igenom en del af England och Skottland, åren 1802 och 1803* introduced the Swedish reading public to the sort of centralised, steam-driven facilities that were to be found at Merthyr Tydfil and other centres of the coal technology revolution. It made for sobering reading. The Cyfarthfa works, Svedenstierna wrote, produced in excess of 10,000 tons a year; the forges at Leufsta, Sweden's largest *bruk*, managed just 1,200 tons.[65]

Svedenstierna returned to the theme in his *Några underrättelser om Engelska jernhandteringen* ('Some information on English iron making') of 1813. This was a scientific treatise, not a travel journal. In it, Svedenstierna made an ambitious attempt to comprehend the enormity of what had happened in Britain. After an opening survey of metallurgy in the British Isles and rival iron producing countries since the middle ages, Svedenstierna turned his attention to the transformations of the modern era. The 1760s, he wrote, had signalled a 'new beginning',

[64] Sven Rinman, *Bergwerks-lexicon* (2 vols., Stockholm, 1788–89), II, p. 413.

[65] E.T. Svedenstierna, *Resa, igenom en del af England och Skottland, åren 1802 och 1803* (Stockholm, 1804), pp. 86ff. An English translation is available as *Svedenstierna's tour of Great Britain 1802–3: the travel diary of an industrial spy* (Newton Abbot, 1973).

with the spread of coke smelting, the take-up of cast iron blowing cylinders, and advances in the use of steam power. Stress was laid on system integration and the continuous improvement of facilities—Svedenstierna spoke of 'daily' innovation.[66]

Några underrättelser, together with a lengthy analysis of the puddling process, published separately by Svedenstierna in 1817, gave Swedish scientists and policy makers a painfully detailed explanation for the parlous state of their nation's iron export. But what remedy could be found? Svedenstierna initiated the debate, but he had not, at the time of his death in 1825, settled on a solution. That was left to a new generation of engineers led by Gustaf Ekman. For Ekman, what distinguished the British mode of iron making from Swedish methods was not the fuel employed but the systemic character it displayed. Ekman visited Britain on three occasions between 1828 and 1833. He returned with the conviction that the British *system* could be emulated.

Gustaf Ekman was not the first to advocate the adoption of British techniques. Experiments in wood-fired puddling had been conducted at Klosters *bruk* as early as 1811, but without satisfactory results. A more sustained attempt to adapt puddling to Swedish conditions was made at Schebo *bruk* between 1817 and 1824 under the supervision of the *Jernkontoret* official Carl David af Uhr, but once more without success. It was not until the late 1820s that Erik Adolf Zethelius, *brukspatron* at Nyby, assisted by the English engineer Samuel Owen, mastered puddling and rolling. Yet the breakthrough at Nyby was not followed up. In 1848 there were only nine puddling furnaces in the whole of Sweden.[67] Doubts persisted about the quality of the iron so made.

Ekman and his peers—A.G. Tamm, Teofron Munktell, Johan Bolinder, and Ludvig Rinman—were less concerned with the seemingly fruitless task of domesticating those elements of the British system, such as puddling, that were most closely associated with the combustion of mineral coal. Their originality lay in a willingness to adapt the organizational principles that underlay British success. They were assisted in this by their recognition that there were pockets of the British iron industry that remained loyal to charcoal as a fuel, yet still used the newer, more streamlined models of production.

---

[66] E.T. Svedenstierna, *Några underrättelser om Engelska jernhandteringen* (Stockholm, 1813), pp. 14f; Svedenstierna, 'Om Puddlingsprocessen'.

[67] Otto Stjernquist 'Om puddling i Sverige och Finland', *Med Hammare och Fackla 1998*, pp. 29–65; Attman, *Fagerstabrukens historia*, p. 167.

Charcoal technology continued to be used at forges that catered for specialist users, such as wire drawers and tinplate manufacturers, who demanded bar iron of particular ductility. Indeed, it would be a mistake to suppose that industrial modernity had imposed flat uniformity on the British iron industry. High-volume production brought no limitation on the range of products. Specialised producers catering for niche markets adopted the new procedures of puddling and rolling but retained, where appropriate, elements of the old charcoal regime. The forge at Hampton Loade in Shropshire stands as an example.[68] In the early 1820s it produced eight different brands of bar iron—quite apart from hoops, strips and sheets. Its premier brand ('Double SC Crown') was a puddled iron that came in two forms. In one, the bars were rolled; in the other, they were drawn out under a traditional forge hammer, having been reheated in an old-style chafery hearth. Other bars, such as 'Charcoal Gun Bars', were finished at a rolling mill but were of iron that had been prepared in a charcoal hearth.[69] This was an environment, in other words, in which coal-charcoal hybrids flourished.

Installations like Hampton Loade were essentially puddling forges that retained some elements of the old regime in order to produce iron designed for very specific ends such as the rolling of gun barrels. But there were other forges that continued to work with charcoal throughout. Pentyrch in south Wales was one such. Fidelity to charcoal was not a sign of unreflecting conservatism. On the contrary, Pentyrch underwent wholesale change in the early nineteenth century. In the 1790s the traditional Walloon method had been employed there; three finery hearths were in use, each attended by a three-man forge crew. Output per hearth was in the region of three tons weekly—a fairly standard figure for the time. But by the 1820s everything, other than the choice of fuel, had changed. Output had grown four-fold in the space of a generation. In the early 1830s (when a consistent series of data is available) forge crews were making over twelve tons of refined metal per week. That this represented a new standard is suggested by output figures from other works that retained charcoal refining (Table 4.2).

---

[68] Norman Mutton, 'The forges at Eardington and Hampton Loade', *Transactions of the Shropshire Archaeological Society*, LVIII (1965–68), 235–43.

[69] Shropshire Records Unit, 5686/2, 'Description of the various qualities of Iron made & the process'.

Table 4.2. Average weekly output of charcoal-refined iron per forge crew in selected early nineteenth-century forges (in tons).

1827–28	Clydach, Monmouthshire	11.85
1830–31	Pentrych, Glamorgan	12.18
1832–33	Old Park, Shropshire	11.30

Sources. Clydach: National Library of Wales, Maybery 3574, 3588, 3598, 3600, 3606a, 3618, 3628. Pentrych: National Museum of Wales, Department of Industry, 89.76I/55. Old Park: John Rylands Library, Manchester, BOT2/8/4.

The hike in productivity was made possible by a more effective use of fuel. The open-sided finery was replaced by an enclosed, oven-like hearth in which the melting down of pig iron was considerably hastened, quickening the pace of work.[70] In addition, fuel savings permitted an expansion in the number of hearths. The three Walloon-style hearths, employing nine finers, had given way to five hearths of the new model by the late 1820s, employing some twenty to twenty-five forgemen. Evidently, the composition of forge crews had changed. They were now four or five strong, with a more complex internal hierarchy than the three-man teams of old. Master forgemen appear rather more in the role of supervisory workers, having charge of three or four hearths at a time.[71]

There was a clear affinity between the revitalised charcoal sector and the coal-fuelled forge sector. Indeed, they were sometimes housed together. The Botfield family ran twin ironworks at Old Park and Stirchley in Shropshire in the 1830s. Their blast furnaces, puddling forges and rolling mills epitomised the new coal-based, integrated technology. Yet alongside the twenty-five puddling furnaces were a handful of charcoal hearths. The forgemen who worked at them occupied the same functional position as puddlers. They received 'finers metal' as their input, and the blooms that they made were subject to the pounding of a shingling hammer. It was this functional affinity that led Ekman and his collaborators to proselytize on behalf of British procedures: the hot blast, or devices like the heavy shingling hammers that the Swedes came to know as 'mumbling hammers'. The effects became apparent in the 1830s with a reform of the German forging method.

[70] John Percy, *Metallurgy: iron and steel* (1864), pp. 581–86.

[71] This conclusion is based upon comparing the utilization of plant suggested in the accounting data for Pentrych from the 1830s and 1840s (NMW, 89.76I/55) and the numbers of forgemen recorded in the censuses of 1841 and 1851.

The 'New German Forging Method'

German forging had been introduced to Sweden in the 1500s, and its essential features had changed little in the two succeeding centuries.[72] The forges were small, and simple in their lay-out: iron was refined and drawn out into bars at one and the same hearth, using a water-powered hammer. The forge crew, as defined in a *Bergscollegium* directive of the seventeenth century, was three-strong: a master, a forgehand, and an apprentice. That remained the case 150 years later, for a new decree in 1823 did little more than restate the longstanding *Bergscollegium* regulations. Productivity was correspondingly static. The output levels achieved at the forges of Gammelbo in the 1730s were still commonplace in the 1820s. In the 1830s, however, there were marked improvements as ironmasters responded to the urging of Ekman and his colleagues. New items of plant appeared. Just as importantly, a new division of labour, clearly British in inspiration, was adopted. A fourth worker was added to the forge crew—the *räckardräng* ('drawing-out apprentice'), who occupied an intermediate position in the forge hierarchy between the forgehand and the apprentice. The presence of an additional worker allowed full advantage to be taken of the potential for speed-up that the enclosed hearth, the heating of the air blast, and the use of robust blowing cylinders presented. Under the traditional German system the forge crew had moved from the hearth to the forge hammer and back again, so that when the hearth was in use the hammer was at rest and *vice versa*. With the 'New German Forging Method' the hearth and hammer were in simultaneous use.

The new system was pioneered at forges in the eastern county of Gästrikland. At Hofors, one of a set of forges owned by Tore Petré, the revised production methods were put in place at the end of the 1830s. The effect was striking. After a century or more of immobility, output per hearth rose by 50 per cent in the space of a decade. Equally striking was the improvement in the quality of the output. By the early 1840s many Gästrikland forges had abandoned the making of staple bar iron in favour of *ståljern* ('steel iron') for the Sheffield market. The contracts that forgemen entered into with Tore Petré were clear on this point.

[72] The following paragraphs are based on Göran Rydén, *Hammarlag och Hushåll. Om relationen mellan smidesarbetet och smedshushållen vid Tore Petrés brukskomplex 1830–1850* (Uppsala, 1991), chapters 5 and 6.

Workers were to reserve the purest iron from the centre of the bloom for *ståljern*; the residue could be used for ordinary bar iron.

The 'New German Forging Method' allowed for improvements in both quantity and quality, enabling Swedish makers to improve their position on the strategic Sheffield market. But it proved to be the last flourish of an old system, not the harbinger of a new. By the mid 1840s the tireless Gustaf Ekman had devised a production system that emulated the British model in a far more thorough-going fashion. He no longer sought to push at the boundaries of the German forging paradigm; he stepped beyond it.

The 'Lancashire Method'

Ekman had resigned as *Jernkontoret*'s roving consultant (*Översmedsmästare*) in 1836 to take charge of the *bruk* at Lesjöfors in Värmland. Here he experimented with a variant of the charcoal forging methods he had observed at Ulverstone in Lancashire.[73] What Ekman had seen was a hybrid technological system in which coal-fired and charcoal-fired methods were intermixed. Coke-smelted pig iron was given a preliminary melting to purge it of silicon, just as it might be prior to puddling. The resultant finers metal was then refined in an enclosed charcoal hearth. The blooms, after being shingled under a 'mumbling hammer', were conveyed to a coal-fired reheating furnace and then rolled into bars. Some elements of this sequence—the enclosed hearths and the shingling hammers—were readily transferable to the Swedish context. Indeed, they had been incorporated into the 'New German Forging Method'. Yet the rolling of bars, the element that could give Swedish iron making a truly industrial speed and scale, was not readily transplanted. It was a problem that Ekman wrestled with for several years.

The fundamental difficulty lay in constructing a reheating furnace that could serve a rolling mill with a continuous supply of iron. The chafery hearth used in the traditional Walloon method was too small for the task. At Ulverstone a coal-fired furnace, a replica of those used in puddling forges, had been employed, but such a solution would not

[73] Gustaf Ekman, *Gustaf Ekman. Svenska järnhanteringens nydanare för 100 år sedan* (Stockholm, 1944); Göran Rydén, 'Gustaf Ekman, Jernkontoret och lancashiresmidet—Ett inlägg i synen på teknisk utveckling', *Polhem: Tidskrift för teknikhistoria*, 1994/2, 132–64.

suffice in coal-poor Sweden. A wood-burning equivalent had to be found. If it was not, German forging, which even in its improved form continued to rely on the forge hammer to shape bars, would mark the limit of technological progress. By resolving the technical difficulties that surrounded the reheating furnace Ekman opened the way for fully industrialised iron making in Sweden. The system he devised was christened the 'Lancashire method' in honour of the inspirational prototype at Ulverstone.[74]

'Lancashire' forging was puddling transposed to a Nordic setting. Each hearth was attended by a two-man team, just as if it were a puddling furnace. The two melters would work for six hours before being relieved by a fresh forge crew—a rotation redolent of the shift system practised in British puddling forges. The parallels were not exact, however. Quite apart from working shorter shifts than puddlers, Lancashire forgemen handled pig iron rather than finers metal. (Because Swedish pig iron had been smelted with charcoal it was free of silicon and needed no preliminary treatment.) Nevertheless, as in British forges the shingling of the blooms was the task of a specialised worker, and the shingled slabs were brought to a welding heat—once more by specialised workers—in a reheating furnace that mimicked those to be found in British puddling forges. Finally, the bars were shaped in rolling mills that featured a division of labour every bit as complex as that found in their British counterparts. At Smedjebacken, Sweden's biggest mill in the mid-1850s, each shift included a master roller, two carriers, two receivers, two lifters, two heaters, two heating apprentices, two charcoal carriers, two straighteners, two straightener apprentices, and lastly two greasers.[75] In Sweden, as in Britain, new technology had brought about a complete rupture with the past.

The 'Lancashire method' began to spread through Värmland in the late 1840s, sweeping eastwards through the other *Bergslagen* districts. Gästrikland, the stronghold of 'New German Forging', proved the most resistant to the new order of things. It was not until 1853 that the first Lancashire forge appeared there. Even so, once it was established that the new system could produce *ståljern*, resistance slowly crumbled. By the early 1860s most Gästrikland forges had switched to the 'Lancashire method'—a transformation signalled by the doubling of regional output

[74] Rydén, 'Gustaf Ekman'.
[75] *Jernkontorets Annaler* (1859), p. 106.

Illustration 4.7. A Lancashire Forge.

Courtesy of Jernkontorets bruksbildsamling.
Caption: This photograph of the mill at Brattfors in Gästrikland, although taken in 1904, shows the technology introduced to Sweden in the mid-nineteenth century. Lifters and receivers stand poised ready to whisk iron back and forth through the rolls. To the right, furnacemen wear traditional white smocks, the garb once worn by Walloon forgemen.

in the course of a decade from 10,000 tons annually to nearly 20,000 tons.[76] The quest for modernity, initiated by Svedenstierna in the early nineteenth century, had been fulfilled.

Walloon iron making transformed

The triumph of 'Lancashire forging' left only one sector of the Swedish iron industry unmodernised—the *Vallonbruk* of Uppland. In 1850 the Uppland works produced 'Orground' iron in the same way that they had in the 1730s. And this premium material was marketed, as it had been in the eighteenth century, through the narrowest of channels. The export of 'Orground' in the early nineteenth century was monopolised by the Stockholm house of Tottie & Arfwedson; its sale in Britain was handled by Joseph Sykes & Sons in Hull. As the location of Sykes & Sons suggests, the iron was bound for Sheffield, the world centre of steel production.[77]

In the post-Napoleonic era 'Orground' iron retained its pre-eminence as the raw material for high-grade steel. When the French metallurgist Frèdèric Le Play visited Sheffield between 1836 and 1843 he found, just as he expected, that 'Orground' abounded in the the non-phosphoric, slag-free qualities that steel makers required. It was 'sound'; it had 'body'. The prices paid on the Sheffield market reflected this. Iron from Leufsta and Carlhom headed the list, commanding £35 per ton. Other 'Orground' brands followed.[78]

Yet the list presented by Le Play revealed that change was at hand. Non-'Orground' irons appeared there. There were examples of *ståljern* made by the 'New German Forging Method'. More troublingly, Russian and English marks featured on the list. The presence of non-Swedish irons was not in itself alarming. Russian iron, such as the 'Sable' mark

[76] Göran Rydén, 'Lancashiresmide och omvandling. Teknisk utveckling, strukturomvandling och produktionstillväxt i Gästrikland under 1800-talet', *Med Hammare och Fackla* (1998), pp. 105–51; Artur Attman, *Svenskt järn och stål 1800–1914* (Stockholm, 1986).

[77] Attman, 'Vallonjärnets avsättning på världsmarknaden', pp. 198ff.

[78] Frèdèric Le Play, 'Mèmoire sur le Fabrication de l'Acier en Yorkshire', *Annales des Mines*, 4me Serie, Tome III (1843), here quoted from its Swedish translation, 'Om ståltillverkningen i Yorkshire, samt jemförelse mellan de förnämsta ståltillverkningsorter i Europa', *Tidskrift för Svenska Bergshandteringen för år 1845*, pp. 1–123. See also W. Ekman, 'Vallonjärnet—en kvalitetsprodukt med världsrykte', in *Forsmark och vallonjärnet*, pp. 120–49, and K.C. Barraclough, 'Svenskt järn och Sheffieldstål', also in *Forsmark och vallonjärnet*, pp. 221–82.

Pris-kurant för Svenska, Norska, Ryska och Engelska jernsorter, som begagnas till ståltillverkning i Yorkshire.

Svenska och Norska stämplar.	*Pris*)*			
	För ett Eng. ton.		För ett Sk:pd St. v.	
	P. St.	sh.	Rgs R.	sk.
Löfsta och Carlholm (Upsala Län)	35	0	85	26
Gimo och Rånäs (dito)	31	0	75	37
Österby (dito)	30	0	73	16
Forsmark (Stockholms Län)	28	0	68	21
Strömsberg och Ullfors (Upsala Län)	28	0	68	21
Gysinge (Gefleborgs Län)	27	0	66	–
Wattholma (Upsala Län)	26	0	63	27
Harg (Stockholms Län)	26	0	63	27
Skeho och Ortala (dito)	25	0	61	5
Öster-Rüsöer (nära Nedenæs i Norge)	24	10	59	42
Ellkarleö (Upsala Län)	21	–	51	16
Sörfors (Wester-Norrlands Län)	21	–	51	16
Hedäker (Westerås Län)	18	10	45	11
Bäckafors (Elfsborgs Län)	18	10	45	11
Söderfors (Upsala Län)	18	0	44	–
Norberg (Gefleborgs Län)?	17	10	42	37
Hedvigsfors (dito)	17	10	42	37
Dådran (Fahlu Län)	16	10	40	16
Rishyttan (dito)	16	0	39	5
Cathrineberg (Gefleborgs Län)	15	10	37	43
Thurbo o Wikmanshyttan (Fahlu L.)	15	10	37	43
Avesta (rättare Svartnäs o. Korså, d:o)	15	0	36	32
Ludvika (ditto)	15	0	36	32
Svanå (Westerås Län)	15	0	36	32
Åmoth (Gefleborgs Län)	15	0	36	32
Strömsbacka och Svabensverk (d:o)	15	0	36	32
Tjärnäs Nedre och Robertsholm (d:o)	15	0	36	32
Hamarby (dito)	15	0	36	32
Storfors (Carlstads Län)	15	0	36	32
Qvarntorp (dito)	15	0	36	32
Fredriksberg (Carlstads Län)	15	0	36	32
Fagersta (Westerås Län)	14	10	35	21

*) Ett P. St. beräknadt till 18 Rdr 12 sk. Rgs.

Illustration 4.8. Irons used in the Sheffield steel trade, 1843.

Courtesy of Jernkontoret.
Source: Frèdèric Le Play, 'Om ståltillverkningen i Yorkshire, samt jemförelse mellan de förnämsta ståltillverkningsorter i Europa', *Tidskrift för Svenska Bergshandteringen för år 1845.*

of the Demidovs, had been used for making steel of a lesser quality since the eighteenth century. Nor was English iron necessarily a threat. Walloon-forged iron from Backbarrow in Lancashire was never likely to be made in significant volumes. But there was every chance that the other English brands named might be, for they were types of puddled iron.

This opening for puddled iron came with the sharp rise in steel production in mid nineteenth-century Britain. The 20,000 tons made at the beginning of the 1840s climbed to over 100,000 tons in the early 1870s.[79] So colossal an increase could not be sustained on the back of 'Orground' iron alone. Although the production of 'Orground' iron had increased from between 5,000 and 6,000 tons in 1800 to about 10,000 tons at mid-century, largely through the spread of Walloon forging into the counties surrounding Uppland, it had not done so to an extent that could keep the cementation furnaces of Sheffield adequately supplied. Other sources of high-quality iron were required. *Ståljern* from Gästrikland was one possible substitute; puddled iron from Britain was another. It was this last prospect that terrified Swedish experts.

Ludvig Rinman rang the tocsin after his tour of Britain in 1849. Puddled iron, he reported, had made major strides since Henry Cort's day. Not only could it be made cheaply, it could be made to a very high specification. Rinman was particularly impressed by the ability of puddlers to vary the level of carbon in the finished product. The best of them could control the fining process far more accurately than most Swedish forgemen. Bars from the Low Moor works in Yorkshire, Rinman noted, were already widely used in Sheffield. They made a fine blister steel; and when crucible steel was to be made, Low Moor iron was considered worthy of being mixed with 'Orground'.[80] If other British makers were to follow the example of Low Moor the *Vallonbruk* would lose the one market that had, for a century and half, been absolutely secure.

A response was required in Uppland, for the *Vallonbruk* continued to follow working patterns that had been fixed for generations. The

---

[79] Barraclough, 'Svenskt järn', p. 259. These statistics are somewhat ambiguous as blister and crucible steel are lumped together. This cannot be, as the latter is produced from the former.

[80] 'Berättelse till herrar Fullmäktige i Jerncontoret om en resa i England år 1849 af Ludvig Rinman', Jernkontorets Arkiv, Fullmäktiges Arkiv, Handlingar ordnade efter Ämne, Vol F II a:20.

routines at Leufsta in the 1840s were still those of the eighteenth century. The plant remained unchanged; the innovations associated with the New German Forging Method or Lancashire forging had passed *Leufstawerken* by. There were no blowing cylinders, no steam engines, and no rolling mills. Nor was there any alteration in the organisation of labour. The ten-man forge crews still worked as they had done in the 1730s when Jacob Tillman had been the master finer at Leufsta's Upper Forge. Indeed, workers with the characteristic Walloon surnames were still to be found at the *bruk*.

After 1850 this changed. New items of plant were installed and new organisational patterns were developed. This is an area in which very little research has been conducted. The evidence from Forsmark, however, is suggestive. In the first decades of the nineteenth century the *bruk*'s two forges operated in a somewhat sluggish fashion, working for only 30 weeks in the years and making approximately 400 tons annually. From mid-century, however, the number of weeks during which the forges were at work grew to between 35 and 40. There was also an increase in the size of the forge crews until they numbered 14 apiece. Finally, new-style blowing cylinders, a heated blast, and steam power came to Forsmark. The consequences were soon evident. Annual output broke the 600-ton barrier in 1860, then 700 tons in 1870. In 1882, the last year in which 'Orground' iron was made at Forsmark, 914 tons was made.[81]

The forges at Leufsta and Österby were upgraded in a somewhat different way. Or rather, they were replaced with entirely new, purpose-built, steam-driven plant from the 1860s. These embodied the 'best-practice' procedures that Gustaf Ekman had seen in British charcoal forges in the 1830s. At Österby the juxtaposition of the old and the new can still be seen. The water-powered Walloon forge of the eighteenth century stands on one side of the stream; on the other bank is the far larger 'Steam Forge' (*Ånghammaren*), emblazoned with the date 1889.[82]

[81] For a longer discussion see Göran Rydén, 'Vallonbruk, vallonsmeder och vallonsmide—en precisering av ett kunskapsläge', in Anders Florén and Gunnar Ternhag (eds), *Valloner—järnets människor* (Hedemora, 2002), pp. 107–35.

[82] Attman, *Fagerstabrukens historia*, pp. 614–18.

Illustration 4.9. The Steam Forge at Österby *bruk*.

By 1850 the 'Atlantic age' had ended for Baltic iron. Russian bar iron was for the most part traded domestically; Swedish iron remained export-orientated, but it no longer turned upon an Atlantic axis. The international markets that Sweden served were now more varied. In any case, the advent of bulk steel-making methods in the 1860s and 1870s opened up a new technological era, one in which steel made by the Bessemer or Siemens-Martin processes replaced malleable bar iron as the staple input of the industrial world.

But in the century and half between the 1690s and the 1840s the Atlantic had exerted a powerful—indeed, determinant—gravitational pull on Baltic iron. For the most part, this pull was channelled through the British Isles. The British market was a vortex into which iron from across the northern hemisphere was dragged. That meant perforce that Sweden, as early modern Europe's leading iron exporter, was bound in with British developments. From Britain, Swedish (and later Russian) iron was spun out into the wider Atlantic world. Sometimes Baltic iron issued forth from British ports in a form that would have been recognisable to the forgemen who had wrought it. Hans Hansson Palt, for example, would have had no difficulty in identifying the voyage iron he had made at Gammelbo *bruk* in the 1730s if, by some extraordinary quirk, he had found himself transported to the Cross River delta where it was traded for slaves. For the most part, however, when Baltic iron entered the Atlantic world it did so embodied in a multiplicity of implements, fastenings and fittings that were bagged up for export to Africa, the Caribbean, or British North America. And many of these incorporated steel made from 'Orground' iron, the most prized sort of Baltic iron. 'Orground' iron had been enormously difficult to capture, but when British merchants succeeded in monopolising it, as they did in the early eighteenth century, they gave British imperialism its cutting edge. Literally so, for the sword blades, bayonets, axes, hoes, planes, chisels and the like that advanced British global hegemony, whether militarily or economically, traced their origin to the *Vallonbruk* of Uppland.

So, just as the Atlantic economy dictated the commodity flows that coursed through the Baltic, the Baltic contributed immeasurably to the growth of the Atlantic economy. Yet the eminence of Baltic iron was a source of unease in the mid-eighteenth century, both for those who produced it and those who consumed it. Swedes wondered if the upward trajectory of their exports could be sustained without incurring excessive social and ecological costs. At the same time, the British were fearful of being dependent upon a hostile power for so strategic a material as iron (not to mention naval stores). It was this that led to the complex and often confused debates, both in Sweden and Britain between the 1730s and 1750s about the future direction of iron production and exchange in the Atlantic world. After much dissension, Swedes endeavoured to shore up their position on the international market by capping production. The British pursued a still more audacious path, by seeking to build a transatlantic iron industry. Mining, smelting, refining and processing would stretch across three thousand miles of ocean. Yet such a project ran counter to the political realities of the revolutionary age. America's nascent capitalist class would not submit to imperial discipline in so constricting a form. The American Revolution came as a rebuke to such mercantilist arrogance.

No sooner had Britain's imperial crisis of the 1770s put paid to the transoceanic division of labour that some of her ironmasters had longed for than technological upheaval set the British iron industry on an entirely different path. For the first time in over a century British ironmasters were able to satisfy home demand without Baltic assistance. Puddling and rolling burst the ecological bounds that had stunted the growth of the British forge sector, plunging the Swedish and Russian iron industries into crisis. Desperate to compensate for their eviction from the British market, Baltic exporters now switched their attention to the American Republic. This was to turn on its head the commercial relationship that had obtained between Sweden, Britain and north America in the mid-eighteenth century. In the 1730s and 1740s British ironmasters and manufacturers had looked to charcoal-rich America for succour: colonial pig iron would be a means of combating import penetration by Baltic producers. Now the roles were reversed. Swedish exporters looked to American consumers to substitute for their lost British clientele. For the Swedes, if not the Russians, the strategy was a successful one. The Atlantic connection endured, even though it was no longer articulated through Britain.

The distinctively Atlantic dimension of Swedish iron making only came to an end in the middle of the nineteenth century, when fresh competition from British ironmasters and import substitution in the United States reduced the importance of America as a destination for iron from *Bergslagen*, just as it had done for Siberian iron. After 1850 the Atlantic ocean, in its guise as vector of an 'iron system', receded. The iron Atlantic, in its eighteenth-century heyday, had broken upon the pink granite skerries that guarded Gothenberg and coursed through the Stockholm archipelago. In 1750 it had lapped against the quays of St Petersburg as surely as it did the shores of the Chesapeake. In 1850 it did so no longer.

CONCLUSION

To write a history that views the Baltic—and Sweden in particular—from the perspective of the Atlantic may seem perverse. After all, Sweden's experience of the Atlantic was largely indirect. The American colony of New Sweden (*Nya Sverige*) had been founded amid high hopes in 1638, but Swedish control of the lower Delaware valley was short-lived.[1] The territory was surrendered to the Dutch in 1655, who lost it in turn to the English. Swedish-speaking Lutheran communities lived on in Pennsylvania, Delaware and New Jersey, but they were just one, increasingly minor element in a colonial scene dominated by English and Welsh Quakers, Scots-Irish Presbyterians, German Pietists of different stripes, and Africans whose spiritual beliefs stood apart from those of their north European masters.[2] Swedish endeavour on the African coast was equally fleeting. Louis De Geer, no less, was among the promoters of the Swedish West India Company (*Guineakompaniet*), established in 1649 to force an entry into the booming slave trade, but the forts that were established along the Gold Coast in the 1650s did not endure. The last of them was seized by the Danes in 1663.[3] Sweden's imperial presence in the Atlantic was not renewed until the 1780s, when Gustav III, exploiting the diplomatic fallout from the American Revolution, acquired the tiny Caribbean island of St Barthélemy.

This was a sorry record, one that compared unfavourably not just with the colonial giants of the Atlantic world, such as Spain or Portugal, but with Nordic neighbours such as Denmark. The Danes not only ousted

---

[1] Stellan Dahlgren and Hans Norman, *The rise and fall of New Sweden: Governor Johan Risingh's journal 1654–55 in its historical context* (Uppsala, 1988).

[2] Although note the argument presented in Terry G. Jordan and Matti Kaups, *The American backwoods frontier: an ethnic and ecological interpretation* (Baltimore, 1989) that the lower Delaware valley was the 'cultural hearth' from which the characteristic practices of American frontier life emerged. And these, it is claimed, were brought across the Atlantic by Savo-Karelians who settled in New Sweden in the 1640s and 1650s. Their expertise with the felling axe, their ecologically wasteful forms of agriculture, their notched log style of domestic architecture, their preference for hard liquor over beer, and their zest for personal violence all endured in frontier culture. In such a way the Swedish imprint on American life was profound.

[3] György Nováky, *Handelskompanier och kompanihandel. Svenska Afrikakompaniet 1649–1663. En studie i feodal handel* (Uppsala, 1990).

the Swedes from their African bridgehead, they maintained a presence in the Caribbean throughout the eighteenth century; the modest but steady slave trade of Copenhagen's *Vestindisk-Guineiske Kompagni* linked the two theatres. Sweden had nothing to compare even with this. The Swedes were not alone in experiencing territorial disappointment, of course. The Dutch lost substantial colonial possessions in both north and south America in the seventeenth century, but they remained a significant trading power in the Atlantic long after their settlements in Pernambuco and the Hudson valley had passed into other hands.[4] Likewise, the wreck of Britain's north American empire in the 1780s did not seriously diminish British maritime prowess. The same could not be said of the Swedes. Swedish shipping played virtually no part in transatlantic commerce until the very end of the eighteenth century.

Given all of this, it might be thought fanciful to add a 'Swedish' Atlantic to the list of conceptual Atlantics already in circulation. Yet Sweden in the eighteenth century had an unmistakeable Atlantic dimension, as this book makes clear. That Sweden's 'Atlanticism' was articulated via Britain, her principal trading partner, did not make it any the less vital, merely singular. And it would to be perverse to deny Atlantic credentials to Stockholm when historians are willing to consider entirely landlocked locations such as the upper Mississippi valley as lying within the ocean's orbit, or to see a city like Lima, whose citizens gazed out across the Pacific, as an important node of the Atlantic slave trade. Conversely, the 'Atlanticity' that can be ascribed to Sweden cannot plausibly be extended to Russia, even though Russian exports of iron, timber, and timber products exceeded those of her Baltic rival by the mid-eighteenth century. To be sure, goods left the quays of St Petersburg and Riga for Atlantic destinations in great number, but the commercial filiations that joined Russian producers to Atlantic consumers had none of the delicacy or precision that characterised their Swedish equivalents. Forgemen at Gammelbo, let it be remembered, adjusted their daily practice in direct response to signals from African markets; their counterparts in the *Vallonbruk* were taught to answer to the demands of steel makers in Birmingham and Sheffield. Workers in the Urals, by contrast, made iron bars without the least regard for the shapes or sizes that Atlantic consumers required. Besides, the mind-set of the Russian court had

[4] Pieter Emmer and Wim Klooster, 'The Dutch Atlantic, 1600–1800: expansion without empire', *Itinerario*, 2 (1999), 48–69.

no pronounced Atlantic orientation. Russian imperial ambitions were continental rather than oceanic. The colonial frontier was driven east and south: into Siberia or toward the Black Sea. Russia was connected to the Atlantic world, but it did not belong to it.

To grant Sweden—but not Russia—a place in the new Atlantic History is therefore allowable. But like all episodes in Atlantic history, Sweden's 'Atlantic moment' had a distinctive chronology. Bar iron from *Bergslagen* appeared on west European markets from the 1620s, but it was not until the 1660s and 1670s, when London became their foremost destination, that exports took on an Atlantic inflection. Significantly, the escalation of exports to London coincided with the great leap forward in Britain's Atlantic commerce brought about by the Navigation Acts of the 1650s and 1660s. By 1700 Swedish iron played an indispensable role in the projection of British power in the western ocean. The merchant fleet and Royal Navy alike owed their seaworthiness to bolts and hoops of Swedish iron, not to mention Swedish tar. It mattered not whether the vessels were built in Shadwell or Salem, Massachusetts; their timbers were shaped by adzes and axes that had been edged with steel of Swedish ancestry. And the British-made hardware and 'toys' that circulated around the ocean either embodied Baltic iron or had been crafted with tools that did.

Bar iron from *Bergslagen* retained a central role in Britain's industrialising economy for the remainder of the eighteenth century, until puddling and rolling changed the entire foundation of malleable iron production. The coal technology revolution in Britain dealt Swedish iron a body blow, but the response of Swedish exporters was to redouble their Atlantic orientation not to abandon it. Now, in the early nineteenth century, as they directed their exports towards markets in the United States, Swedish producers and merchants became direct participants in an Atlantic iron trade rather than adjuncts to the British, as formerly. As a result, Swedish iron production kept a distinctively Atlantic flavour into the 1840s, long after Russia's iron industry had turned in on itself to serve Eurasian markets.

But to bring yet another part of the globe into the Atlantic's embrace is to invite difficulties. 'Atlanticity' is now being ascribed to so many communities, cultures, and commodities that the capacity of the Atlantic History paradigm to bear ever more historiographical freight is open to question. It is in danger of slumping beneath the explanatory demands made upon it. As the study of Atlantic History has progressed, so it has become clear that the integrative, centripetal forces that pulled together peoples, things, and ideas from the continents that bordered the ocean did *not* make the Atlantic basin a place apart. On the contrary, the early modern Atlantic was being drawn into a tighter relationship with other zones of the world economy. Whilst Mexican silver was shipped across the Atlantic to fill the royal treasury in Spain, it also exited Mexico to the west, leaving Acapulco for Manila, where it was exchanged for silk, porcelain, and lacquer wares. And much of the silver that was docked at Cadiz was trans-shipped for Asia. (Sweden's tiny East India Company made its own contribution to the drain of bullion to the east: her ships invariably tied up at Cadiz to take on board the silver *piastres* needed to buy tea at Canton).[5] America and Africa, in other words, combined to bring Europe closer to Asia.

The number of European vessels that ventured east of the Cape of Good Hope was, of course, relatively small in comparison with those that rode the Atlantic, but those that did navigate the Indian Ocean found it flush with shipping of all sorts. Indeed, historians of the Indian Ocean insist that the dynamic interconnectedness that Atlantic historians see as peculiar to their ocean in the early modern era was far from unique. Similar sorts of maritime interaction were to be found in the Arabian sea, along the east coast of Africa, in the Bay of Bengal, and—pre-eminently—amid the sea lanes that linked the Indian Ocean to the South China Sea.[6] This thriving commerce only tended to register in European minds when Bohea tea was served in middle-class parlours, ideally in 'china' cups, and to register principally as an Asian contribution to the business of the western hemisphere. Yet the trade of the Indian Ocean also intruded into the Atlantic when Gujarati textiles or cowries from the Maldives were traded on the Guinea coast. The instinctive reflex of Atlantic historians is to treat the commodity chains that originated in south Asia as tributaries of an Atlantic system. But might not the Atlantic be viewed with equal justice as an arm of Asian commerce?

This is to say that the Atlantic History paradigm that has enlivened historical discussion for the last twenty years is too enclosed—too

---

[5] Christian Koninckx, *The first and second charters of the Swedish East India Company (1731–1766): a contribution to the maritime, economic, and social history of north-western Europe in its relationship with the Far East* (Kortrijk, 1980), pp. 120, 126–27, 189–98.

[6] See Sushil Chaudhury and Michel Morineau (eds), *Merchants, companies and trade: Europe and Asia in the early modern era* (Cambridge, 1999); Om Prakash, *European commercial enterprise in pre-colonial India* (Cambridge, 1998).

enclosed in the sense that the Atlantic basin was part of a global system and is best appreciated in its global setting. That Atlantic History had epochal extra-Atlantic ramifications is something that historians of China are keen to point out. The New World provided Europeans with material resources that were denied to east Asians. The Americas supplied immense quantities of food and fibre to European consumers—resources that could not have been found within Europe itself: rice from Carolina or cotton from Georgia. The bounty of the New World, in other words, generated for the most part by coerced labour, allowed a growth in Europe's non-agricultural workforce and a growth in European industrial production. And it did so without precipitating a catastrophic decline in European living standards. China did not benefit from the same windfall. Continued growth implied a movement onto marginal land or an over-exploitation of lands already used to raise food or industrial raw materials. Diminishing returns set in and in the nineteenth century China succumbed to ecological degradation and Malthusian crisis.[7] Atlantic History is therefore a planetary rather than a hemispheric phenomenon: it underlay the 'great divergence' of East and West.

Yet if the Atlantic History paradigm is in one sense too closed off, it is in another sense a little too open. There is tendency among many historians of an 'Atlanticist' bent to play down the obstacles to economic and social interaction within the Atlantic theatre, opting for concepts and metaphors that bring collegiality or adaptive collaboration to mind ('community', 'conversation', 'Atlantic creole'). Yet obstacles there were. The ocean was ringed by institutions and structures designed expressly to thwart interchange between rival political systems. Smugglers might subvert them and religious visionaries try to disregard them, but they were there nonetheless. British commercial success during the lifetime of Graffin Prankard—roughly speaking the years between the Glorious Revolution and the Seven Years' War—was built upon fiscal centralisation, naval might, and the ruthless exclusion of foreign shipping from colonial trade routes. It proved a happy combination, allowing for a massive growth in Atlantic trade, of which the *Baltick Merchant*'s annual expedition to South Carolina was one small expression. The record of

[7] Kenneth Pomeranz, *The great divergence: China, Europe, and the making of the modern world economy* (Princeton NJ, 2000), chapters 5 and 6.

competitor states that could not boast the same fiscal efficiency (France) or protectionist rigour (the United Provinces) was inferior.[8]

So, systems of power placed limits on what was possible, as Swedes well knew. After the loss of *Nya Sverige* and the surrender of *Guineakompaniet*'s African forts Sweden's direct link to the western ocean was lost. Henceforth her 'Atlanticism' was contingent upon British mediation. Swedish commentators were uncomfortably conscious of this and sought to override British influence by imitating Britain's own protectionist code. The *Produktplakat* of the 1720s drew unashamedly on the Navigation Acts, and the boost it gave to Swedish shipping was appreciable. But nothing could dislodge the grip that British merchants had established on Swedish iron. Although British ministers fretted at what they saw as over-dependence on Baltic supplies of iron and naval stores, the truth was that Sweden's iron makers stood in a subordinate position to British merchant capital.

The concept of 'conversation' has been profitably employed to explain how Madeira wine, one of the most successful Atlantic commodities of the eighteenth century, arose out of a dialogue between its producers and consumers—the product was refined in terms of its flavour, appearance and packaging in response to shifts in consumer taste around the ocean's circumference.[9] The concept is less successfully applied to Swedish iron. The major 'Orground' brands underwent changes in the ways they were manufactured and marketed, but these changes did not stem from a conversation; they arose from exchanges that had a curt, peremptory tone to them. And if most of Sweden's iron exports were contracted for by British merchants the chances of exploring alternative markets were limited. Much to his dismay, Reinhold Angerstein found few items of Swedish provenance on the quayside at Cadiz when he visited the port in 1752. The iron was mostly Spanish and the steel '*Azero de Venetia*'—Alpine steel shipped from Venice.[10] Angerstein made hopeful estimates of the markets that Swedish commodities could win

[8] David Ormrod, *The rise of commercial empires: England and the Netherlands in the age of mercantilism, 1650–1770* (Cambridge, 2003).

[9] David Hancock, 'Commerce and conversation in the eighteenth-century Atlantic: the invention of Madeira wine', *Journal of Interdisciplinary History*, XXXIX, 2 (1998), 197–219.

[10] Anders Florén and Göran Rydén, 'Sketches of Spain: the journey of Reinhold Rücker Angerstein, 1752', in K. Benson, M. Mörner, and I. Söhrman, (eds), *Spanish-Swedish relations from the mid-seventeenth century to the early nineteenth century* (Göteborg, 2002), pp. 291–305.

in the Iberian Atlantic—markets that could be fed via the fleets that sailed to Havana, Vera Cruz, Cartagena, and Buenos Aires—but for as long as Sweden's iron exports remained under the mercantile control of British factors Angerstein's hopes were to be unfulfilled. And so it proved. British hegemony in Stockholm was immovable.

The iron that left Stockholm's *Jernvåg* had, of course, a profound impact on the British Isles, the point of departure for a still wider Atlantic world. It did so, as we know, by providing an essential raw material for the makers of hardware and other ferrous articles, and by providing it in bulk. But it was not just a question of quantity; it was the *quality* of Swedish iron that played a quite decisive role in the British Atlantic. By monopolising the supply of 'Orground' iron in the early eighteenth century British merchants facilitated a growth in English steel making. Indeed, the existence of such an industry in Sheffield and elsewhere depended *entirely* on the command that British factors were able to establish over iron from *Leufstawerken* and *Strömbergswerken*. Had it been otherwise, English steel making must have perished and the hardware trades languished for want of steel. A comparison with France, Britain's great adversary, is instructive. French industrialists, *savants* and state officials were intent on developing a steel sector to match Britain's, but misplaced economic nationalism led them to forswear the use of Swedish 'Orground' iron.[11] Réaumur, whose *L'art de convertir le fer forgé en acier* (1722) was a major contribution to metallurgical science, saw no call for Baltic iron.

> I have said before that, if it should be necessary to resort to the irons of Sweden, the inconvenience would not be great, as it can be had in our ports at approximately the same price as our own. But we are far from this necessity. I have tried the irons of several forges in Berry and have been very successful with them. I am experienced with good iron from Nivernais...'[12]

Réaumur went on to list a further six French provinces whose forges could supply iron worthy of conversion to steel. But it was not so. By

[11] J.R. Harris, *Industrial espionage and technology transfer: Britain and France in the eighteenth century* (Aldershot, 1998), pp. 205–21.

[12] [René Antoine Ferchault de Réaumur], *Reaumur's memoirs on iron and steel* (1722: Chicago, 1956), p. 120. Réaumur (1683–1757) was best known as an entomologist whose *Mémoires pour servir a l'histoire des insectes*, published in six volumes between 1734 and 1742, was the model for Charles De Geer's own eight-volume study of the same title, published between 1752 and 1778.

spurning the appropriate inputs, French steel makers could do no more than stumble forward. A refusal to acknowledge Sweden's potential sapped at France's Atlantic endeavours; a willingness to exploit Baltic resources strengthened Britain's.[13]

As the case of 'Orground' iron suggests, the growth of imperial power in the Atlantic depended upon trade—more specifically, upon trade as contemporaries understood the term. Trade, as Daniel Defoe insisted, was an amalgam of 'Dealing and Manufacturing'; it comprised commodity chains that stretched from wharf to warehouse to workshop and on again.[14] This book has applied Defoe's dictum to the international iron trade. Historiographically, the iron trade has long been disregarded, yet—as has been demonstrated here—Baltic iron, the most widely traded of irons, was an integral part of the eighteenth-century Atlantic world. To be fully appreciated, however, the iron trade must be understood on the terms that Defoe proposed, as the unity of making and exchange. Only by tracing the passage of metallic matter from mine to furnace bank, from casting house to forge, and on into the commercial maelstrom of the Atlantic can the pervasive presence of Baltic iron be revealed. It was there in the guns and blades that provided the Atlantic with its violence; it was to be found in the scientific instruments or burnished gewgaws that lent finesse to the Atlantic world. Baltic iron, in one form or another, penetrated to the furthest reaches of the ocean. Little of this was apparent to the Walloon forgemen who heaved spark-scattering blooms of iron onto the floor of the Great Forge at Leufsta, but as they dragged the blooms to the hammer block they set in train a world-changing process.

---

[13] Pierrick Pourchasse, 'Problems of French trade with the North in the eighteenth century', paper presented at the annual conference of the Economic History Society, 1 April 2006: http://www.ehs.org.uk/ehs/conference2006/Assets/IIDPourchasse.doc

[14] Daniel Defoe, *Plan of the English commerce. Being a complete prospect of the trade of this Nation, as well home as foreign* (2nd edn., London, 1737), p. 3.

DRAMATIS PERSONAE

The ALLOWAY family of Bridgwater in Somerset were merchants and mariners, trading in both European and American waters in the late seventeenth and early eighteenth centuries. Their involvement in the herring fishery of England's south-west and the cod fishery of Newfoundland led to their investing in the manufacture of salt, that essential preserving agent. Graffin Prankard (q.v.) married into the family in 1708—they were Quakers, like him—and followed them into the salt business. Nathaniel Alloway, Prankard's cousin, was master of the *Baltick Merchant*.

Reinhold Rütker ANGERSTEIN (1718–1760), an ironmaster's son, was appointed to the *Bergscollegium* in 1738. Between 1752 and 1758 he was engaged in a more or less continuous tour of Europe, visiting the German lands, Hungary, Italy, Spain, France, Britain and the Netherlands. At the time of his death Angerstein held the post of *Assessor*; only the *Bergscollegium*'s president and his two counsellors outranked him.

John BANNISTER (d. 1743) had day-to-day management of the Crowley family firm after the death of John Crowley in 1728 and was a partner of Crowley Hallett's (q.v.) in the firm Harrison, Bannister & Hallett. His appearance before the 1737 committee of the House of Commons on the iron trade featured a tendentious warning against the dangers of hardware manufacturing being tolerated in the North American colonies.

Anders BERCH (1711–1774), professor of economics at Uppsala University from 1741, wrote the first economics text-book in Swedish, *Inledning til almänna hushålningen, innefattande grunden til politie, oeconomie och cameral wetenskaperne* [An introduction to common householding, including the foundations of political, economical and cameral science] (1747).

Christer BERCH (1733–1792), the son of Anders Berch and his successor as professor of economics at Uppsala. Berch's 'grand tour' in 1757–1761, included a spell in Britain.

Willem de BESCHE (1573–1629), Dutch entrepreneur, was Louis De Geer's chief associate in opening up Sweden's iron and copper industries to foreign investment in the 1620s. Like his compatriot, De Besche owned or rented ironworks in Uppland and several other counties.

Johan (or Jean) BOLINDER (1813–1899), engineer and businessman. Bolinder was recruited to *Jernkontoret* in 1838 and sent on a mission to Britain in 1842–43. He was instrumental in transplanting British production techniques to Sweden, where he established in his own engineering firm in 1850.

Carl BONDE (1581–1652), Swedish nobleman and servant to the Crown, was one of the founding fathers of the *Bergscollegium.*

Auguste-Henri de BONNARD (1781–1857), mining engineer and geologist, graduated from the Ecole des Mines de Paris in 1801 and undertook a tour of British coalfields during the brief period of peace that followed the Treaty of Amiens in 1802. After his return to France he enjoyed a distinguished career as secretary to the Conseil général des Mines.

The BOTFIELD family were important practitioners of coal-based technologies in the Shropshire iron industry. Thomas Botfield (1738–1801) became a partner in Lightmoor furnace in 1758, just as coke smelting became standard practice on the east Shropshire coalfield. He founded a new works at Old Park in 1790 that was soon equipped with a steam-driven rolling mill.

Gustaf BROLING (1766–1838), *Bergscollegium* agent, visited Britain in 1797–99. His well-received travel account included a full description of crucible steel making in Sheffield. Broling established his own crucible plant in 1808.

Robert CAMPBELL (d. 1758) came to Stockholm from Scotland in the first decade of the eighteenth century. By the 1720s he was one of the city's leading exporters of bar iron.

The CARROLL family of Annapolis, Maryland, were among the largest planters and merchants in the Chesapeake. Charles Carroll (1691–1755) was the principal investor in the Baltimore Company, established in 1731, which was to be one of the largest iron-making concerns in the mid-Atlantic region and one of those most focused on the British market.

Henry Cort (1741–1800) patented the puddling and rolling method in the mid-1780s, but he derived little personal benefit from the technological breakthrough he did so much to bring about. He was unwittingly caught up in the malfeasance of his business partners and bankrupted in 1789. His patents were forfeit to the state, but the government made little effort to collect the royalties due. As a result, British ironmasters were able to exploit Cort's methods for free.

Thomas Coster (1684–1739), MP for Bristol from 1734 until his death, was a member of an eminent industrial dynasty, one responsible for the rise of a coal-fired copper smelting industry in the Forest of Dean (from the 1690s), then in the Neath and Swansea valleys of south Wales (from the 1720s). Related investments in brass manufacture tied the family's fortunes very firmly to the slave-based commerce of the city Thomas Coster represented in Parliament. Indeed, he was part-owner of the *Amoretta*, which shipped slaves to Carolina through the 1730s.

Richard Crawshay (1739–1810) began his career in the 1760s as a dealer in iron hollow wares in London. Soon, he moved into the Baltic trade, importing iron and timber from Sweden. By the 1780s he was probably the capital's foremost iron merchant. Not content with that, he re-launched himself as an ironmaster, taking over the Cyfarthfa works in Wales. Within a decade the works were Britain's largest, where the latest in coal technology was showcased.

The Crowley family was established as a force in the British iron trade by Ambrose Crowley (1658–1713), the son of a Stourbridge ironmonger of the same name. A man of granite authority, Ambrose Crowley established himself as an ironmonger in London in the 1680s and, having fallen out with his suppliers, began to manufacture iron wares on his own account, establishing factories in the North East of England for that purpose. Although born a Quaker, Ambrose Crowley became an Anglican and an ardent Tory. His son and successor, John Crowley (1689–1728), was a Jacobite—a risky allegiance for someone whose business depended on naval contracts from the Hanoverian regime. After John's early death the sprawling Crowley empire was headed by his widow Theodosia (1694–1782).

Abraham Darby (1678–1717), a native of the nail country around Dudley, set up business in Bristol as a brass manufacturer and iron founder at the turn of the eighteenth century. Soon after, he shifted his attention to Shropshire, where he was involved in establishing a brass battery, rolling mill, wire mill and steel furnace near Shrewsbury and taking over the blast furnace at Coalbrookdale. Graffin Prankard was an investor in both projects, albeit fleetingly. Darby was an important innovator in brass making, but it was in ferrous metallurgy that he made his most enduring contribution, using coke to smelt iron at Coalbrookdale.

The De Geer family was of Walloon origin but Louis De Geer (1587–1652), the founder of the family's fortunes, first made his mark in Amsterdam where he flourished as a merchant. In the 1620s he moved again, this time to Sweden, where he established himself as the country's leading merchant, mine owner and industrialist. The Uppland *bruk* at Leufsta, Österby and Gimo, and the cannon foundry at Finspång, south of Stockholm, were his principal bases. In the early eighteenth century the family concerns were dominated by two of Louis De Geer's grandsons, the brothers Carl De Geer (1660–1730) and Jean Jacques De Geer (1666–1738). They consolidated and extended the De Geer domains in the county of Uppland, taking additional shares in the Dannemora mine and acquiring new ironworks. Carl De Geer died childless, but Jean Jacques had three sons—Louis De Geer (1705–1758), Charles De Geer (1720–1778) and Antoine De Geer (1721–1756)—under whose guidance the family continued to dominate iron making in Uppland and in Sweden as a whole.

The Demidov family dominated Russian iron making in the eighteenth century. The family's fortunes were founded by Nikita Demidovich Antufyev (1656–1725), a blacksmith from the metalworking city of Tula in western Russia who was one of Peter the Great's metallurgical pioneers on the Siberian frontier. His son, Akinfy Nikitich Demidov (1678–1745), added to his father's industrial empire, so that he controlled 25 ironworks and copper works at the time of his death. The sable mark that was imprinted on the family's iron was an internationally recognised brand.

Caleb Dickinson (1716–1783), Bristol merchant and landowner. Dickinson was apprenticed to his co-religionist Graffin Prankard and married his master's daughter on the completion of his term in 1738. He also inherited estates in Somerset and a share in a sugar plantation in Jamaica, yielding a wealth that allowed him to save his father-in-law from bankruptcy during the crisis of Prankard's affairs in 1739–1740.

William DONNE, the Bristol ironmonger, claimed to have 'two furnaces in Virginia & two slitting mills in Eng[lan]d' when he gave evidence to a committee of the House of Commons in 1737. One of the slitting mills was that at Congresbury in Somerset where he processed Russian iron bought from Graffin Prankard. He was also a partner in the 'Bristol' ironworks in Virginia, which supplied pig iron to the forges of the Knight family (q.v.). The second colonial furnace has not been identified.

Gustaf EKMAN (1804–1876), the son of a Gothenberg merchant, joined *Jernkontoret* in 1827 after graduating from Uppsala. His first-hand observation of British forge techniques, made on three separate trips between 1828 and 1833, led him to devise the Lancashire forging method.

The gun-maker Joseph FARMER (d. 1741) described himself as 'a manufacturer of steel wares at Birmingham' when he appeared before the House of Commons committee on the iron trade in 1737 as an advocate of importing American bar iron. Farmer had visited the plantations in 1718–1719 and become convinced of their potential as iron making territories. He acted on his conviction, helping to establish the Principio Company in Maryland in the 1720s.

James FARMER (1715–1773) took over the gun-making business of his father Joseph (q.v.) in the 1740s. In partnership with his brother-in-law Samuel Galton, he was one of Birmingham's leading manufacturers.

Jacob FEIFF (1679–1736), a Stockholm merchant born into a family of Scottish origin, handled the export of iron from a variety of *bruk*, including Gammelbo. He was also an ironmaster in his own right, renting Wattholma *bruk*, for example, in the early 1730s.

The FINCH family of Dudley were major putting-out employers in the West Midlands nail trade. They also maintained a London warehouse.

Edward Finch (c. 1697–1771) was a younger son of the Earl of Nottingham. He was MP for Cambridge from 1727 to 1768, but for most of his early career he served as a diplomat in the Baltic. He was envoy to the Polish court from 1725 to 1727, to the Swedish court from 1728 to 1739, and to the Russian court from 1740 to 1742.

The Freeths were prominent as Birmingham tool makers over several generations. Sampson Freeth features regularly in Graffin Prankard's accounts from the 1730s, buying between four and seven tons of iron annually.

John Galton (1671–1743), like Graffin Prankard, his fellow Bristol merchant, married into the Alloway family (q.v.). His son Samuel (1720–1799) moved to Birmingham to set in partnership with his kinsman James Farmer (q.v.), the gun manufacturer.

The Grill family were merchants of Dutch descent, active in both Amsterdam and Stockholm. The firm headed by Carlos Grill (1681–1736) and his nephew Claes Grill (1705–1767) was one of the leading iron export houses in the Swedish capital. They traded mainly with Amsterdam, where the sale of iron was handled by Anthony Grill (1705–1783), the twin brother of Claes. The family also owned important ironworks, such as Söderfors, purchased in 1748, and Österby, acquired from Charles De Geer in 1758. Members of the family were also very active in the Swedish East India Company.

Fredrik Gyllenborg (1698–1759) was an important ironmaster and president of the *Bergscollegium* from 1750.

Crowley Hallett (1711–1767), grandson of Ambrose Crowley (q.v.), was a London ironmonger and sometime partner of Josias Wordsworth (q.v.).

William Hill (1741–1816), ironmaster and politician. Like many Scots-Irish of his generation, Belfast-born Hill found his way to the Carolina backcountry, where he had acquired more than 5,000 acres of land by the outbreak of the Revolution. The revolutionary war saw him engage in munitions production at his new-built Aera furnace and perform military service as an officer in the famous militia brigade of Thomas Sumter. As an activist in state politics after the war, William

Hill was instrumental in promoting river navigation and canal schemes that would open up the backcountry.

Homfray family. Francis Homfray (1674–1737) served Ambrose Crowley (q.v.) as a nail warehouseman in the Stourbridge district in the 1690s. In 1702 he branched out on his own account and became an important figure in the West Midland nail trade in his own right. The business, which comprised forges and mills on the river Stour, continued to thrive after his death under the management of his wife Mary (d. 1758) and his son Francis (1725–1798). The third generation was instrumental in the coal technology revolution in South Wales in the 1780s and 1790s. Samuel Homfray (1762–1822) and his brother Jeremiah (1759–1833) established the Penydarren works at Merthyr Tydfil, one of the first to employ the puddling process.

Benjamin Huntsman (1704–1776), the inventor of crucible steel, began his working life as a clockmaker and continued in that trade into the 1740s. By the 1750s, however, he had specialised as a steel refiner, whose distinctive product enjoyed a European celebrity.

The London Quaker, Thomas Hyam, was Graffin Prankard's banker. Bills of exchange drawn on Hyam were the means by which payment was made to Prankard's suppliers in Holland and the Baltic.

Gabriel Jars (1732–1769) was a star pupil at the Ecole des Ponts et Chaussées, the main centre of training for engineers in Enlightenment France. He was sent to investigate English technology in the wake of the Seven Years' War. His mission was wide-ranging, but his instructions directed him to pay particular attention to English steel making and to the manufacture of files, which were thought superior to those of any other European country.

Francis Jennings (1692–1754), an Ulsterman who naturalised as *Frans Jenning* in Sweden, was Stockholm's most important iron exporter in the second quarter of the eighteenth century. Like many export merchants, Jennings acquired *bruk* of his own, in this case Forsmark, bought in 1751.

Henric Kahlmeter (1693–1750), *Bergscollegium* agent, spent four years in Britain in the 1720s.

Pehr Kalm (1716–1779), a naturalist who trained under Linnaeus at Uppsala, became the first professor of 'Oeconomia' at Åbo in his native Finland.

John Kettle of Birmingham was Graffin Prankard's single most important customer. His two cementation steel furnaces consumed a major part of the 'Orground' iron that Prankard imported.

The Knight family dominated the iron trade in the English Midlands in the middle decades of the eighteenth century. Richard Knight (1659–1745) and his sons ran a variety of furnaces and forges in the Severn and Stour valleys, with much of the output being marketed through their kinsman and partner Abraham Spooner (q.v.). Because the capacity of their forges regularly outran the supply of pig iron from their furnaces, the Knights were keen to experiment with American pig when it appeared on the British market in the 1730s.

Pierre-Guillaume-Frèdèric Le Play (1806–1882), engineer, metallurgist and social scientist, was trained at the Ecole des Mines in Paris. An inveterate traveller and investigator, Le Play was familiar with both the Ural ironworks of the Demidov family (q.v.) and the *Vallonbruk* of Uppland.

Sampson Lloyd (1699–1779) was the son of Quaker ironmonger in Birmingham. Ambrose Crowley (q.v.) was a maternal uncle. By the 1730s Lloyd had expanded his father's business by integrating backwards into iron slitting at the 'town mill' in Birmingham. He was also to invest in forges, such as that at Powick in Worcestershire. In later life Sampson Lloyd engaged in banking; his bank was the ancestor of the present-day Lloyds TSB Group.

Charles Marescoe (*c.* 1633–1670) was one of the most important Baltic merchants in Restoration London. He was a major importer of Swedish tar, iron and copper, and an exporter of colonial sugar and tobacco to Amsterdam and Hamburg. His brother-in-law and partner Peter Joye (1636–1721) continued in the Baltic trade into the eighteenth century, supplying iron to the East India Company in Queen Anne's reign.

Daniel Mather ran a manufactory at Toxteth Park, near Liverpool, where 'tongs, pliers, wires, small hammers and other tools for watchmakers, which he claims to be the best in England', were made. Mather published a catalogue of horological tools in 1775.

Teofron Munktell (1805–1887), engineer and entrepreneur, toured England in 1835. The knowledge of British practice that he gained was put to use at his own engineering works at Eskilstuna.

Anders Nordencrantz (1697–1772), parliamentarian and writer on economic affairs. In his youth Nordencrantz travelled widely, spending eighteen months in London studying English political and economic conditions. On his return to Sweden he was elected to the *Riksdag*, where he served as a dedicated Hat (although he was later to switch to the Caps). In 1728 he was appointed Swedish consul in Lisbon, allowing him to refine his views on international commerce. Returning once more to Sweden, he became ironmaster and a key figure in establishing *Jernkontoret* in 1747.

Henry Norris (fl. 1725–1736) was an eminent London Baltic merchant and a bitter rival of Graffin Prankard. In the early 1730s he was the most influential figure in the marketing of 'Orground' iron in Britain, which allowed him to discriminate against those who, like John Kettle, were associates of Prankard. It was, Prankard wailed, 'very hard on me to See ['Orground' iron] Pass by me here & up into ye Markett & Sold by a Person that wont Sell it on any reasonable terms or really not at all to my best Chapp [Kettle]' (letter to Francis Jennings, 16 August 1732).

Eric Odelstierna (1661–1704), metallurgist and *Bergscollegium* agent. Odelstierna made two lengthy tours of Europe. During the second, begun in 1690, he made the earliest Swedish observation of English steel making.

The English engineer Samuel Owen (1774–1854), who founded a mechanical workshop in Stockholm in 1809, played a crucial role in conveying technological novelties, such as rolling mills and steam engines, from Britain.

Axel Oxenstierna (1583–1654), Swedish statesman. As chancellor from 1612, Oxenstierna was the architect of *stormaktstiden* in both its political and economic aspects.

The PEMBERTON family of Birmingham were prominent Quaker merchants who resided in the fashionable Square on the northside of the town. They dealt in hardware, like their kinsman and co-religionist Sampson Lloyd (q.v.). Another branch of the family was based in Philadelphia, playing a leading role in the commercial and political life of that city.

Tore PETRÉ (1793–1853), parliamentarian and Gästrikland's leading ironmaster during the introduction of the 'New German Forging Method' to the county.

Christopher POLHEM (1661–1751) was a self-taught engineer who came to fame after repairing the astronomical clock at Uppsala Cathedral. He was then employed at the Falu copper mine where he erected pumping and lifting gear. In 1700 Polhem founded Stjernsund *bruk* where both bar iron and metal wares were made with the aid of mechanical devices of his own design. The *brukspatron* of Stjernsund was a prolific writer who addressed the economic development of Sweden as well as scientific matters.

Graffin PRANKARD (d. 1756), Quaker merchant and industrialist, was active in the commerce of Bristol between the reign of Queen Anne and his insolvency in 1740. In addition to dealing in Baltic iron and timber, he sold cast iron wares from Coalbrookdale (where he was briefly a partner of Abraham Darby's) and salt produced at his own works in Worcestershire.

Bengt Andersson QVIST (1726–1799) was appointed to the *Bergscollegium* in 1755, and promoted to the senior post of *Assessor* in 1782. Qvist made an extensive grand tour in the second half of the 1760s. He used the observations he made in Britain to set up Sweden's first crucible steelworks on his return.

Sven RINMAN (1720–1792) was the most renowned Swedish engineer and metallurgist of the later eighteenth century. An official of the *Bergscollegium*, he rose through its ranks to achieve the position of *Bergsråd*, the second highest office available, in 1782. Rinman was an able publicist, whose *Bergwerks Lexicon* (1788–89), an encyclopaedia of metallurgical knowledge, was his most notable work. He was the founder of a dynasty, for his son CARL RINMAN (1763–1826) and grandson LUDVIG RINMAN (1815–1890) were also important figures in Swedish metallurgy.

Samuel SCHRÖDER (1720–1779), who was ennobled as Schröderstierna in 1770, was an official of the *Bergscollegium*. He travelled widely through Holland, Britain, Germany, and France in 1748–1751. On his return home he was given oversight of bar iron manufacturing in Sweden.

The SHALLARD family owned the steel cementation furnace at Keynsham, near Bristol. John Shallard (d. 1735) operated the furnace in partnership with his son William, who outlived his father by just one year. On William's death in 1736 the business passed to his teenage son Christopher (1718–1754). Reinhold Angerstein (q.v.) visited the site shortly after Christopher Shallard's death; he found his widow 'so big and fat that she could have concealed the whole steel furnace'.

Samuel SHORE (1676–1751), merchant and steel maker of Sheffield. A Dissenter of modest beginnings, Shore established two cementation furnaces in the Sheffield district in Queen Anne's reign. By the late 1720s, when Shore was trading in partnership with his son Samuel (1707–1785), he felt sufficiently powerful to join with Graffin Prankard in an endeavour to monopolise the import of 'Orground' iron into the British Isles.

Abraham SPOONER (1691–1788) was described as an 'Importer of Iron as well as Ironmonger' when he gave evidence before a committee of the House of Commons in 1737, who imported 'Orground' iron and several types of Russian iron. Graffin Prankard could testify to that, for Spooner was his deadly rival in the struggle to monopolise the supply of 'Orground' iron in the Midlands. The Birmingham ironmonger was supplied with Baltic iron via Henry Norris (q.v.).

Eric VON STOCKENSTRÖM (1703–1790), ironmaster and *Bergscollegium* official, was one of the driving forces behind the establishment of *Jernkontoret* in the 1740s.

Peter STUBS (1756–1806) established a filesmith's business in his native Warrington in the 1770s that became internationally renowned. The 'PS' mark stamped on his files was sought after not just in Britain but in America, where Sheffield-made counterfeits also circulated.

Eric Thomas SVEDENSTIERNA (1765–1825), an official first of the *Bergscollegium* and then of *Jernkontoret*, made the grand tour that was customary for men in his position at the start of the nineteenth century. It fell to him to formulate a response to the crisis that struck Swedish iron making in the Napoleonic age.

Georg SWEBILIUS (d. 1735) was the general manager of *Leufstawerken* from 1722 until his death.

Adolf Gustaf TAMM (1805–1851) was appointed to *Jernkontoret* in 1827 and embarked on an investigative tour of Germany, France and Britain shortly afterwards. His findings were published as *Anteckningar öfver främmande länders jernhandtering, gjorde under en resa i Tyskland, Frankrike och England, åren 1830 och 1831* (Stockholm, 1832).

Daniel TILAS (1712–1772), polymath and parliamentarian, enjoyed a long and distinguished career with the *Bergscollegium*. He was ennobled in 1766.

The firm of TOTTIE & ARFWEDSON, founded by Carl Christopher Arfwedson (1735–1826) and his cousin Anders Tottie in 1771, dominated the export of bar iron from Stockholm in the last decades of the eighteenth century.

Eric TOUSCHER succeeded Georg Swebilius (q.v.) as *directeur* at *Leufstawarken* in 1735.

Carl David af UHR (1770–1849), ironmaster and the author of an important work on charcoal making.

Olof WIJK (1786–1856), iron exporter and parliamentarian from Gothenberg.

Josias WORDSWORTH (d. 1749) was a London iron merchant and Baltic trader; he was also a considerable industrialist in the North East of England and a major exporter of hardware. He held the contract to supply bar iron to the Navy Board in 1733–36, 1740, 1742–43 and 1746. He also sold iron to the East India Company.

Samuel WORDSWORTH, who was born about 1701, moved to Stockholm soon after the ending of the embargo that extinguished direct trade between Sweden and Britain in 1717–1719. In the 1730s he handled the export of 'Orground' iron to Samuel Shore (q.v.) of Sheffield, usually in association with Samuel Worster (q.v.), another British-born merchant.

Samuel WORSTER (d. 1746), a merchant of English origin, most likely from London, settled in Stockholm in the second decade of the eighteenth century.

John WYKE (1720–1787) was born near Prescot, Lancashire, where his father was a watch spring maker and where Wyke was to set up his own business as a watch maker and tool manufacturer. At the end of the 1750s Wyke moved to Liverpool where he became a prominent figure in the town's affairs. He supplied tools not only to the watch trade in his native region, but to more far flung industrial innovators and intellectuals like Josiah Wedgwood and James Watt.

Eric Adolf ZETHELIUS (1781–1864), *brukspatron* at Nyby and Surahammar, was one of the earliest proponents of puddling in Sweden.

GLOSSARY

Bergscollegium
The Swedish Board of Mines, founded in 1649. This body was responsible for overseeing the extraction and processing of Sweden's mineral wealth. It comprised a central administration in Stockholm and twelve regional bureaux or *Bergmästardömen* (q.v).

bergsdeputation
The standing committee of the *Riksdag* (q.v.), established in 1723, with special responsibility for formulating policy on the iron trade.

Bergslagen
The mining district of central Sweden. Historical geographers dispute its exact definition, but most would agree that it swept in a broad arc north and west of Mälaren, the huge lake system that extends inland from Stockholm.

Bergmästardöme
One of the twelve regional offices of the *Bergscollegium* (q.v.). The chief officer (*bergmästare*) presided over a local 'mining court' that enforced the *Bergscollegium*'s decrees.

bergsman (pl. *bergsmän*)
Peasant miners who combined agriculture with ore mining and charcoal making. In the late medieval period *bergsmän* were responsible for all aspects of iron making. From the early seventeenth century, however, they were restricted to the mining and smelting of ore. The making of malleable iron was entrusted to *brukspatroner* (q.v.).

Bertram steel
See shear steel.

blister steel
The product of the cementation furnace (q.v.). See also shear steel.

BLOOM
See *smältan.*

BOKARE
An ore crusher at the blast furnace.

BRUK (pl. *bruk*)
An iron-making estate, with one or more forges, and sometimes a blast furnace, annexed to large tracts of woodland for the production of charcoal.

BRUKSPATRON (pl. *brukspatroner*)
An ironmaster; the proprietor of a *bruk.*

CAST STEEL
See crucible steel.

CEMENTATION FURNACE
The standard plant used in steel making in eighteenth-century Britain. Bars of iron were subject to a prolonged heating whilst in contact with charcoal dust. The charcoal acted as a source of carbon, which infused into the iron, producing its hard alloy, steel.

CHAFERY
See Walloon forging.

COMPANIGRUVA (pl. *companigruvor*)
One of those areas of the Dannemora mine (q.v.) that was worked by several *bruk* on a rota basis.

CRUCIBLE STEEL
Blister steel (q.v.) that had been melted down to achieve a uniform distribution of carbon through the metal and to allow the skimming off of residual slag. The outcome of this technique, developed in the 1740s by Benjamin Huntsman, was also known as cast steel or Huntsman steel.

DALER KOPPARMYNT
'Copper dollars': the monetary unit in which most business was conducted in early eighteenth-century Sweden. Each *daler* was made up of 32 *öre.*

DANNEMORA
The enormous iron ore mine in the county of Uppland. Its phosphorous-free ores were used to make 'Orground iron' (q.v.).

DEALS
Lengths of sawn timber that were a staple item of trade between Sweden and Britain

DEPUTATION (pl. *deputationer*)
A standing committee of the *Riksdag* (q.v.). See also *bergsdeputation*.

ENGLISH-GERMAN STEEL
See shear steel.

EGNA KOL
See *kol*.

FINERY
See Walloon forging.

FRIHETSTIDEN
'The Age of Liberty': the period of constitutional government that followed the death of Charles XII in 1718. During *Frihetstiden* the power of the monarchy was countered by that of the four estates: the aristocracy, the clergy, the burghers and the peasantry. 'The Age of Liberty' was terminated by a royal coup in 1772.

GERMAN FORGING
The most widely practised forge technique in eighteenth-century Sweden. A single hearth was used both to fine the metal and to reheat it during hammering. The distinction found in Walloon forging (q.v.) between the finery and the chafery—and between finers and hammermen—was unknown. A single forge crew undertook all elements of the process.

GERMAN STEEL
Either steel that originated in the German-speaking lands (Westphalia, Carinthia, etc.) *or* English-made shear steel (q.v.), which was made by a technique believed to be of German origin.

GEUSE
The form in which iron from Swedish blast furnaces was sometimes cast. The universal practice in Britain was for the liquid iron to be tapped into a series of branching channels that brought to mind a sow feeding her young, hence pig iron. In Sweden, by contrast, at works where the Walloon forging technique was used, cast iron (*tackjärn*) was allowed to flow into a single long depression, the *geuse*.

GOUJAR
A charcoal carrier at the forge.

HAMMERMAN
A specialised worker at the chafery (q.v.) in Walloon forging (q.v.).

HAYFORD STEEL
See shear steel.

HIELPEKARL
Auxiliary forge worker.

HUNTSMAN STEEL
See crucible steel.

HYTTDRÄNG
Assistant furnace keeper.

JERNBÄRARE
An 'iron carrier'; a worker at a *jernvåg* (q.v.).

JERNKONTORET
The Ironmasters' Association, established as a quasi-state body in 1747 to supply credit to *brukspatroner*. It also acted as a clearing house for information about technology and markets.

JERNVÅG
'Iron Weigh'. This was the compound in *stapelstäder* (q.v.) through which bar iron had to pass before being exported.

KOL
Charcoal. A legal distinction was drawn between two types of charcoal used in Swedish iron making: *egna kol* ('own charcoal') supplied by a *bruk*'s tenants, and *köpekol* ('bought coal') purchased from freeholders.

KÖPEKOL
See *kol*.

LANCASHIRE FORGING
A charcoal-fired refining technique introduced in Sweden in the 1840s that emulated the rapid throughput of puddling and rolling (q.v.). It paved the way for the industrialisation of Swedish iron making.

LASS
A measure of iron ore. It varied from mine to mine, but usually fell in the range 400 to 600 kilograms.

LEUFSTAWERKEN
The industrial complex in Uppland that comprised Leufsta *bruk*, with its blast furnace and four forges, Åkerby *bruk*, and the furnace at Toboborg, to which *bruk* at Hillebola and Carlholm were later added.

LOOP
See *smältan*.

MALM SKUTARE
An ore carrier at the blast furnace.

MANUFAKTURPOLITIK
In the 1720s the Swedish state reaffirmed its traditional interest in the promotion of import-substituting industries through the so-called *Manufakturpolitik*. Credit and legal privileges were extended to a number of new enterprises, especially in the textiles sector.

MASMÄSTARE
Furnace keeper.

MILLE
A traditional Walloon unit of weight used in the *Vallonbruk*, equal to 510 kilograms.

MÄSTERRÄCKARE
Master hammerman. Also known as a *räckarmästare*.

MÄSTERSMÄLTARE
Master finer.

MÄSTERSVEN
Finer's assistant.

NEW GERMAN FORGING
The forging technique developed in the 1830s in Sweden. It was an amplified and accelerated version of the traditional 'German' forging method (q.v.) that had been in use since the 1500s.

NEWCASTLE STEEL
See shear steel.

ÖRE
See *daler kopparmynt*.

'ORGROUND' IRON
The high-quality bar iron made at the *Vallonbruk* (q.v.) of Uppland, so called because much of it had traditionally passed through the Baltic port of Öregrund on its way to Stockholm.

OSMUND IRON
A type of fined iron produced in Sweden and parts of Germany in the medieval and early modern periods. It was the principal sort of iron made by *bergsmän* (q.v.) before the reform of the Swedish iron industry in the seventeenth century. Physically, it was produced in lumps that were exported by the barrel.

PRODUKTPLAKAT
The edict of 1724 that remodelled Swedish commercial policy. Foreign vessels entering Swedish ports were restricted to carrying goods from their country of origin. The intention was to promote Swedish shipping and shipbuilding.

PUDDLING AND ROLLING
The coal-fired method of refining iron patented by Henry Cort in 1783–1784. It became the standard method of producing malleable iron in the industrial world in the nineteenth century.

RÄCKARDRÄNG
Hammerman's assistant.

RÄCKARMÄSTARE
See *mästerräckare*.

RIKSDAG
The national diet of Sweden. It was composed of four estates: the nobles, the clergy, the burghers, and the peasantry.

SCHAMPLUNER
Bar iron made to specific dimensions.

SHEAR STEEL
Blister steel (q.v.) that had been cut into short lengths and forge-welded into a single mass. The procedure was intended to counteract the uneven distribution of carbon in blister steel, which was a potential source of weakness, by pounding together high-carbon and low-carbon portions of the bar. Shear steel was known by variety of different names: 'Hayford' steel (after Denis Hayford, its supposed inventor); Bertram steel (after another early practitioner); spur steel (after the mark stamped on one of the hardest brands); German steel (because of the supposed geographical source of the technique); English-German steel (to distinguish it from 'true' German steel); and Newcastle steel (after the port through which most of it was shipped). See also crucible steel.

SKEPPUND
The unit of weight used to measure iron in Sweden. Confusingly, the value of the *skeppund* varied according to the form of the iron being measured and the place where it was being weighed. A *skeppund* of pig iron weighed over 194 kilograms, but a *skeppund* of bar iron, weighed at the forge, was a little less than 150 kilograms. The difference reflected the weight loss incurred during the forging process; one *skeppund* of pig iron was assumed to yield one *skeppund* of bar iron. The measure used

at the point of the export was different again: a *skeppund stapelstadvikt* was just 136 kilograms.

SPUR STEEL
See shear steel.

SKEPPSBRON
The main quay in Stockholm and the residential quarter for many of the city's merchant elite in the eighteenth century.

SMÄLTAN
The mass of decarburised iron and liquid slag, known as a 'loop' in English, that was pulled from the hearth once the fining process was complete. The *smält* was then pounded beneath the forge hammer to expel the slag and to shape the metal into a roughly squared block: a *smältstycke* or what British forgemen called a 'bloom'. See also Walloon forging.

SMÄLTARDRÄNG
Apprentice finer.

SMÄLTSTYCKE
See *smältan*.

STÅLJERN
'Steel iron': the high-quality iron that could be made using the New German forging method (q.v.). It was intended for the Sheffield market.

STAPELSTAD (pl. *stapelstäder*)
A 'staple' town, one of the twenty-four designated centres through which Sweden's international trade was channelled.

STORMAKTSTIDEN
'The Age of Greatness': the period of military and imperial success that opened with Sweden's entry into the Thirty Years' War in the late 1620s and closed with her defeat in the Great Northern War (1699–1721).

STRÖMSBERGSWERKEN
The industrial complex comprising the *bruk* at Ullfors, Wessland and Strömsberg.

TACKJÄRN
Pig iron or cast iron.

TOURNEIJ
The four-hour shift worked by forgemen at the *Vallonbruk*.

UPPSÄTTARE
An ore charger at the blast furnace

VALLONBRUK
'Walloon ironworks': *bruk* (q.v.) settled by immigrants from the southern Netherlands during the seventeenth century at which the Walloon method of forging (q.v.) was practised

VOYAGE IRON
Bars of iron designed for export to Africa. They were made to quite precise specifications, reflecting the instructions of slave traders on the Guinea coast.

WALLOON FORGING
The method of making malleable bar iron that originated in the Walloon region in the late middle ages. It was the technique employed in Britain and in the *Vallonbruk* (q.v.) of Sweden during the eighteenth century. Walloon forges featured two separate hearths: the finery (in which pig iron was decarburised or 'fined') and the chafery (in which the fined iron was reheated before being drawn out into bars at the forge hammer). This division of labour distinguished the Walloon method from German forging (q.v.).

BIBLIOGRAPHY

1. *Unpublished Sources*

Britain

Aberystwyth—National Library of Wales (NLW)
Bedford MSS
EL Chappell MSS
Maybery MSS
Tredegar MSS

Birmingham—Birmingham City Archives (BCA)
Boulton & Watt MSS (MII)
Galton MSS (405/1–2)
Samuel Smith note books 1794 (1513)

Bradford—West Yorkshire Archive Service
Spencer Stanhope MSS (SpSt)

Bristol—Bristol Record Office
Isaac Hobhouse letters (8029)

Cardiff—Cardiff Central Library (CCL)
William Lewis, 'Mineral and Chemical History of Iron' (MS 3.250)

Cardiff—Glamorgan Archives (GA)
'B. Haddock's Memorandum Book Jany 1st 1818' (CL/MS 1.170)
Dowlais ironworks MSS (D/D G)
Melingriffith forge accounts (D/D X 809)
Pentyrch cashbook 1790–93 (D/D Xn 3)

Cardiff—National Museums and Galleries of Wales (NMW)
Melingriffith MSS (Department of Industry 89.76 and 1994.120)

Carmarthen—Carmarthenshire Record Office
Trostre MSS

Cwmbran—Gwent Record Office
Richard Crawshay's letter book 1788–97 (D2.162)

Dudley—Dudley Archives
Cradley forge stock and yield accounts 1805–12 (Z121)

Edinburgh—National Archives of Scotland
Carron Company MSS (GD58)

Edinburgh—National Library of Scotland
Cadell of Grange MSS (5381)

Glasgow—Mitchell Library
Adam Montgomerie letter book 1699–1702 (SR352)

Gloucester—Gloucestershire Record Office
Blagden Hale MSS (D1086/F116–28)

Hull—Brynmor Jones Library, University of Hull
Maister day book 1714–15 (DP/82)

Hull—Hull City Archives
Broadley MSS (DFB)

Ipswich—Suffolk Record Office
John Crowley inventory 1728 (HAI/GD/5)

Ironbridge—Ironbridge Gorge Museum
Richard Ford letter book 1733–36 (1992.11941)

Leeds—West Yorkshire Archive Service
Kirkstall forge MSS (K/F)

London—British Library
Crowley Hallett & Co inventory 1751 (Oriental and India Office Collections, MSS Eur F 218/113–15)
India Office Records, L/AG series
Crowley 'law book' (Add MS 34555)

London—Corporation of London Record Office
Joseph Baker letter book 1746–49

London—House of Lords Record Office
House of Lords main papers 1750

London—The National Archives (TNA)
PRO, ADM 106 series, Navy Board minutes and correspondence
PRO, BT 6 series, Board of Trade miscellanea
PRO, C 11/822/3, Court of Chancery proceedings: Remnant and Legas versus Wordsworth, Hallett and Smith
PRO, C 107/1–15, business papers of James Rogers.
PRO, CO 5 series, Board of Trade papers and correspondence for the West Indies and America
PRO, CO 388 series, Board of Trade correspondence
PRO, CO 390 series, Board of Trade miscellanea
PRO, CUST 3 series, ledgers of imports and exports
PRO, CUST 14 series, ledgers of imports and exports for Scotland
PRO, CUST 15 series, ledgers of imports and exports for Ireland
PRO, E 190 series, port books

London—National Maritime Museum
Chatham dockyard correspondence (CHA)

London—Religious Society of Friends Library
Lloyd MSS (TEMP MSS 210)

London—Royal Society of Arts
Correspondence files

London—Science Museum Library (SML)
Weale MSS (MS 371)

Manchester—John Rylands Library
Botfield MSS (BOT)
Richard Dalton letter books 1735–50 (B 5/4)

Manchester—Manchester Archives
Peter Stubs MSS (L24/1)

Newcastle upon Tyne—Northumberland Record Office
Bebside survey 1771 (ZMD 66/46)
Thomlinson correspondence (ZAN M13)

Newcastle upon Tyne—Tyne and Wear Archives Service
Plan of Swalwell (DX104/1)

Sheffield—Sheffield Archives
Osborn MSS (OR)
Spencer Stanhope MSS (SpSt)
Wharncliffe Muniments

Shrewsbury—Shropshire Records Unit
Gilbert Gilpin correspondence (1781/6)
Hampton Loade forge MSS (5686/2)

Taunton—Somerset Archives (SA)
Graffin Prankard MSS (DD/DN 423–452)

Worcester—Worcestershire Record Office (St Helens Branch)
Stour works general accounts (MSS 810:399)

Denmark

Copenhagen—Rigsarkivet
Øresunds Toldkammer 1737, microfilms S.15.085 and S.15.086

Sweden

Stockholm—Riksarkivet (RA)
Bergskollegiets arkiv
Börstorpssamlingen
Handlanden Peter Westmans arkiv
Jernkontorets Arkiv
Leufsta Arkivet

Stockholm—Krigsarkivet (KrA)
Ekaterinburg: 0406:19:031:001 Plan Der Festung Catrinenburg und darein befindliche Wercken, Fabriquen und gebeyden.
St. Petersburg: 0406:19:015:026 Afritning på residensstaden och festningen St Peterborg anvijsandes huru mycket war byggt år 1721 och huruledes då dessein war den wijdare at fullföra…

Stockholm—Jernkontorets Bibliotek (JK)
Bruksbildssamlingen
'Dagbok öfver Resan genom England åren 1753, 1754 och 1755 af Reinhold R. Angerstein'.

Stockholm—Stockholms Stadsarkiv (SSA)
Tolagsjournaler, 1720–1754, Statskamrerarens arkiv

Stockholm—Kungliga Biblioteket (KB)
'Dagbok rörande Handel, Näringar och Manufakturer m.m. Uti Danmark, Holland, England, Frankrike och Tyskland. Under verkstälde resor, åren 1748–1751 förd af Samuel Schröder' (X:303)
'Dagbok öfver en Resa till Roslags Bergslagen hållen 1753 af Christer Berch' (M 172)

Uppsala—Uppsala Universitetsbibliotek (UUB)
Georg Biurman's *Charta öfwer Upland och Södertörn* (1742), Kart- och Bildenheten
'Christer Berch relation öfver sin resa, 1757–1761' (X407b)
'Daniel Tilas, Diarum pro Anno 1733' (X 300)

Uppsala—Landsarkivet (ULA)
Bergmästarämbetet i Gävleborgs, Uppsala och Stockholms län arkiv
Österby bruksarkiv
Forsmarks bruksarkiv
Vattholma bruksarkiv

Leufsta Bruk—Leufstaarkivet
Leufsta bruksarkiv
Gimo bruksarkiv

Gammelbo—Bruksarkivet
Gammelbo bruksarkiv

United States

Winterthur DE—Winterthur Museum & Library
Samuel Rowland Fisher journal 1767–1768 (microfilm 296)
Andrew Clow & Co MSS (Joseph Downs collection C363)

Literature

Adamson, Rolf, 'Swedish exports to the United States, 1783–1860', *Scandinavian Economic History Review*, XVII (1969), 58–114.
Ågren, Maria (ed.), *Iron making societies: early industrial development in Sweden and Russia, 1600–1900* (Oxford, 1998).
Ahonen, Kalevi, *From sugar triangle to cotton triangle: trade and shipping between America and Baltic Russia, 1783–1860* (Jyväskylä, 2005).
Albion, R.G., *Forests and sea power: the timber problem of the Royal Navy* (Cambridge MA, 1926).
Alder, Ken, *Engineering the Revolution: arms and Enlightenment in France, 1763–1815* (Princeton NJ, 1997).
Almquist, Johan Axel, *Bergskollegium och bergslagsstaterna, 1637–1857: administrativa och biografiska anteckningar* (Stockholm, 1909).

Alpern, Stanley A., 'What Africans got for their slaves: a master list of trade goods', *History in Africa*, XXII (1995), 5–43.
Anon., *A brief state of facts relative to the new method of making bar iron, discovered and brought to perfection by Mr Henry Cort* (nd).
——, *American husbandry; containing an account of the soil, climate, production and agriculture, of the British colonies in North-America and the West-Indies* (1775).
——, 'A description of the Province of South Carolina, drawn up at Charles-Town in Sept. 1731', *The Gentleman's Magazine*, XX (August 1732), 894–96.
——, *The state of the trade and manufactory of iron in Great-Britain considered* (1750).
Appleby, Joyce, *Economic thought and ideology in seventeenth-century England* (Princeton NJ, 1978).
Armitage, David, 'Three concepts of Atlantic history', in David Armitage and Michael J. Braddick (eds), *The British Atlantic world, 1500–1800* (Basingstoke, 2002), pp. 11–27.
Arrighi, Giovanni, *The long twentieth century: money, power, and the origins of our times* (London, 1994).
Ashton, T.S., *Iron and steel in the Industrial Revolution* (Manchester, 1924; revised edition 1963).
——, *An eighteenth-century industrialist: Peter Stubs of Warrington, 1756–1806* (Manchester, 1939).
Ashworth, William J., *Customs and excise: trade, production, and consumption in England 1640–1845* (Oxford, 2003).
Åström, Sven-Erik, *From cloth to iron: the Anglo-Baltic trade in the late seventeenth century. Part 1: The growth, structure and organization of the trade* (Helsinki, 1963).
——, 'Swedish iron and the English iron industry about 1700: some neglected aspects', *Scandinavian Economic History Review*, XXX (1982), 129–41.
Atkinson, M. and C. Baber, *The growth and decline of the south Wales iron industry, 1760–1880: an industrial history* (Cardiff, 1987).
Attman, Artur, *Fagerstabrukens historia. Del II. Artonhundratalet* (Uppsala, 1958).
——, *Svenskt järn och stål 1800–1914* (Stockholm, 1984).
——, 'Vallonjärnets avsättning på världsmarknaden 1800–1914', in *Forsmark och vallonjärnet* (Stockholm, 1987), pp. 188–219.
Awty, B.G., 'Charcoal ironmasters of Cheshire and Lancashire, 1600–1785', *Transactions of the Historical Society of Lancashire and Cheshire*, CIX (1975), 71–124.
——, 'The continental origins of Wealden ironworkers, 1451–1544', *Economic History Review*, XXXIV (1981), 524–39.
Awty, Brian G., 'Hayford, Denis (*c.* 1635–1733)', *Oxford Dictionary of National Biography*, Oxford University Press, 2004 [http://www.oxforddnb.com/view/article/47478].
Aye, Efiong U., *Old Calabar through the centuries* (Calabar, 1967).
Bailey, De Witt and Douglas A. Nie, *English gunmakers: the Birmingham and provincial gun trade in the eighteenth and nineteenth century* (London and Melbourne, 1978).
Bailey, F.A. and T.C. Barker, 'The seventeenth-century origins of watchmaking in south-west Lancashire', in J.R. Harris (ed.), *Liverpool and Merseyside: essays in the economic and social history of the port and its hinterland* (London, 1969), pp. 1–15.
Bailyn, Bernard, *Atlantic history: concept and contours* (Cambridge MA, 2005).
Barbot, John, *A description of the coasts of North and South Guinea* (1746).
Barraclough, K.C., *Steelmaking before Bessemer. Volume 1. Blister steel: the birth of an industry* (London, 1984).
——, *Steelmaking before Bessemer. Volume 2. Crucible steel: the growth of technology* (London, 1984).
——, 'Svenskt järn och Sheffieldstål', in *Forsmark och vallonjärnet* (Stockholm, 1987), pp. 221–82.
Battye, Kathleen M., 'Scythe makers and other metal workers in the parish of Norton, 1533–1750', *Tools and Trades*, XIV (2005), 46–77.
Baugh, Daniel A., 'Maritime strength and Atlantic commerce: the uses of "A Grand Marine Empire"', in Lawrence Stone (ed.), *An imperial state at war: Britain from 1689 to 1815* (London, 1994), pp. 185–223.

Bedingfield, G., 'Congresbury's mills and the iron industry', *Bristol Industrial Archaeology Society Journal*, 31 (1998), 28–30.

Behagg, Clive, 'Mass production without the factory: craft producers, guns and small firm innovation, 1790–1815', *Business History*, XL, 3 (1998), 1–15.

Behre, Göran, Lars-Olof Larsson and Eva Österberg, *Sveriges historia 1521–1809. Stormaktsdröm och småstadsrealitet* (Stockholm, 2001).

Behrendt, Stephen D., 'Markets, transaction cycles, and profits: merchant decision making in the British slave trade', *The William and Mary Quarterly*, 3rd series LVIII, 1 (2001), 171–204.

Behrendt, Stephen D. and Eric J. Graham, 'African merchants, notables and the slave trade at Old Calabar, 1720: evidence from the National Archives of Scotland', *History in Africa*, XXX (2003), 37–61.

Benson, Adolph B. (ed.), *The America of 1750: Peter Kalm's travels in North America. The English version of 1770* (New York, 1937).

Berch, Anders, *Inledningen til den almänna hushålningen. Innefattande grunden til politie, oeconomie och cameral wetenskaperne* (Stockholm, 1747).

Berg, Maxine, 'New consumer industries in eighteenth-century England: products, markets and metal goods in Birmingham and Sheffield', in René Leboutte (ed.), *Proto-industrialization: recent research and new perspectives* (Geneva, 1996), pp. 211–36.

——, *Luxury and pleasure in eighteenth-century Britain* (Oxford, 2005).

Berg, Torsten and Peter Berg (eds), *R.R. Angerstein's illustrated travel diary, 1753–1755: industry in England and Wales from a Swedish perspective* (London, 2001).

Berlin, Ira, 'From Creole to African: Atlantic Creoles and the origins of African-American society in mainland North America', *The William and Mary Quarterly*, 3rd series LIII (1993), 251–88.

——, *Many thousands gone: the first two centuries of slavery in North America* (Cambridge MA, 1998).

Bettey, J.H., 'From Quaker traders to Anglican gentry: the rise of a Somerset dynasty', *Proceedings of the Somerset Archaeology and Natural History Society*, CXXXV (1991), 1–9.

——, 'Graffin Prankard: an eighteenth-century Bristol merchant', *Southern History*, XII (1990), 34–47.

——, 'The capture of the *Baltick Merchant* 1740', *Mariner's Mirror*, LXXVI, 1 (1990), 36–39.

Bezís-Selfa, John, *Forging America: ironworkers, adventurers, and the Industrious Revolution* (Ithaca NY, 2004).

Bialuschewski, Arne, 'Between Newfoundland and the Malacca Strait: a survey of the golden age of piracy', *The Mariner's Mirror*, XC, 2 (2004), 167–86.

Bining, Arthur C., *British regulation of the colonial iron industry* (Philadelphia PA, 1933).

——, *Pennsylvania iron manufacture in the eighteenth century* (Harrisburg PA, 1938).

Bisset, James, *A poetic survey round Birmingham; with a brief description of the different curiosities and manufactories of the place... accompanied by a magnificent directory* (Birmingham, 1800).

Blackburn, Robin, *The making of New World slavery: from the baroque to the modern, 1492–1800* (London, 1997).

Blanchard, Ian, 'Eighteenth-Century Russian economic growth: state enterprise or peasant endeavour?', *Jahrbücher für Osteuropas*, XLV (1997), 541–51.

——, 'Russia and international iron markets, ca. 1740–1850', http://www.esh.(ed.)ac.uk/Research%20IB/Rus_Ind/R_market.pdf

Blom, Philipp, *Encyclopédie: the triumph of reason in an unreasonable age* (London, 2004).

Boëthius, Bertil, 'Hammarkommissionerna på 1680-och 1720-talet. En studie över deras ställning i bergskollegiets brukspolitiska system', in *En bergsbok till Carl Sahlin* (Stockholm, 1921), 193–211.

Boëthius, Bertil and Åke Kromnow, *Jenrkontorets historia, Del I–III* (Stockholm, 1947–68).
Bowen, H.V., 'Sinews of trade and empire: the supply of commodity exports to the East India Company during the late eighteenth century', *Economic History Review*, LV, 3 (2002), 466–86.
Boyer, Charles S., *Early forges and furnaces in New Jersey* (Philadelphia PA, 1931).
Bradshaw, Brendan and John Morrill (eds), *The British problem, c. 1534–1707: state formation in the Atlantic archipelago* (Basingstoke, 1996).
Braudel, Fernand, *Civilization and capitalism: 15th–18th century. Volume 3: The perspective of the world* (Berkeley CA, 1992).
——, *Afterthoughts on Material Civilization and Capitalism* (Baltimore MD, 1977).
——, *La Méditerranée et le monde méditerranéen à l'époque de Philippe II* (Paris, 1949).
Braund, Kathryn E. Holland, *Deerskins and duffels: the Creek Indian trade with Anglo-America, 1685–1815* (Lincoln NE, 1993).
Breen, T.H., 'An empire of goods: the anglicization of colonial America, 1690–1776', *Journal of British Studies*, XXV (1986), 467–99.
Brewer, John, *The sinews of power: war, money and the English state, 1688–1783* (London, 1989).
Brockliss, Laurence and David Eastwood (eds), *A union of multiple identities: the British Isles, c. 1750–c. 1850* (Manchester, 1997).
Broling, Gustaf, *Anteckningar under en resa i England åren 1797, 1798, och 1799*, 3 volumes (Stockholm, 1811–1817).
Brooke, E.H., *Chronology of the tin plate works in Great Britain* (Cardiff, 1944).
Bruce, Kathleen, *Virginia iron manufacture in the slave era* (New York, 1931).
Brydon, G. MacLaren, 'The Bristol iron works in King George County', *Virginia Magazine of History and Biography*, XLII, 2 (1934), 97–102.
Campbell, Robert, *The London tradesman* (1747).
Cannadine, David, 'British history: past, present—and future?', *Past and Present*, 116 (1987), 169–91.
Cappon, Lester J. (ed.), *Atlas of early American history: the revolutionary era 1760–1790* (Princeton NJ, 1976).
Carney, Judith A., *Black rice: the African origins of rice cultivation in the Americas* (Cambridge MA, 2001).
Castells, Manuel, *The information age: economy, society and culture*, 3 volumes (Oxford, 1996–98).
Chaplin, Joyce E., *An anxious pursuit: agricultural innovation and modernity in the Lower South, 1730–1815* (Chapel Hill NC, 1993).
Chaudhuri, K.N., *Trade and civilisation in the Indian Ocean: an economic history from the rise of Islam to 1750* (Cambridge, 1985).
Chaudhury, Sushil and Michel Morineau (eds), *Merchants, companies and trade: Europe and Asia in the early modern era* (Cambridge, 1999).
Cheney, Robert C., 'Roxbury eight-day movements and the English connection, 1785–1825', *The Magazine Antiques* (April 2000), 606–15.
Cherry, Gordon E., *Birmingham: a study in geography, history and planning* (Chichester, 1993).
Child, Sir Josiah, *A new discourse of trade* (1693), reprinted in Lars Magnusson (ed.), *Mercantilism. Volume III: a science of trade* (London, 1995).
Childs, W.R., 'England's iron trade in the fifteenth century', *Economic History Review*, XXIV (1981), 25–47.
Cleere, Henry and David Crossley, *The iron industry of the Weald* (Cardiff, 1995).
Clemens, P.G.E., 'The rise of Liverpool, 1665–1750', *Economic History Review*, XXIX (1976), 211–25.
Coclanis, Peter A., *The shadow of a dream: economic life and death in the South Carolina Low Country, 1670–1920* (New York, 1989).
——, 'The hydra head of merchant capital: markets and merchants in early South Carolina', in David R. Chesnutt and Clyde N. Wilson (eds), *The meaning of South Carolina history: essays in honor of George C. Rogers, Jr.* (Columbia SC, 1991), pp. 1–18.

——, (ed.), *The Atlantic economy during the seventeenth and eighteenth centuries: organization, operation, practice, and personnel* (Columbia SC, 2005).

Colley, Linda, *Britons: forging the nation, 1707–1837* (New Haven CT, 1992).

Coulton, Barbara, 'Tern Hall and the Hill family: 1700–75', *Shropshire History and Archaeology*, LXVI (1989), 97–105.

Cowan, Thomas, 'William Hill and the Aera ironworks', *Journal of Early Southern Decorative Arts*, XIII (1987), 1–32.

Cox, Nancy, 'Imagination and innovation of an industrial pioneer: the first Abraham Darby', *Industrial Archaeology Review*, XII, 2 (1990), 127–44.

Cracraft, James, *The Petrine revolution in Russian architecture* (Chicago IL, 1988).

Cranstone, David, *Derwentcote steel furnace: an industrial monument in County Durham* (Lancaster, 1997).

Cross, A.G., *"By the banks of the Thames": Russians in eighteenth-century Britain* (Newtonville MA, 1980).

——, *By the banks of Neva: chapters from the lives and careers of the British in eighteenth-century Russia* (Cambridge, 1997).

Curtin, Philip D., *Economic change in precolonial Africa: Senegambia in the era of the slave trade* (Madison WI, 1975).

Daff, Trevor, 'The introduction of furnace blowing cylinders', *Steel Times* (May 1973), 401–02.

Daggett, Charles, Evelyne Jay and Frederic Osada, 'The *Griffin*, an English East Indiaman lost in the Philippines in 1761', *International Journal of Nautical Archaeology and Underwater Exploration*, XIX, 1 (1990), 35–41.

Dahlgren, Erik W., *Louis De Geer 1587–1652. Hans lif och verk* (Uppsala, 1923).

——, *Järnvräkeri och järnstämpling. Ett bidrag till den svenska järnhandelns historia* (Stockholm, 1930).

Dahlgren, Stellan and Hans Norman, *The rise and fall of New Sweden: Governor Johan Risingh's journal 1654–55 in its historical context* (Uppsala, 1988).

Dalrymple, Sir John, *Address & proposals from Sir John Dalrymple, Bart. on the subject of the coal, tar, and iron, branches of trade* (1784).

Dane, E. Surrey, *Peter Stubs and the Lancashire hand tool industry* (Altrincham, 1973).

Davies, K.G., *The Royal African Company* (London, 1957).

Davis, Ralph, *The rise of the English shipping industry in the seventeenth and eighteenth centuries* (Newton Abbot, 1962).

Day, Joan, *Bristol brass: a history of the industry* (Newton Abbot, 1973).

de Bonnard, A.H., 'Memoire sur les procédés employés en Angleterre pour le traitement du fer par le moyen de la houille', *Annales des arts et manufactures*, XXIII (1805), 113–51 and 225–54, and XXIV (1806), 44–62.

de Hennin, G.W., *Description of Ural and Siberian factories in 1735* (Washington D.C., 1992).

de Vries, Jan, *European urbanization 1500–1800* (London, 1984).

——, 'Between purchasing power and the world of goods: understanding the household economy in early modern Europe', in John Brewer and Roy Porter (eds), *Consumption and the world of goods* (London, 1993), pp. 85–132.

——, 'The industrial revolution and the industrious revolution', *Journal of Economic History*, LIV (1994), 248–71.

de Vries, Jan and Ad van der Woude, *The first modern economy: success, failure, and perseverance of the Dutch economy, 1500–1815* (Cambridge, 1997).

Defoe, Daniel, *A tour through the whole island of Great Britain* [1724–26], (ed.) P.N. Furbank and W.R. Owens (New Haven CT, 1991).

——, *Plan of the English commerce. Being a complete prospect of the trade of this Nation, as well home as foreign* (2nd edn., London, 1737).

——, *The complete English tradesman* (1738).

Devine, T.M., *The tobacco lords: a study of the tobacco lords of Glasgow and their trading activities, c. 1740–1790* (Edinburgh, 1975).
Dicken, Peter, Philip F. Kelly, Kris Olds and Henry Wai-Chung Yeung, 'Chains and networks, territories and scales: towards a relational framework for analysing the global economy', *Global Networks*, I, 2 (2001), 89–112.
Directory of Sheffield (1787).
Dixon, Simon, *The modernisation of Russia 1676–1825* (Cambridge, 1999).
Donnan, Elizabeth (ed.), *Documents illustrative of the slave trade to America*, 4 volumes (New York, 1965).
Douhan, Bernt, *Arbete, kapital och migration. Valloninvandringen till Sverige under 1600-talet* (Uppsala, 1985).
Downer, Silas ['A Son of Liberty'], *A discourse at the dedication of the Tree of Liberty* [1768], in Bruce Frohnen (ed.), *The American Republic: primary sources* (Indianapolis IN, 2002), pp. 143–44.
Dresser, Madge, *Slavery obscured: the social history of the slave trade in an English provincial port* (London, 2001).
Duncan, J.G., 'A Scottish trading house in eighteenth-century Gothenburg: Carnegy and Shepherd', *Northern Scotland*, XI (1991), 1–9.
Edelson, Max S., *Plantation enterprise in colonial South Carolina* (Cambridge MA, 2006).
Edgar, Walter B. (ed.), *The letterbook of Robert Pringle. Volume 1: April 2, 1737–September 25, 1742* (Columbia SC, 1972).
Eggeby, Eva and Klas Nyberg, 'Stad i Stagnation 1720–1850', in Lars Nilsson (ed.), *Staden på Vattnet. Del I: 1252–1850* (Stockholm, 2002), pp. 187–276.
Eklund, Åsa, 'Iron production, iron trade and iron markets. Swedish iron on the British market in the first half of the eighteenth century' (Licenciate thesis, Department of Economic History, University of Uppsala, 2001).
Eklund, Åsa, Chris Evans and Göran Rydén, 'Baltic iron and the organisation of the British iron market in the eighteenth century', in Patrick Salmon and Tony Barrow (eds), *Britain and the Baltic: studies in commercial, political and cultural relations* (Sunderland, 2003).
Ekman, Gustaf, 'Bihang om Valsverk', in A.G. Tamm, *Anteckningar öfver främmande länders jernhandtering, gjorde under en resa i Tyskland, Frankrike och England, åren 1830 och 1831* (Stockholm, 1832).
——, *Gustaf Ekman. Svenska järnhanteringens nydanare för 100 år sedan* (Stockholm, 1944).
Ekman, Wilhelm, 'Vallonjärnet—en kvalitetsprodukt med världsrykte' in *Forsmark och vallonjärnet* (Stockholm, 1987), pp. 121–49.
Elliott, J.H., *Empires of the Atlantic world: Britain and Spain in America 1492–1830* (New Haven CT and London, 2006).
Ellis, Steven G. and Sarah Barber (eds), *Conquest and union: fashioning a British state, 1485–1735* (London, 1995).
Eltis, David, Stephen D. Behrendt, David Richardson and Herbert S. Klein (eds), *The transatlantic slave trade: a database on CD-ROM* (Cambridge, 1999).
Emmer, Pieter and Wim Klooster, 'The Dutch Atlantic, 1600–1800: expansion without empire', *Itinerario: European Journal of Overseas History*, 2 (1999), 48–69.
Engerman, Stanley, 'Mercantilism and overseas trade, 1700–1800', in Roderick Floud and Donald McCloskey (eds), *The economic history of Britain since 1700. Volume 1: 1700–1860* (2nd edn, Cambridge, 1994).
Englund, Peter, *The battle that shook Europe: Poltava and the birth of the Russian empire* (London, 2003).
Ericsson, Birgitta, Karl-Gustaf Andersson and Per-Arne Karlsson, 'Privilegiegivningen till järnbruk och järnmanufaktur i Sverige under frihetstiden', in *Industri og Bjergværksdrift. Privilegier i Norden i det 18. århundrade* (Oslo, 1985), 245–319.
Evans, Chris, 'Manufacturing iron in the North-East in the eighteenth century: the case of Bedlington', *Northern History*, XXVIII (1992), 178–96.

——, 'Iron puddling: the quest for a new technology in eighteenth-century industry', *Llafur*, VI, 3 (1994), 44–57.
——, 'The corporate culture of the British iron industry 1650–1830', in Göran Rydén (ed.), *The social organization of the European iron industry 1600–1900* (Stockholm, 1997), pp. 121–46.
——, 'A skilled workforce during the transition to industrial society: forgemen in the British iron trade, 1500–1850', *Labour History Review*, LXIII, 2 (1998), 143–59.
——, 'Work and workloads during industrialization: the experience of forgemen in the British iron industry, 1750–1850', *International Review of Social History*, XLIV (1999), 197–215.
——, 'Global commerce and industrial organization in an eighteenth-century Welsh enterprise: the Melingriffith Company', *Welsh History Review*, XX (2001), 413–34.
——, 'The Industrial Revolution in iron in the British Isles', in Chris Evans and Göran Rydén (eds), *The Industrial Revolution in iron: the impact of British coal technology in nineteenth-century Europe* (Aldershot, 2005), pp. 15–28.
Ferguson, Leland, *Uncommon ground: archaeology and early African America* (Washington D.C., 1992).
Flavell, Neville, 'The economic development of Sheffield and the growth of the town, *c.* 1740–1820', (Ph.D thesis, University of Sheffield, 1996).
Flinn, M.W. (ed.), *The law book of the Crowley ironworks* (Publications of the Surtees Society, CLXVII, 1957).
——, 'Industry and technology in the Derwent valley of Durham and Northumberland in the eighteenth century', *Transactions of the Newcomen Society*, XXIX (1953–55), 255–62.
——, 'The growth of the English iron industry 1660–1760', *Economic History Review*, XI (1958), 144–53,
——, *Men of iron: the Crowleys in the early iron industry* (Edinburgh, 1962).
——, (ed.), *Svedenstierna's tour of Great Britain 1802–3: the travel diary of an industrial spy* (Newton Abbot, 1973).
Florén, Anders, *Disciplinering och konflikt. Den sociala organiseringen av arbetet. Jäders bruk 1640–1750* (Uppsala, 1987).
——, 'Social organization of work and labour conflicts in proto-industrial iron production in Sweden, Belgium and Russia', in Catharina Lis, Jan Lucassen and Hugo Soly (eds), *Before the unions: wage earners and collective action in Europe, 1300–1850* (International Review of Social History, supplement 2, 1994), pp. 83–113.
——, *Vallonskt järn. Industriell utveckling i de södra Nederländerna före industrialiseringen* (Uppsala, 1998).
——, 'The making of the forgeman. Social relations and bar iron production in Sweden *c.* 1650–*c.* 1750', in Hans-Jürgen Gerhard, Karl Heinrich Kaufhold and Ekkehard Westermann (eds), *Europäische Montanregion Harz* (Bochum, 2001), pp. 193–211.
Florén, Anders, and Göran Rydén, *Arbete, hushåll och region. Tankar om industrialiseringsprocesser och den svenska järnhanteringen* (Uppsala, 1992).
——, 'A journey to the market society: a Swedish pre-industrial spy in the middle of the eighteenth century', in Ragnar Björk and Karl Molin (eds), *Societies made up of history: essays in historiography, intellectual history, professionalisation, historical social theory and proto-industrialisation* (Uppsala, 1996), pp. 265–307.
——, 'The social organisation of work at mines, furnaces and forges', in Maria Ågren (ed.), *Iron-making societies: early industrial development in Sweden and Russia, 1600–1900* (Oxford, 1998), pp. 61–138.
——, 'Sketches of Spain: the journey of Reinhold Rücker Angerstein, 1752', in K. Benson, M. Mörner, and I. Söhrman (eds), *Spanish-Swedish relations from the mid-seventeenth century to the early nineteenth century* (Göteborg, 2002), pp. 291–305.
Forde, Daryll (ed.), *Efik traders of Old Calabar* (Oxford, 1956).
Frank, Andre Gunder, *ReOrient: global economy in the Asian age* (Berkeley CA, 1998).

French, Christopher J., '"Crowded with Traders and a Great Commerce": London's dominion of English overseas trade, 1700–1775', *The London Journal*, XVII (1992), 27–35.

Fuchs II, Ronald W., '"At Elk Ridge Furneis as you See, William Williams He Mad Me": the story of an eighteenth-century Maryland iron furnace', *Journal of Early Southern Decorative Arts*, XXII, 2 (1996), 40–59.

Gadd, Carl-Johan, *Den Agrara Revolutionen 1700–1870* (Stockholm, 2000).

Gale, W.K.V., 'Wrought iron: a valediction', *Transactions of the Newcomen Society*, XXXVI (1963), 1–11.

Gallay, Alan, *The Indian slave trade: the rise of the English empire in the American south, 1670–1717* (New Haven CT and London, 2003).

Gaynor, James M. and Nancy L. Hagedorn, *Tools: working wood in eighteenth-century America* (Williamsburg VA, 1993).

[Gee, Joshua], *The trade and navigation of Great-Britain considered* (1729).

Gemery, H.A. and J.S. Hogendorn, 'Technological change, slavery and the slave trade', in C. Dewey and A.G. Hopkins (eds), *The imperial impact: studies in the economic history of Africa and India* (London, 1973), pp. 243–59.

Gereffi, Gary, Miguel Korzeniewicz and Roberto P. Korzeniewicz, 'Introduction: global commodity chains', in Gary Gereffi and Miguel Korzeniewicz (eds), *Commodity chains and global capitalism* (Westport CT, 1994), pp. 1–14.

Goldenberg, Joseph A., *Shipbuilding in colonial America* (Charlottesville VA, 1976).

Goldstone, Jack A., 'Efflorescences and economic growth in world history: rethinking the 'Rise of the West' and the Industrial Revolution', *Journal of World History*, XIII, 2 (2002), 323–89;

Goody, Jack, *The domestication of the savage mind* (Cambridge, 1977).

Gordon, Robert B., *American iron 1607–1900* (Baltimore MD, 1996).

Goucher, Candice, 'Iron is iron 'til it rust: trade and ecology in the decline of West African iron-smelting', *Journal of African History*, XXII (1981), 179–89.

Haan, Richard L., 'The "trade do's not flourish as formerly": the ecological origins of the Yamassee War of 1715', *Ethnohistory*, XXVIII, 4 (1982), 341–58.

Hair, P.E.H., 'Antera Duke of Old Calabar—a little more about an African entrepreneur', *History in Africa*, XVII (1990), 359–65.

Hammersley, G.F., 'The charcoal iron industry and its fuel', *Economic History Review*, XXVI (1973), 593–613.

Hancock, David, *Citizens of the world: London merchants and the integration of the British Atlantic community, 1735–1785* (Cambridge, 1995).

——, 'Commerce and conversation in the eighteenth-century Atlantic: the invention of Madeira wine', *Journal of Interdisciplinary History*, XXXIX, 2 (1998), 197–219.

——, 'The British Atlantic world: co-ordination, complexity, and the emergence of an Atlantic market economy, 1651–1815', *Itinerario: European Journal of Overseas History*, XXIII (1999), 107–26.

Hardy, Stephen G., 'Colonial South Carolina's rice industry and the Atlantic economy: patterns of trade, shipping, and growth', in Jack P. Greene, Rosemary Brana-Shute, and Randy J. Sparks (eds), *Money, trade, and power: the evolution of colonial South Carolina's plantation society* (Columbia SC, 2001), pp. 108–40.

Harris, J.R., *The British iron industry 1700–1850* (London, 1988).

——, *Industrial espionage and technology transfer: Britain and France in the eighteenth century* (Aldershot, 1998).

Hartley, E.N., *Ironworks on the Saugus: the Lynn and Braintree ventures of the Company of Undertakers of the Ironworks in New England* (Norman OK, 1957).

Hayman, Richard, 'The Cranage brothers and eighteenth-century forge technology', *Historical Metallurgy*, XXXVIII, 2 (2004), 113–120.

Heckscher, E.F., *Sveriges ekonomiska historia från Gustav Vasa. Andra delen. Det moderna Sveriges grundläggning* (Stockholm, 1949).

Hey, David, *The rural metalworkers of the Sheffield region: a study of rural industry before the Industrial Revolution* (Leicester, 1972).
——, *The fiery blades of Hallamshire: Sheffield and its neighbourhood, 1660–1740* (Leicester, 1991).
Hildebrand, K-G., *Fagerstabrukens historia. Del I. Sexton- och Sjuttonhundratalen* (Uppsala, 1957).
——, 'Foreign markets for Swedish iron in the eighteenth century', *Scandinavian Economic History Review*, VI (1958), 3–52.
——, *Svenskt järn. Sexton- och sjuttonundratal. Exportindustri före industrialismen* (Stockholm, 1987).
——, 'Gammalt och nytt i det svenska järnets historia. En översikt över fem årtionden', in *Dædalus 1997. Svenskt järn under 2500 år. Från gruvpigor och smedsdrängar till operatörer* (Stockholm, 1997), pp. 1–30.
Himmelfarb, Gertrude, *The roads to modernity: the British, French, and American Enlightenments* (New York, 2004).
Högberg, Staffan, *Utrikeshandel och sjöfart på 1700-talet. Stapelvaror i svensk export och import 1738–1808* (Lund, 1969).
Hopkins, Eric, *Birmingham: the first manufacturing town in the* world (London, 1989).
Hopkins, Thomas K. and Immanuel Wallerstein, 'Commodity chains: construct and research', in Gary Gereffi and Miguel Korzeniewicz (eds), *Commodity chains and global capitalism* (Westport CT, 1994), pp. 17–20.
Hopkinson, G.G., 'The charcoal iron industry in the Sheffield region, 1500–1775', *Transactions of the Hunter Archaeological Society*, VIII (1961), 122–51.
Hoppit, Julian, 'Political arithmetic in eighteenth-century England', *Economic History Review*, XLIX, 3 (1996), 516–40.
Hoppus, Edmund, *Practical measuring made easy* (7th edn., 1765).
Hounshell, David A., *From the American System to mass production, 1800–1932: the development of manufacturing technology in the United States* (Baltimore MD, 1984).
Hudson Jr., Hugh D., *The rise of the Demidov family and the Russian iron industry in the eighteenth century* (Newtonville MA, 1986).
——, 'Religious persecution and industrial policy in the reign of Anna I: V.N. Tatishchev and the Old Believers reconsidered', *Jahrbücher für Osteuropas*, L (2002), 22–36.
Hudson Jr., Hugh D., Bruce J. DeHart and David M. Griffiths, 'Proletarians by fiat: the compulsory Ural metallurgical work force, 1630–1861', *International Labour and Working-Class History*, XLVIII (1995), 94–111.
Hughes, Lindsey, *Russia in the age of Peter the Great* (New Haven CT, 1998).
Hulme, E.W., 'Statistical history of the iron trade of England and Wales, 1717–1750', *Transactions of the Newcomen Society*, IX (1928–29), 12–35.
Hussey, David, *Coastal and river traffic in pre-industrial England: Bristol and its region, 1680–1730* (Exeter, 2000).
Hutton, William, *An history of Birmingham, to the end of the year 1780* (Birmingham, 1781).
Hyde, C.K., *Technological change and the British iron industry 1700–1870* (Princeton NJ, 1977).
Ince, Laurence, 'Homfray family (per. 1702–1833)', *Oxford Dictionary of National Biography* (Oxford University Press, 2004), [http://www.oxforddnb.com/view/article/47499].
——, *The Knight family and the British iron industry, 1695–1902* (Birmingham, 1991).
Inikori, Joseph E., *Africans and the Industrial Revolution in England: a study in international trade and economic development* (Cambridge, 2002).
Iron and steel on the European market in the 17th century: a contemporary Swedish account of production forms and marketing (Stockholm, 1982).

Isacson, Maths, 'Bergskollegium och den tidigindustriella järnhanteringen' *Dædalus 1998. Människa, teknik, industri* (Stockholm, 1998), 43–58.
Israel, Jonathan, *The Dutch Republic: its rise, greatness, and fall 1477–1806* (Oxford, 1995).
Jackson, Gordon, *Hull in the eighteenth century: a study in economic and social history* (Oxford, 1972).
Jackson, R.G. and R.H. Price, *Bristol clay pipes: a study of their makers and their marks* (Bristol, 1974).
Johansen, H.C., 'How to pay for Baltic products?', in Wolfram Fischer, R. Marvin McInnis and Jürgen Schneider (eds), *The emergence of a world economy 1500–1914. Part 1: 1500–1850* (Wiesbaden, 1986), pp. 123–42.
John, A.H. (ed.), *Minutes relating to Messrs Samuel Walker & Co., Rotherham, iron founders and steel refiners, 1741–1829* (Council for the Preservation of Business Archives, 1951).
Johnson, B.L.C., 'The Stour valley iron industry in the late seventeenth century', *Transactions of the Worcestershire Archaeological Society*, XXVII (1950), 35–46.
——, 'The charcoal iron industry in the early eighteenth century', *The Geographical Journal*, CXVII (1951), 167–77.
——,'The Foley partnerships: the iron industry at the end of the charcoal era', *Economic History Review*, VI (1952), 322–40.
——, 'The iron industry of Cheshire and Staffordshire, 1688–1712', *Transactions of the North Staffordshire Field Club*, LXXXVIII (1953–54), 32–55.
Johnson, Keach, 'The genesis of the Baltimore ironworks', *Journal of Southern History*, XIX, 2 (1953), 157–79.
——, 'The Baltimore Company seeks English markets: a study of the Anglo-American iron trade, 1731–1755', *The William and Mary Quarterly*, 3rd series XVI (1959), 37–60.
Jones, A.C. and C.J. Harrison, 'The Cannock Chase ironworks, 1590', *English History Review*, XCIII (1978), 795–810.
Jones, D.W., *War and economy in the age of William III and Marlborough* (Oxford, 1988).
Jordan, Terry G. and Matti Kaups, *The American backwoods frontier: an ethnic and ecological interpretation* (Baltimore MD, 1989).
Kaplan, Herbert H., *Russian overseas commerce with Great Britain during the reign of Catherine II* (Philadelphia PA, 1995).
Karlsson, Per-Arne, *Järnbruken och ståndssamhället. Institutionell och attitydmässig konflikt under Sveriges tidiga industrialisering 1700–1770* (Stockholm, 1990).
Kea, R.A., 'Firearms and warfare on the Gold and Slave Coasts from the sixteenth to the nineteenth centuries', *Journal of African History*, XII, 2 (1971), 196–97.
Kent, H.S.K., *War and trade in northern seas: Anglo-Scandinavian economic relations in the mid-eighteenth century* (Cambridge, 1973).
King, P.W., 'The Vale Royal company and its rivals', *Transactions of the Historical Society of Lancashire and Cheshire*, CXLII (1992), 1–18.
——, 'Early statistics for the iron industry: a vindication', *Historical Metallurgy*, XXX, 1 (1996), 23–46.
——, 'The cartel in Oregrounds iron: trading relationships in the raw material for steel', *Journal of Industrial History*, VI, 1 (2003), 25–48.
King, Peter, 'The production and consumption of bar iron in early modern England and Wales', *Economic History Review*, LVIII, 1 (2005), 1–33.
Kirby, D.G., 'The Royal Navy's quest for pitch and tar during the reign of Queen Anne', *Scandinavian Economic History Review*, XXII (1974), 97–116.
Klein, Herbert S., *African slavery in Latin America and the Caribbean* (New York, 1986).
——, *The Atlantic slave trade* (Cambridge, 1999).
Koepp, C.J., 'The alphabetical order: work in Diderot's *Encyclopédie*', in S.L. Kaplan and C.J. Koepp (eds), *Work in France: representations, meaning, organization, and practice* (Ithaca NY, 1986).

Koninckx, Christian, *The first and second charters of the Swedish East India Company (1731–1766): a contribution to the maritime, economic, and social history of north-western Europe in its relationship with the Far East* (Kortrijk, 1980).

Krawczynski, Keith, 'Hill, William', *American National Biography Online* [http://www.anb.org/articles/01/01–01286.html]

Latham, A.J.H., *Old Calabar, 1600–1891: the impact of the international economy upon a traditional society* (Oxford, 1973).

Law, Robin and Kristin Mann, 'West Africa in the Atlantic community: the case of the Slave Coast', *The William and Mary Quarterly*, 3rd series LVI (1999), 307–34.

Law, Robin (ed.), *The English in West Africa, 1681–1683. The local correspondence of the Royal African Company of England 1681–1699. Part 1* (Oxford, 1997).

Lawson, John, *A new voyage to Carolina* (1709).

Le Play, Frèdèric, 'Om ståltillverkningen i Yorkshire, samt jemförelse mellan de förnämsta ståltillverkningsorter i Europa', *Tidskrift för Svenska Bergshandteringen* (1845), 1–123.

Liedman, Sven-Eric, *Den synliga handen. Anders Berch och ekonomiämnena vid 1700-talets svenska universitet* (Värnamo, 1986).

Lilja, Sven, *Tjuvehål och stolta städer. Urbaniseringens kronologi och geografi i Sverige (med Finland) ca 1570-tal till 1810-tal* (Stockholm, 2000).

Lindqvist, Svante, *Technology on trial: the introduction of steam power technology into Sweden, 1715–1736* (Uppsala, 1984).

Lindroth, Sten, *Svensk lärdomshistoria. Frihetstiden* (Stockholm, 1978).

Linebaugh, Peter and Marcus Rediker, *The many-headed hydra: sailors, slaves, commoners, and the hidden history of the revolutionary Atlantic* (Boston MA, 2000).

Littlefield, Daniel C., *Rice and slaves: ethnicity and the slave trade in colonial South Carolina* (Baton Rouge LA, 1981).

Lloyd, H., *The Quaker Lloyds and the Industrial Revolution* (London, 1975).

Lotareva, R., 'Factory-towns of Russia and C.-N. Ledoux', in *Claude-Nicolas Ledoux and Russian architecture* (Ekaterinburg, 2001), 43–55.

Lovejoy, Paul E. and David Richardson, 'Trust, pawnship, and Atlantic history: the institutional foundations of the Old Calabar slave trade', *American Historical Review*, CIV, 2 (1999), 333–55.

——, 'Letters of the Old Calabar slave trade, 1760–1789', in Vincent Caretta (ed.), *Genius in bondage: literature of the early Black Atlantic* (Louisville KY, 2001), pp. 89–115.

——, '"This horrid hole": royal authority, commerce and credit at Bonny, 1690–1840', *Journal of African History*, XLV (2004), 363–92.

MacLeitch, Gail D., '"Red" labor: Iroquois participation in the Atlantic economy', *Labor: Studies in Working-Class History of the Americas*, I, 4 (2004), 69–90.

Magnusson, Lars, *Mercantilism: the shaping of an economic language* (London, 1994).

——, *Sveriges ekonomiska historia* (Stockholm, 1996).

——, *Merkantilism. Ett ekonomiskt tänkande formuleras* (Stockholm, 1999).

——, *Äran korruptionen och den borgerliga ordningen. Essäer från svensk ekonomihistoria* (Stockholm, 2001).

Malinovskij, Konstantin V., 'En stad växer fram. Sankt Petersburg 1703–1740', in *Vattenstäder Sankt Petersburg—Stockholm* (Stockholm, 1998), 79–108.

Malone, Joseph P., *Pine trees and politics: the naval stores and forest policy in colonial New England, 1691–1775* (London, 1964).

Marshall, J.D. (ed.), *The autobiography of William Stout of Lancaster 1665–1752* (Manchester, 1967).

Marx, Karl, *Capital: a critique of political economy, vol. III* (Harmondsworth, 1981).

Massey, Joseph, *Representation concerning the knowledge of commerce as a national concern* (1760).

McCusker, John J. and Russell R. Menard, *The economy of British America, 1607–1789* (Chapel Hill NC and London, 1985).
McIntyre, Sylvia, 'The mineral water trade in the eighteenth century', *Journal of Transport History*, VII, 1 (1973), 1–19.
Melkersson, Martin, *Staten, ordningen och friheten. En studie av den styrande elitens syn på statens roll mellan stormaktstiden och 1800-talet* (Uppsala, 1997).
Menard, Russell R., 'The Africanization of the Lowcountry labor force, 1670–1730', in Winthrop D. Jordan and Sheila L. Kemp (eds), *Race and family in the colonial* South (Jackson MI, 1987), pp. 81–108.
Mitchell, B.R. with Phyllis Deane, *Abstract of British historical statistics* (Cambridge, 1962).
Molander, Bo, 'Forsmarks stångjärnsstämpel. Vallonbrukens stångjärnsstämplar i över 250 år', in *Forsmark och vallonjärnet* (Stockholm, 1987), pp. 70–99.
Montelius, S., G. Utterström and E. Söderlund, *Fagerstabrukens historia. Del V. Arbetare och arbetareförhållanden* (Uppsala, 1959).
Morgan, Gwenda and Peter Rushton, *Eighteenth-century criminal transportation: the formation of the criminal Atlantic* (Houndsmill, 2004).
Morgan, Kenneth, *Bristol and the Atlantic trade in the eighteenth century* (Cambridge, 1993).
——, 'The organization of the colonial American rice trade', *The William and Mary Quarterly*, 3rd series, LII (1995), 433–52.
——, 'Slave sales in colonial Charleston', *English Historical* Review, CXIII (1998), 905–27.
——, *Slavery, Atlantic trade and the British economy, 1660–1800* (Cambridge, 2000).
Morgan, Philip D., *Slave counterpoint: Black culture in the eighteenth-century Chesapeake and Lowcountry* (Chapel Hill NC, 1998).
Mörner, Magnus, 'Vår östersjökust nedbränd 1719–1721. Terror avpassad för att framtvinga fred', *Militärhistorisk tidskrift* (2004), 161–219.
Morton, G.R. and N. Mutton, 'The transition to Cort's puddling process', *Journal of the Iron and Steel Institute*, CCV (1967), 722–28.
Mott, R.A. (ed. Peter Singer), *Henry Cort, the great finer, creator of puddled iron* (London, 1983).
——, 'Abraham Darby (I and II) and the coal-iron industry', *Transactions of the Newcomen Society*, XXXI (1957–59), 49–93.
——, 'The Coalbrookdale group Horsehay Works: Part I', *Transactions of the Newcomen Society*, XXXI (1957–59), 271–87 and 'Part II', XXXII (1959–60), 43–56.
Moxon, Joseph, *Mechanick exercises, or the doctrine of handy-works* (3rd edition, 1703).
Mulcahy, Matthew, '"Melancholy and Fatal Calamities": disaster and society in eighteenth-century South Carolina', in Jack P. Greene, Rosemary Brana-Shute, and Randy J. Sparks (eds), *Money, trade, and power: the evolution of colonial South Carolina's plantation society* (Columbia SC, 2001), pp. 278–98.
Müller, Leos, *Consuls, corsairs, and commerce: the Swedish consular service and long-distance shipping, 1720–1815* (Uppsala, 2004).
——, *The merchant houses of Stockholm, c. 1640–1800: a comparative study of early-modern entrepreneurial behaviour* (Uppsala, 1998).
Mutton, Norman, 'The forges at Eardington and Hampton Loade', *Transactions of the Shropshire Archaeological Society*, LVIII (1965–68), 235–43.
Myrdal, Janken, *Jordbruket under feodalismen 1000–1700* (Stockholm, 1999).
Nash, R.C., 'South Carolina and the Atlantic economy in the late seventeenth and eighteenth centuries', *Economic History Review*, XLV (1992), 677–702.
——, 'Urbanization in the colonial South: Charleston, South Carolina, as a case study', *Journal of Urban History*, XIX, 1 (1992), 3–29.
Nergård, M-B., *Mellan krona och marknad. Utländska och svenska entreprenörer inom svensk järnhantering från ca 1580 till 1700* (Uppsala, 2001).

Neve, Richard, *The city and country purchaser's and builder's dictionary: or, the complete builder's guide* (3rd edition, 1736).

Newman, S.J., 'Russian foreign trade, 1680–1780: the British contribution', (University of Edinburgh Ph.D thesis, 1985).

Nisser, Marie, 'Forsmark—ett av vallonbruken kring Dannemora gruvor', in *Forsmark och vallonjärnet* (Stockholm, 1987), pp. 15–69.

Norman, Hans, 'Befolkningsförhållandena vid två uppländska vallonbruk. En studie av Forsmark och Söderfors 1775–1855', in Anders Florén and Gunnar Ternhag (eds), *Valloner—järnets människor* (Hedemora, 2002), pp. 166–84.

Norrby, Jonas, *Jennings* (Köping, 1991).

Northrup, David, *Trade without rulers: pre-colonial economic development in south-eastern Nigeria* (Oxford, 1978).

Nováky, György, *Handelskompanier och kompanihandel. Svenska Afrikakompaniet 1649–1663. En studie i feodal handel* (Uppsala, 1990).

O'Brien, Patrick, 'European economic development: the contribution of the periphery', *Economic History Review*, XXXV (1982), 1–18.

——, 'Inseparable connections: trade, economy, fiscal state, and the expansion of empire', in P.J. Marshall (ed.), *The Oxford history of the British Empire. Volume II: the eighteenth century* (Oxford, 1998), pp. 53–77.

Ogilvie, Sheilagh and Markus Cerman (eds), *European proto-industrialization* (Cambridge, 1996).

Ogilvie, Sheilagh, 'The European economy in the eighteenth century', in T.C.W. Blanning (ed.), *The eighteenth century: Europe 1688–1815* (Oxford, 2000), pp. 91–130.

Omberg, Ture, *Bergsmän i hyttelag. Bergsmansnäringens utveckling i Linde och Ramsberg under en 100-årsperiod från mitten av 1700-talet* (Uppsala, 1992).

Ormrod, David, *The rise of commercial empires: England and the Netherlands in the age of mercantilism, 1650–1770* (Cambridge, 2003).

Owen Jansson, Sam, *Måttordenboken* (Stockholm, 1995).

Pearson, Edward A., '"A Countryside Full of Flames": a reconsideration of the Stono Rebellion and slave rebelliousness in the early eighteenth-century South Carolina Lowcountry', *Slavery and Abolition*, XVII (1996), 22–50.

Percy, John, *Metallurgy: iron and steel* (London, 1864).

Petander, Karl, *De nationalekonomiska åskådningarna i Sverige. Sådane de framträda i litteraturen* (Stockholm, 1912).

Pietschmann, Horst, 'Introduction: Atlantic history—history between European and global history', in Horst Pietschmann (ed.), *Atlantic history: history of the Atlantic system 1580–1830* (Göttingen, 2002), pp. 11–54.

Pocock, J.G.A., 'British history: a plea for a new subject', *Journal of Modern History*, XLVII, 4 (1975), 601–28.

——, 'The limits and divisions of British history: in search of an unknown subject', *American History Review*, LXXXVII (1982), 311–36.

——, *Virtues, commerce, and history: essays on political thought and history, chiefly in the eighteenth century* (Cambridge, 1985).

Pollard, S. and R.S.W. Davies, 'The iron industry 1750–1850', in C.H. Feinstein and S. Pollard (eds), *Studies in capital investment in the United Kingdom 1750–1920* (Oxford, 1988), pp. 73–104.

Pomeranz, Kenneth, *The great divergence: China, Europe, and the making of the modern world economy* (Princeton NJ, 2000).

Postlethwayt, Malachy, *The universal dictionary of trade and commerce: with large additions and improvements* (4th edn, London, 1774).

Pourchasse, Pierrick, 'Problems of French trade with the North in the eighteenth century', paper presented at the annual conference of the Economic History Society, 1 April 2006: http://www.ehs.org.uk/ehs/conference2006/Assets/IIDPourchasse.doc

Prakash, Om, *European commercial enterprise in pre-colonial India* (Cambridge, 1998).
Price, Jacob M., 'Multilateralism and/or bilateralism: the settlement of British trade balances with "The North"', *Economic History Review*, XIV (1961), 254–74.
——, 'Economic function and the growth of American port towns in the eighteenth century', in *idem*, *The Atlantic frontier of the Thirteen Colonies and States: essays in eighteenth-century commercial and social history* (Aldershot, 1996), 123–86.
Price, Jacob M. and Paul G.E. Clemens, 'A revolution of scale in overseas trade: British firms in the Chesapeake trade, 1675–1775', *Journal of Economic History*, XLVII, 1 (1987), 1–43.
Raistrick, A., 'The South Yorkshire iron industry, 1698–1756', *Transactions of the Newcomen Society*, XIX (1938–39), 51–86.
Raistrick, A., and E. Allen, 'The south Yorkshire ironmasters, 1690–1750', *Economic History Review*, IX (1938), 168–85.
Réaumur, René Antoine Ferchault de, *Réaumur's memoirs on iron and steel* (1722; Chicago IL, 1956).
Rees, Jane and Mark Rees, *Christopher Gabriel and the tool trade in eighteenth-century London* (Ipswich, 1997).
Rehder, J.E., 'The change from charcoal to coke in iron smelting', *Historical Metallurgy*, XXI (1987), 37–43.
Reimer, S. and A. Hughes (eds), *Geographies of commodity chains* (London, 2003).
Renting, Arnold, *I skuggan av Lövsta bruk. Järnbruksrörelsens inverkan på agrarsamhället in norra Uppland 1630–1930* (Stockholm, 1996).
Richards, W.A., 'The import of firearms into West Africa in the eighteenth century', *Journal of African History*, XXI (1980), 43–59.
Richardson, David, 'West African consumption patterns and their influence on the eighteenth-century English slave trade', in H.A. Gemery and J.S. Hogendorn (eds), *The uncommon market: essays in the economic history of the Atlantic slave trade* (New York, 1979), pp. 303–30.
——, 'The British slave trade to colonial South Carolina', *Slavery and Abolition*, XII, 3 (1991), 125–72.
——, *Bristol, Africa and the eighteenth-century slave trade to America. Volume 1: The years of expansion 1698–1729* (Bristol Record Society Publications, XXXVIII, 1986).
——, *Bristol, Africa and the eighteenth-century slave trade to America. Volume 2: The years of ascendancy 1730–1745* (Bristol Record Society Publications, XXXIX, 1987).
——, *Bristol, Africa and the eighteenth-century slave trade to America. Volume 3: The years of decline, 1746–1769* (Bristol Record Society Publications, XLIII, 1991).
——, *Bristol, Africa and the eighteenth-century slave trade to America. Volume 4: The final years* (Bristol Record Society Publications, XLVIII, 1996).
——, 'Slavery and Bristol's "Golden Age"', *Slavery and Abolition*, XXVI (2005), 35–45,
Richter, Daniel K., 'Native peoples of North America and the eighteenth-century British empire', in P.J. Marshall (ed.), *The Oxford history of the British Empire. Volume II: the eighteenth century* (Oxford, 1998), pp. 347–71.
Riden, Philip, 'The final phase of charcoal iron-smelting in Britain, 1660–1800', *Historical Metallurgy*, XXVIII (1994), 14–26.
——, *A gazetteer of charcoal-fired blast furnaces in Great Britain in use since 1660* (Cardiff, 1993).
Rinman, Sven, *Bergwerks-lexicon*, 2 volumes (Stockholm, 1788–89).
——, *Tjänsteberättelser rörande den grövre järnförädlingen 1761–1770*, ed. E. Malmborg (Stockholm, 1935).
Robbins, Michael W., *The Principio Company: iron-making in colonial Maryland* (New York, 1986).
Robertson, John, 'The Enlightenment above national context: political economy in eighteenth-century Scotland and Naples', *Historical Journal*, XL, 3 (1997), 667–97.

Rogers, Nicholas, 'Archipelagic encounters: war, race, and labor in American-Caribbean waters', in Felicity A. Nussbaum (ed.), *The global eighteenth century* (Baltimore MD, 2003), pp. 211–25

Roseveare, Henry (ed.), *Markets and merchants of the late seventeenth century: the Marescoe-David letters 1668–1680* (Oxford, 1987).

Rowlands, Marie B., *Masters and men in the West Midland metalware trades before the Industrial Revolution* (Manchester, 1975).

——, 'Continuity and change in an industrial society: the case of the West Midlands industries', in Pat Hudson (ed.), *Regions and industries: a perspective on the industrial Revolution in Britain* (Cambridge, 1989), pp. 103–31.

Runefelt, Leif, *Hushållningens dygder. Affektlära, hushållningslära och ekonomiskt tänkande under svensk stormaktstid* (Stockholm, 2001).

Rydberg, Sven, *Svenska studieresor till England under Frihetstiden* (Uppsala, 1951).

Rydén, Göran, *Hammarlag och Hushåll. Om relationen mellan smidesarbetet och smedshushållen vid Tore Petrés brukskomplex 1830–1850* (Uppsala, 1991).

——, 'Gustaf Ekman, Jernkontoret och lancashiresmidet—Ett inlägg i synen på teknisk utveckling', *Polhem: Tidskrift för teknikhistoria*, 1994/2, 132–64.

——, 'Lancashiresmide och omvandling. Teknisk utveckling, strukturomvandling och produktionstillväxt i Gästrikland under 1800-talet', *Med Hammare och Fackla* (1998), 105–51.

——, *Production and work in the British iron trade in the eighteenth century: a Swedish perspective* (Uppsala, 1998).

——, '"Det Andra" som det kommersiella och det industriella—Svenska tankar om handel och produktion under 1700-talet', in Maths Isacson and Mats Morell (eds), *Industrialismens tid. Ekonomisk-historiska perspektiv på industriell omvandling under 200 år* (Stockholm, 2002), pp. 37–56.

——, 'Vallonbruk, vallonsmeder och vallonsmide—en precisering av ett kunskapsläge', in Anders Florén and Gunnar Ternhag (eds), *Valloner—järnets människor* (Hedemora, 2002), pp. 107–35.

Ryder, A.F.C., *Benin and the Europeans, 1485–1897* (Harlow, 1969).

Sacks, David Harris, *The widening gate: Bristol and the Atlantic economy, 1450–1700* (Berkeley CA, 1991).

[Salvius, Lars], *Tanckar öfwer den Swenska Oeconomien igenom Samtal yttrade* (Stockholm, 1738).

Samuelsson, Kurt, *De stora köpmanshusen i Stockholm 1730–1815* (Stockholm, 1951).

Savall, Jordi, 'J.S. Bach's Musical Testament', in the booklet acompanying the CD *Die Kunst der Fuge* (AliaVox, AVSA9818, 2001).

Schafer, R.G., 'Genesis and structure of the Foley "Ironworks in Partnership" of 1692', *Business History*, XIII (1971), 19–38.

Schlumbohm, Jürgen, 'Relations of production—production forces—crises in proto-industrialization', in Peter Kriedte, Hans Medick, and Jürgen Schlumbohm, *Industrialization before industrialization* (Cambridge, 1981), pp. 94–125.

Schubert, H.R., *History of the British iron and steel industry from c. 450 B.C. to A.D. 1775* (London, 1957).

Schumpeter, Elizabeth B., *English overseas trade statistics 1697–1808* (Oxford, 1960).

Segal, Ronald, *Islam's Black slaves: the history of Africa's other Black diaspora* (London, 2002).

Sennefelt, Karin, *Den politiska sjukan. Dalaupproret 1743 och frihetstida politisk kultur* (Hedemora, 2001).

Sheridan, Richard B., 'The formation of Caribbean plantation society, 1689–1748', in P.J. Marshall (ed.), *The Oxford history of the British Empire. Volume II: the eighteenth century* (Oxford, 1998), pp. 394–414.

Smith, Adam, *An inquiry into the nature and causes of the wealth of nations* (1776: Indianapolis IN, 1981).

Smith, B.M.D., 'The Galtons of Birmingham: Quaker gun-merchants and bankers, 1702–1831', *Business History*, IX (1967), 132–50.
Smith, Mark M., 'Remembering Mary, shaping revolt: reconsidering the Stono Rebellion', *Journal of Southern History*, LXVII, 3 (2001), 513–34.
Smith, Richard S., 'Sir Francis Willoughby's ironworks, 1570–1610', *Renaissance and Modern Studies*, XI (1967), 90–140.
Smith, S.D., 'The market for manufactures in the thirteen continental colonies, 1698–1776', *Economic History Review*, LI, 4 (1998), 676–708.
——, 'The Atlantic History paradigm', *New England Quarterly* (March 2006), 123–33.
Sparks, Randy J., *The two princes of Calabar: an eighteenth-century Atlantic odyssey* (Cambridge MA, 2004).
St. John [de Crèvecoeur], J. Hector, *Letters from an American farmer* (London, 1782).
Steele, Ian K., *The English Atlantic 1675–1740: an exploration of communication and community* (Oxford, 1986).
Stjernquist, Otto, 'Om puddling i Sverige och Finland', *Med Hammare och Fackla* (1998), 29–65.
Stumpf, Stuart O., 'The merchants of colonial Charleston, 1680–1756', (Ph.D. thesis, University of Michigan, 1971).
Sundström, Lars, *The exchange economy of pre-colonial tropical Africa* (London, 1974).
Svedenstierna, E.T., *Resa, igenom en del af England och Skottland, åren 1802 och 1803* (Stockholm, 1804).
——, 'Tal om den svenska jernhandteringen i äldre och nyare tider', in *Kungliga Vetenskaps Akademins handlingar* (Strängnäs, 1810).
——, *Några underrättelser om Engelska jernhandteringen* (Stockholm, 1813).
——, 'Om Puddlingsprocessen', *Jernkontorets Annaler* (1817, Sednare häftet), 92–153.
Symonds, James (ed.), *The historical archaeology of the Sheffield cutlery and tableware industry 1750–1900* (Sheffield, 2002).
Szostak, Rick, *The role of transportation in the Industrial Revolution* (Montreal, 1991).
Tamm, A.G., *Anteckningar öfver främmande länders jernhandtering, gjorde under en resa i Tyskland, Frankrike och England, åren 1830 och 1831* (Stockholm, 1832).
Thomas, Hugh, *The slave trade: the history of the Atlantic slave trade, 1440–1870* (London, 1997).
Thornton, John K., 'African dimensions of the Stono Rebellion', *American Historical Review*, XCVI (1991), 1101–113.
Thornton, John, *Africa and Africans in the making of the Atlantic world, 1400–1800* (2nd edn, Cambridge, 1998).
Thörnwall, Folke, *Leufsta. Ett gammalt upplandsbruk* (Tierp, 1968).
Timmins, J.G., 'Concentration and integration in the Sheffield crucible steel industry', *Business History*, XXIV (1982), 61–78.
Timmins, Samuel (ed.), *The resources, products, and industrial history of Birmingham and the Midland hardware district* (London, 1866).
Trainor, Richard H., *Black Country elites: the exercise of authority in an industrialized area 1830–1900* (Oxford, 1993).
Unger, W.S., 'Trade through the Sound in the seventeenth and eighteenth centuries', *Economic History Review*, XII, 2 (1959), 206–21.
Uriarte Ayo, Rafael, 'Anglo-Spanish trade through the port of Bilbao during the second half of the eighteenth century: some preliminary findings', *International Journal of Maritime History*, IV (1992), 193–217.
——, 'El hierro vasco y los mercados europeo y colonial durante el Antiguo Régimen', *Revista de Estudios Marítimos del País Vasco*, IV (2003), 313–26.
von Stockenström, Eric, *Tal om svenska Järn-Bruksnäringen, samt om Jern-Contoiret. Hållet för Kongl. Vetenskaps Academien, vid Præsidii Nedläggande Den 11 April 1767* (Stockholm, 1767).
Wahlund, J., *Dannemora grufvor. Historisk skildring* (Stockholm, 1879).

Wakelin, A.P., 'Pre-industrial trade on the river Severn: a computer-aided study of the Gloucester port books *c.* 1640–*c.* 1770' (unpublished CNAA Ph.D. thesis, 1991).

Wallerstein, Immanuel, *The modern world-system: capitalist agriculture and the origins of the European world-economy in the sixteenth century* (New York, 1974).

——, *The modern world-system II: mercantilism and the consolidation of the European world-economy, 1600–1750* (New York, 1980).

Walsh, Lorena S., *From Calabar to Carter's Grove: the history of a Virginia slave community* (Charlottesville VA, 2001).

Wax, D.D., '"The great risque we run": the aftermath of the slave rebellion at Stono, South Carolina, 1739–1745', *The Journal of Negro History*, LXVII, 2 (1982), 136–47.

Wayman, M.L. (ed.), *The ferrous metallurgy of early clocks and watches: studies in post medieval steel* (British Museum Occasional Paper 136, 2000).

Weber, Klaus, 'The Atlantic coast of German trade: German rural industry and trade in the Atlantic, 1680–1840', *Itinerario: European Journal of Overseas History*, XXVI, 2 (2002), 99–119.

Weir, Robert M., '"Shaftesbury's Darling": British settlement in the Carolinas at the close of the seventeenth century', in Nicholas Canny (ed.), *The Oxford history of the British Empire. Volume 1: the origins of empire: British overseas enterprise to the close of the seventeenth century* (Oxford, 1998), pp. 375–97.

Williams, Gomer, *History of the Liverpool privateers* (1897).

Williams, Justin, 'English mercantilism and Carolina naval stores, 1705–1776', *Journal of Southern History*, I, 2 (1935), 169–85.

Wise, Michael J., 'Birmingham and its trade relations in the early eighteenth century', *University of Birmingham Historical Journal*, II, 1 (1949), 53–79.

Wise, M.J. and B.L.C. Johnson, 'The changing regional pattern during the eighteenth century', in *Birmingham and its regional setting: a scientific survey* (Birmingham, 1950), pp. 161–86.

Witt, Cleo, Cyril Weeden and Arlene Palmer Schwind, *Bristol glass* (Bristol, 1984).

Wolff, Christoph, *Johann Sebastian Bach, the learned musician* (Oxford, 2000).

Wood, Peter H., *Black majority: negroes in colonial South Carolina from 1670 through the Stono Rebellion* (New York, 1975).

Wrigley, E.A., *Continuity, chance and change: the character of the Industrial Revolution in England* (Cambridge, 1988).

Wrigley, E.A. and R.S. Schofield, *The population history of England 1541–1871: a reconstruction* (Cambridge, 1981).

Wyke, John, *A catalogue of tools for watch and clock makers by John Wyke of Liverpool, with an introduction and technical commentary by Alan Smith* (Charlottesville VA, 1978).

Zahedieh, Nuala, 'London and the colonial consumer in the late seventeenth century', *Economic History Review*, XLVII, 2 (1994), 239–61.

——, 'Making mercantilism work: London merchants and Atlantic trade in the seventeenth century', *Transactions of the Royal Historical Society*, 6th series, IX (1999), 143–58.

INDEX

Please note that the Swedish letters å, ä and ö come at the end of the alphabet and are indexed accordingly, i.e. Åkerby comes after Wira, and Södermalm follows Stockholm.

Note also that members of the De Geer family are indexed according to the Swedish form of their surname rather than the Dutch usage, i.e. *De Geer, Louis* rather than *Geer, Louis de*.

1.1 The Royal Dockyard, Deptford in the 1750s

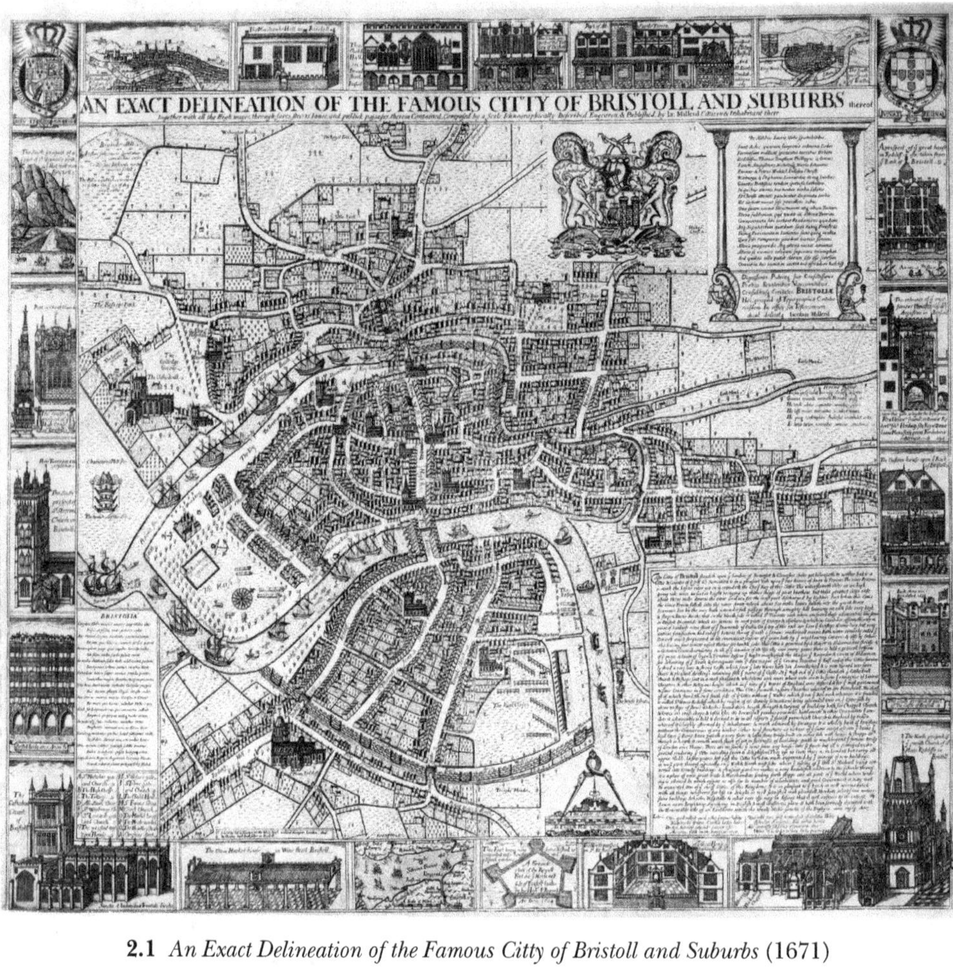

2.1 *An Exact Delineation of the Famous Citty of Bristoll and Suburbs* (1671)

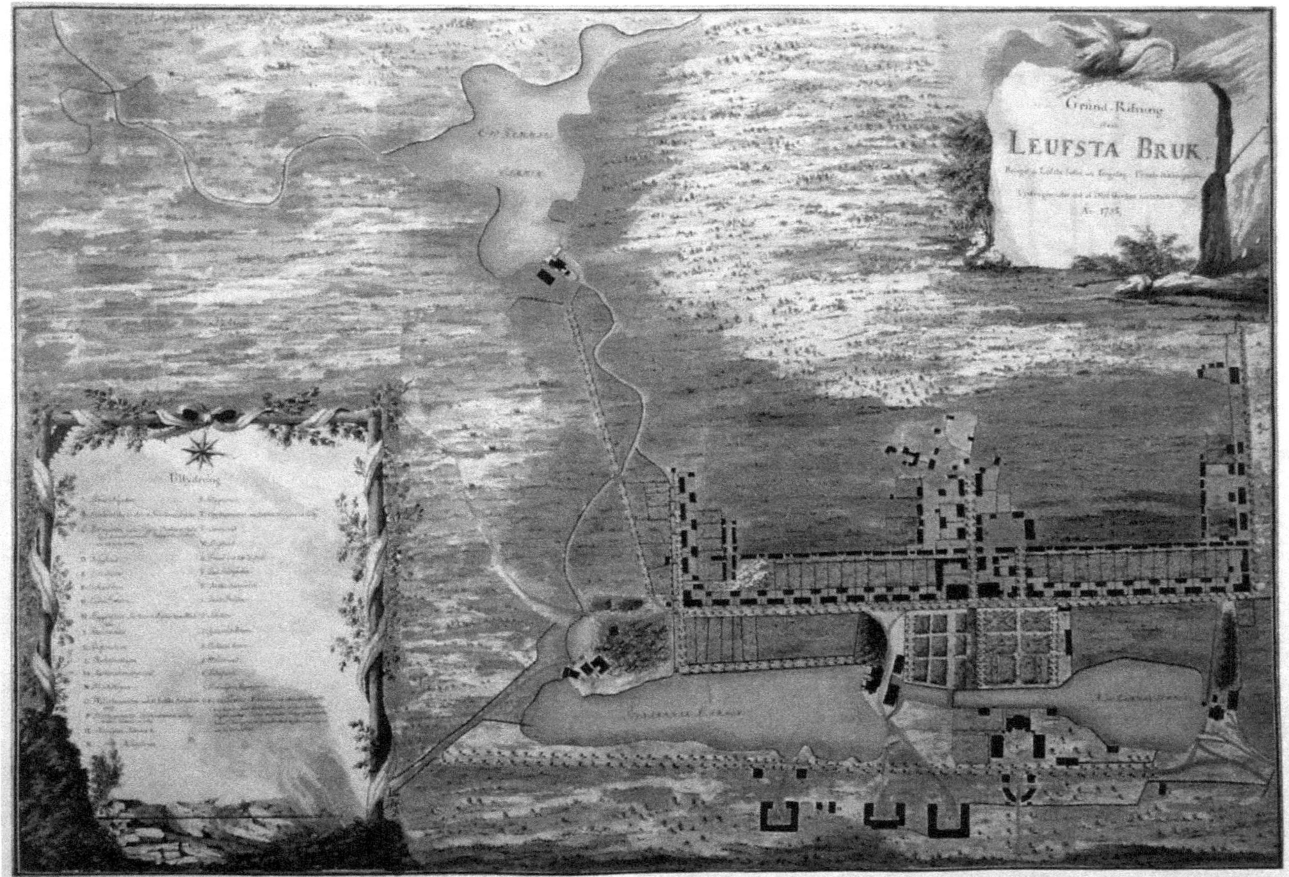

2.4 Leufsta *bruk* in 1735

2.4 Leufsta *bruk* in 1735 (detail)

2.7 Dannemora mine

2.9 A Walloon forge in Leufsta *bruk c.*1790

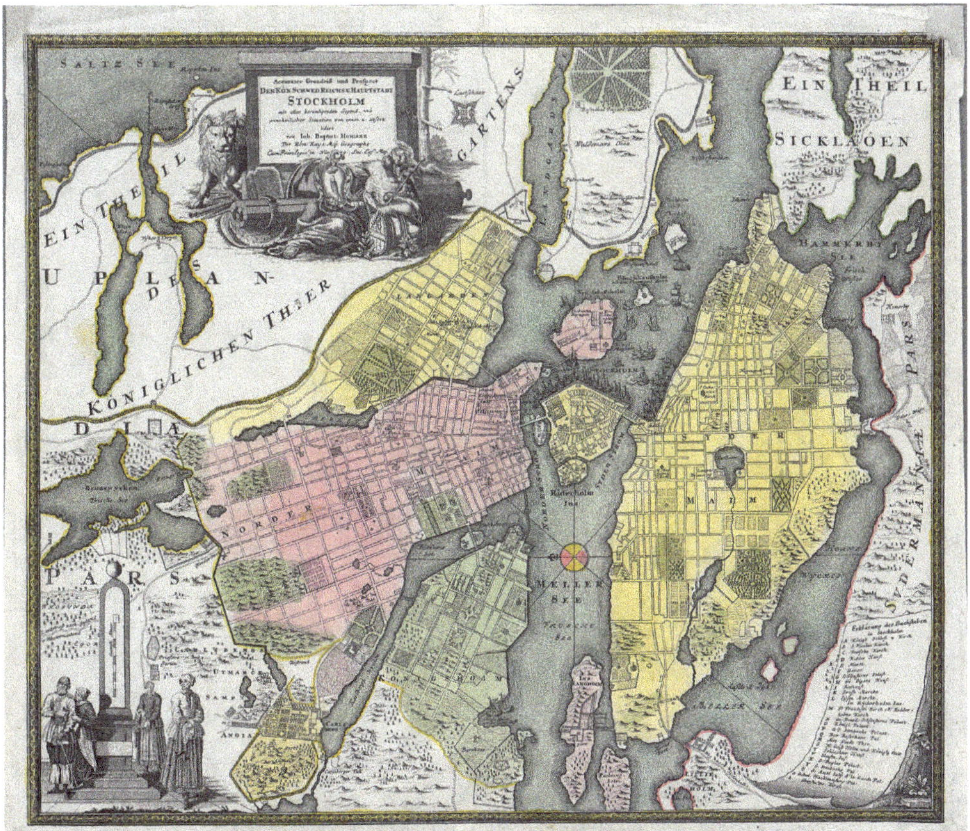

2.11 Stockholm *c.*1720

2.15 The steel furnace at Derwencote

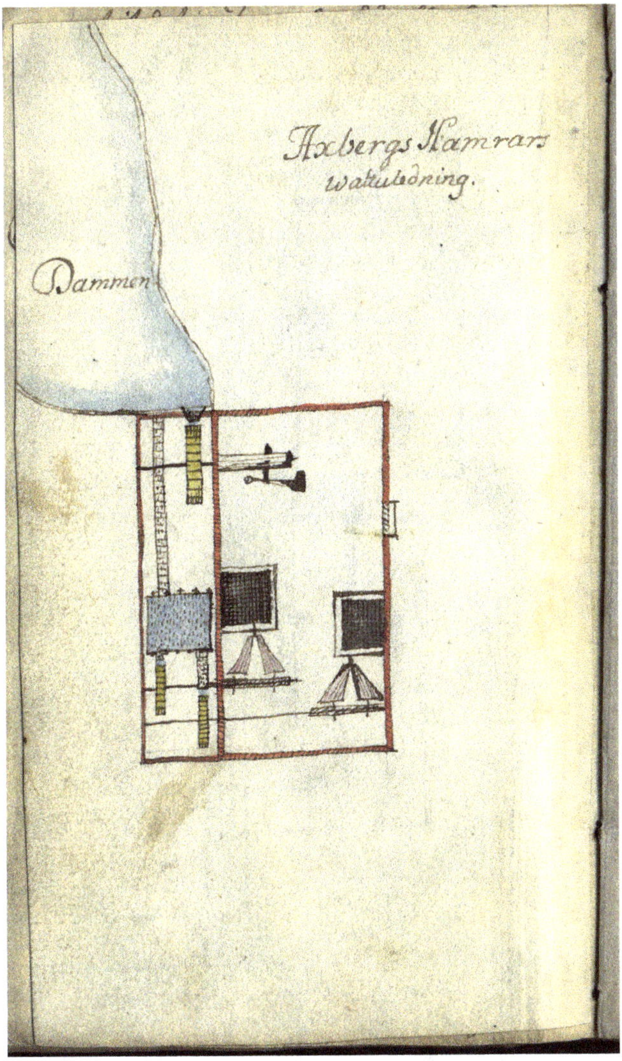

2.23 The German forge at Axberg.

2.24 St Petersburg in 1722

2.24 St Petersburg in 1722 (detail)

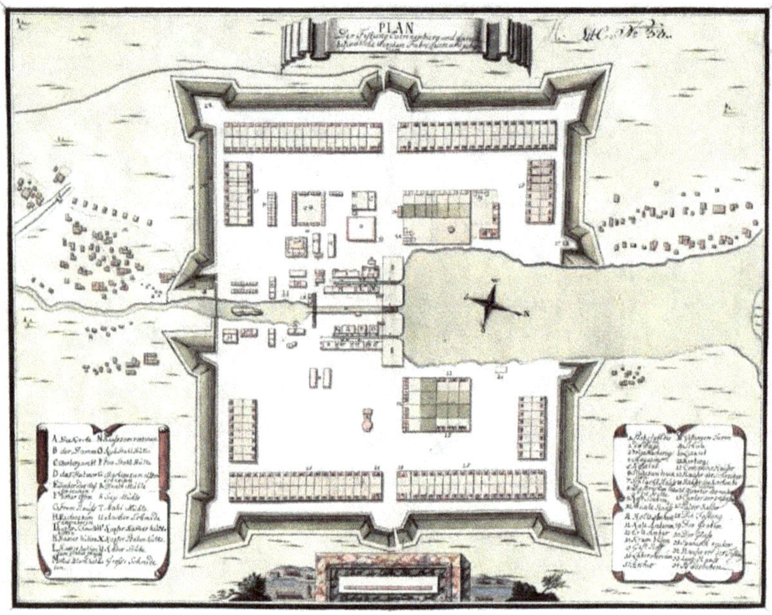

2.25 Ekaterinburg.

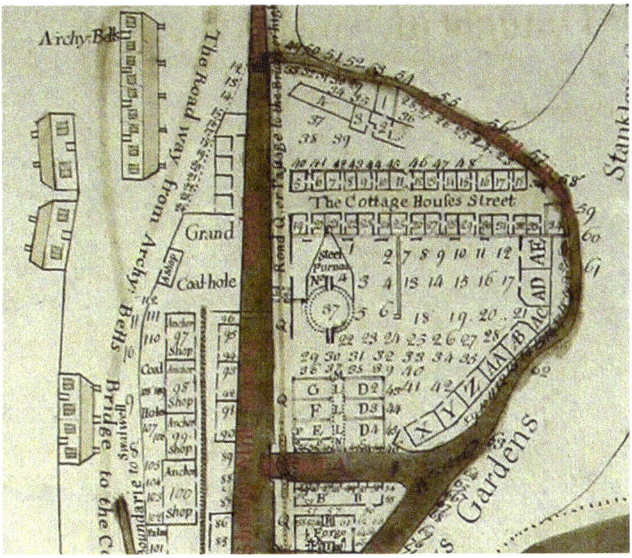

2.26 Swalwell works

2.28 'Indiens going a hunting', 1736

4.3 A loop of puddled iron being turned under the shingling hammer at Cyfarthfa ironworks, 1792

4.5 A rolling mill at Cyfarthfa ironworks, *c.*1830

www.ingramcontent.com/pod-product-compliance
Lightning Source LLC
LaVergne TN
LVHW010639300926
849161LV00005B/986